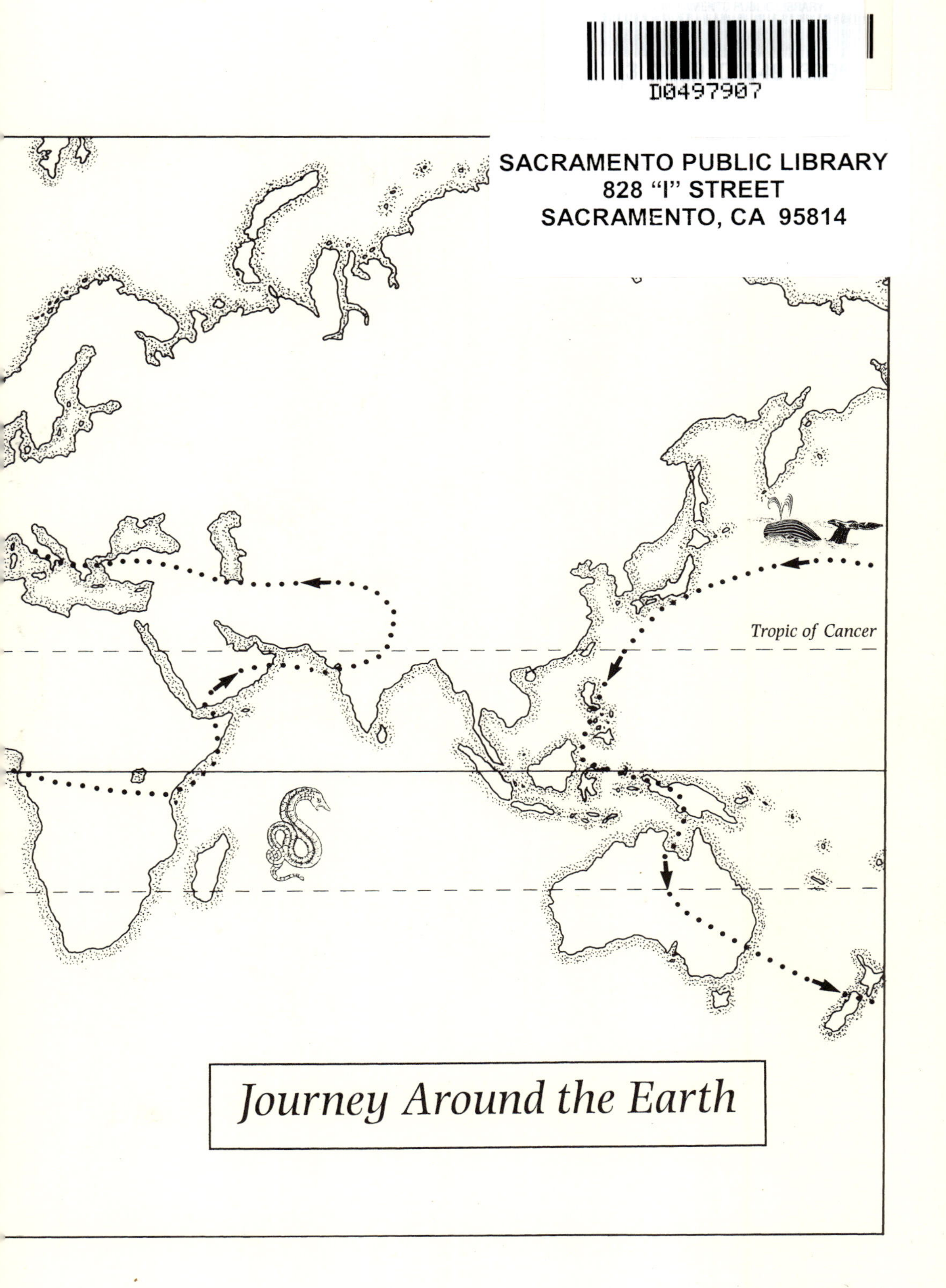

Journey Around the Earth

ALSO BY RICHARD FORTEY

Life: A Natural History of the First Four Billion Years of Life on Earth

Trilobite! Eyewitness to Evolution

Earth

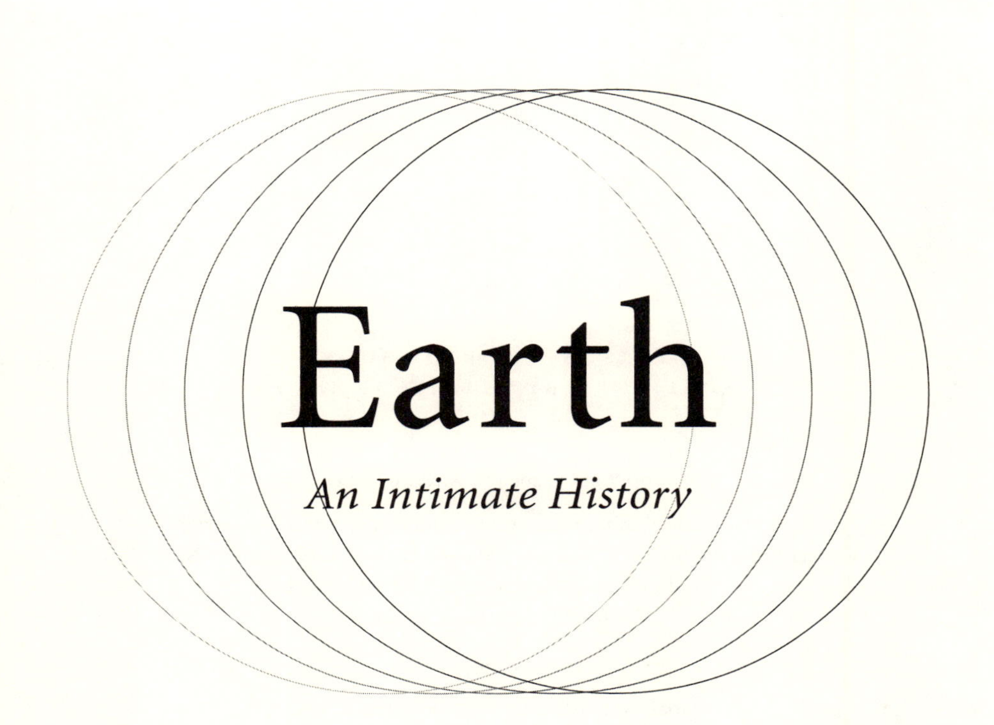

Earth

An Intimate History

RICHARD FORTEY

Alfred A. Knopf  *New York* 2005

THIS IS A BORZOI BOOK
PUBLISHED BY ALFRED A. KNOPF

 Published in the United States by Alfred A. Knopf, a division of Random House, Inc., New York. Distributed by Random House, Inc., New York.

www.aaknopf.com

Originally published in Great Britain as *The Earth* by HarperCollins Publishers, London.

Knopf, Borzoi Books, and the colophon are registered trademarks of Random House, Inc.

Library of Congress Cataloging-in-Publication Data

Fortey, Richard A.
Earth, an intimate history / by Richard Fortey.
p. cm.
Includes bibliographical references and index.
ISBN 0-375-40626-3
1. Historical geology. I. Title.

QE28.3.F66 2004
551.7—dc22 2004046470

Manufactured in the United States of America
First American Edition Published November 3, 2004
Reprinted Once
Third Printing, January 2005

For Jules, with my love

Contents

Acknowledgements

It would have been impossible to write this book had I not been awarded the Collier Chair in the Public Understanding of Science and Technology at the University of Bristol for 2002. I am most grateful to the Institute of Advanced Studies, University of Bristol, and particularly Professor Bernard Silverman, for granting me time, and a quiet attic, to escape the hassles of my normal life. Karine Taylor, secretary to the Institute, was helpful in countless small ways which made my stay in Bristol a pleasure. I thank Paul Henderson of The Natural History Museum in London for facilitating my sabbatical leave at Bristol.

While researching the book I was compelled to explore parts of the geological column that are not my usual habitat, and to extend my expertise into new fields. I profited enormously from the generosity of several geologists who guided me over territory well-known to them, but novel to me. Geoff Milnes, and his daughter Ellen, were charming guides around the eastern Alps, and patiently tolerated my naive questions. Graham Park took me around the Northwest Highlands at a time of the year when most sensible people (even Scottish people) are sat round the fire with a hot toddy. Geoff and Graham also revised the appropriate parts of my manuscript, and corrected mistakes and misapprehensions. I need hardly add that any infelicities or inaccuracies that remain are entirely the responsibility of the writer. Several years ago, Dr. Ajit Varkar of Pune arranged my trip to the Deccan traps in northwestern India, negotiating with drivers and guides alike. I could not have completed this excursion without his help. Paul and Jodi Moore provided unstinting hospitality in Hawai'i for me and my family, and were the best possible company in the tropical evenings (I hope they got the sand out of the carpet). I cannot thank them enough. Sam Gon helped me understand aspects of Hawaiian culture. Several colleagues generously gave me advice in return for no more than a free lunch. I may still owe one or two. I particularly thank Claudio Vita-Finzi and David Price of

University College, London, for advice on the Bay of Naples, and the middle of the earth, respectively. Bernard Wood of the University of Bristol spent an afternoon showing me how rock samples can be crushed so hard as to change their character, which was invaluable for my chapter "Deep Things." I also thank John Dewey, Tony Harris, David Gee, Bob Symes, Joe Cann, and my old Newfoundland friends for advice of one kind or another. Adrian Rushton and Derek Siveter provided moral support in the pub at times of crisis.

I am particularly indebted to Robin Cocks, John Cope and Heather Godwin, who read through the first version of the book, and made several suggestions for its improvement. Heather has always had a crucial role in my writing career, as an infallible arbiter of taste, and excisor of bad jokes. Arabella Pike at HarperCollins has proved again an enthusiastic and supportive editor, and her organisational skills ensured that the whole work came together.

I sincerely thank my wife, Jackie, for once again tolerating the writer at work, and, more specifically, for organising several of the field trips which comprise the core of the book. Jackie and my daughter, Rebecca, did much of the picture research for the colour plates.

Robert Francis provided me with many excellent photographs, which greatly enhanced the attractiveness of my natural history of the earth. I thank Ray Burrows for his skillful drawings. James Secord, Ted Nield, John Cope, David Gee, John Moorby, Jerry Ortner, Graham Park, and Geoff Milnes also provided illustrations, for which I am much indebted.

Preface

For some years I have been thinking about how best to describe the way in which plate tectonics has changed our perception of the earth. The world is so vast and so various that it is evidently impossible to encompass it all within one book. Yet geology underlies everything: it founds the landscape, dictates the agriculture, determines the character of villages. Geology acts as a kind of collective unconscious for the world, a deep control beneath the oceans and continents. For the general reader, the most compelling part of geological enlightenment is discovering what geology does, how it interacts with natural history, or the story of our own culture. Most of us engage with the landscape at this intimate level. Many scientists, by contrast, are propelled by the search for the inclusive model, a general theory that will change the perception of the workings of the world. In most scientific papers, the intimacies of plants or places are hardly given a second glance. Plate tectonics has transformed the way we understand the landscape, for the world alters at the bidding of the plates, but much of the transformation has been expressed in the cool prose of the scientific treatise. The problem is how we can marry these two contrasting modes of perception—the intelligent naturalist's sensitive view of the details of the land with the geologist's abstract models of its genesis and transformation. My solution has been to visit particular places, to explore their natural and human history in an intimate way, thence to move to the deeper motor of the earth—to show how the lie of the land responds to a deeper beat, a slow and fundamental pulse. I have chosen my examples with some care, for they are all places that have figured in unscrambling this complex and richly patterned planet of ours. I have visited them all, so that the reader will also have this particular guide's reactions to the sights, sounds, smells, and ambience of the critical localities. At the same time, I have endeavoured to show how knowledge of the deeper tectonic reality has changed over the last century or so. Great minds have pondered the shape of the world, and have purveyed "theories of everything" that have come and gone. Many past theories have been founded on good reasoning for their time and place, and it would be a complacent scientist today who would claim that present knowledge is as far as it goes. Understanding advances by building upon, and criticising the work of those who went before. It is a messy and complicated business, in which the human heart has as much a part to play as human intellect. This, too, is an ingredient in my story. The most difficult decisions I faced were not what to include, but what to leave out. I am acutely aware that there are areas of science that are merely sketched herein, any one of which would merit a book of its own. Geochemical cycles and their role in earth systems are a case in point. The intercedence of extraterrestrial events in our history is another fascinating field in which many recent advances have been made. The omission of such things in the interest of a coherent narrative was a painful necessity. What my story lacks in omniscience I hope it makes up for in coherence and accessibility.

Earth

1

Up and Down

It should be difficult to lose a mountain, but it happens all the time around the Bay of Naples. Mount Vesuvius slips in and out of view, sometimes looming, at other times barely visible above the lemon groves. In parts of Naples, all you see are lines of washing draped from the balconies of peeling tenements or hastily constructed apartment blocks: the mountain has apparently vanished. You can understand how it might be possible to live life in that city only half aware of the volcano on whose slopes your home is constructed, and whose whim might control your continued existence.

As you drive eastwards from the centre of the city, the packed streets give way to a chaotic patchwork of anonymous buildings, small factories, and ugly housing on three or four floors. The road traffic is relentless. Yet between the buildings there are tended fields and shaded greenhouses. In early March the almonds are in flower, delicately pink, and there are washes of bright daffodils beneath the orchard trees; you can see women gathering them for market. In the greenhouses exotic flowers such as canna lilies can be glimpsed, or ranks of potted plants destined for the supermarket trade. Oranges and lemons are everywhere. Even the meanest corner will have one or two citrus trees, fenced in and padlocked against thieves. The lemons hang down heavily, as if they were too great a burden for the thin twigs that carry them. The soil is marvellously rich: with enough water, crops would grow and grow.

This was an abundant garden in Roman times, and it still is, even if crammed between scruffy apartments and scrap-metal yards. Volcanic soil is rich in minerals; it is correspondingly generous to crops. Outside the city,

The Bay of Naples in the early nineteenth century; Vesuvius in the distance, an Arcadian prospect. An engraving by E. Benjamin from an original painting by G. Arnald.

Vesuvius is more of a continuous presence; the ground rises gently towards its brown summit. New buildings cling on to the side of the mountain, even high up among the low trees and broom bushes that clothe its flanks. The buildings are indistinct, however, hidden by a creamy-yellow haze of petrochemical smog spreading outwards from the frantic centre of Naples towards the mountainside. You pass a road sign to Pompeii, but from the road there is little to distinguish this suburb from any other, for all its fame.

When the road rises into the hills that abut the southern margin of the Bay of Naples, the urban sprawl begins to thin out. The orange groves are more orderly, with the trees neatly planted in rows inside cages made of makeshift wooden struts, draped over the top with nets. The slopes become much steeper than on the volcanic flanks—close terraces piled one upon the other, each banked up with a wall of pale limestone blocks. Medium-sized trees with small grey-green leaves—which appear almost silvery in the afternoon light—cling to the most precipitous terraces. These are olive trees, the definitive Mediterranean survivors, oil producers and suppliers of piquant fruit. Their deep roots can seek out the narrowest cracks. They relish limestone soils, however poor they are in comparison with volcanic loam. The villages in this part of the bay are as you would expect of regular, tourist Italy, with piazzas and ristorante-pizzerias and youths with slick hairstyles on the lookout for a fast buck. Even this long before the summer season there is opportunity for a smooth operator. You find yourself agreeing to hire a cab for a day for €200 to hug the congested roads, when you could travel faster on the excellent Circumvesuviana railway for a tiny fraction of the price. Somehow, you, the visitor, have become the rich volcanic soil primed to yield a good harvest.

Near the tip of the southern peninsula, Sorrento commands a wonderful prospect of Mount Vesuvius across the entire Bay of Naples. From this steep-sided town, Vesuvius looks almost the perfect, gentle-sided cone. It could be a domestic version of Mount Fuji, the revered volcanic mountain in Japan. It can appear blue, or grey, or occasionally stand revealed in its true brown colours. On clear days Vesuvius is starkly outlined against a bright sky: a dark, heavy, almost oppressive presence. Or on a misty morning its conical summit can rise above a mere sketch or impression of the lower slopes, which are obscured in vapour, as if it were cut off from the world to make a house for the gods alone. At night, ranks of lights along Neapolitan roads twinkle incessantly. Vesuvius is often no more than a dark shape against a paler, but still Prussian blue sky. The lights might persuade you that the mountain was still in the process of eruption, with points of white illumination tracking lava flows running down the hillsides. From Sorrento, you can make of Vesuvius what you will, for within a day it will have remade itself.

The Bay of Naples is where the science of geology started. The description of the eruption of Vesuvius and the destruction of Pompeii in A.D. 79 by Pliny the Younger is probably the first clear and objective description of a geological phenomenon. No dragons were invoked, no clashes between the Titans and the gods. Pliny provided observation, not speculation. Not quite two millennia later, in 1830, Charles Lyell was to use an illustration of columns from the so-called Temple of Serapis at Pozzuoli, north of Naples, as the frontispiece to volume 1 of the most seminal work in geology—his *Principles of Geology.* This book influenced the young Charles Darwin more than any other source in his formulation of evolutionary theory: so you could say that the Bay of Naples had its part to play, too, in the most important *bio*logical revolution. Everybody who was anybody in the eighteenth and nineteenth centuries visited the bay, and marvelled at its natural and archaeological phenomena. For geology—a latecomer to the pantheon of sciences—the area is the nearest thing to holy ground that there is. If you were going to choose anywhere to retrace the growth in our understanding about how the earth is constructed, what better place to begin? Where else more appropriate to explain first principles? The long intellectual journey that eventually led to plate tectonics started in this bite out of the western shin of Italy's boot-shaped profile. A voyage around this particular bay is a pilgrimage to the foundations of comprehension about our planet.

Everything about Sorrento is rooted in the geology. The town itself is in a broad valley surrounded by limestone ranges, which flash white bluffs on the hillsides and reach the sea in nearly vertical cliffs—an incitement to dizziness for those brave enough to look straight down from the top. Seen from a distance, the roads that wind up the sides of the hills look like folded tagliatelle. Stacked blocks of the same limestone are used in the walls that underpin the terraces supporting the olive groves. In special places there are springs that spurt out fresh, cool water from underground caverns. These sources are often flanked by niches containing the statue of a saint, or of the Virgin: water is not taken for granted in these parts. There are deep ravines through the limestone hills, probably marking where caves have collapsed. The country backing the Bay of Naples is known as Campania, and the same name, Campanian, is applied to a subdivision of geological time belonging to the Cretaceous period. If you look carefully on some of the weathered surfaces of the limestones, you will see the remains of seashells that were alive in the age of the dinosaurs. I saw some obvious clams and sea urchins, belonging to extinct species, emerging from the cliffs as if they were on a bas-relief. A palaeontologist can identify the individual fossil species, and use them to calibrate the age of the rocks, since the succession of species is a measure of geological time. The implication is clear enough: in Cretaceous times all these hilly regions were beneath a shallow, warm sea. Limy muds accumulated there as sediments, and entombed the remains of the animals living on the sea floor. Time and burial hardened the muds into the tough limestones we see today. They are sedimentary rocks, subsequently uplifted to become land; earth movements then tilted them—but this is to anticipate. What one can say is that the character of the limestone hills is a product of an ancient sea.

The massive limestones continue westwards on to the island of Capri, which is a twenty-minute ferry ride from Sorrento and bounds the southern edge of the Bay of Naples. The island rises sheer from the sea, circumscribed by steep limestone cliffs, and your first thought is how could it support the smallest village, let alone a town. The town of Capri lies at the top of a vertiginous funicular railway running from the harbour. The buildings are ancient and quaint, and, naturally enough, built of the local stone. The blocks themselves are often concealed under stucco. There is a fine medieval charterhouse where the pale limestone is put to good effect in columns supporting cloisters. Almost everything else is fabricated of limestone—walls, floors, piazzas. In the bright Mediterranean light there is an overwhelming sense of whiteness; some of the villas glimpsed on the hillside have the appearance of frosted cakes tucked under umbrella pines. Only dark basalt must have been imported from Vesuvius to make the surfaces of the streets: this volcanic rock is less liable to shatter than limestone. It is not difficult to imagine the racket that iron-rimmed wheels made as they clattered over these roughly matched, large blocks. On the inner side of the island there are truly astonishing vertical limestone cliffs dropping hundreds of metres to the sea. The Roman Emperor Tiberius spent his declining years in a palace on the island, the ruins of which endure. According to the prurient accounts of his chronicler Suetonius, he indulged every kind of sexual perversion in a life of epicene self-gratification. Small boys were favoured. Those who displeased him were liable to be thrown off the monstrous cliffs. There is a subtle undercurrent in the Caprese atmosphere that hints at such darker things. Just offshore there are two enormous and forbidding sea-stacks—masses of limestone isolated from the main cliff by the relentless erosion of the sea. According to Norman Douglas, this was the abode of the Sirens, whose alluring and fatal song Odysseus was able to resist only by being strapped to the mast of his vessel, while his muffled crew rowed onwards to safety. Capri makes you wonder whether an idyllic hilltop haven might eventually also deprave and destroy. One of the grandest villas (now a hotel) overlooking the fearsome cliffs was built by the Krupps dynasty, once the armourers of German ambitions. Unexpectedly, the builder apparently immersed himself in studying the growth of lampreys, a primitive and parasitic fish. On this island there is a seamless continuity with the past—with Hellenic myth and Roman decadence and medieval devotion. The island gardens have seen the ages come and go, perched high upon the hardened sediments of a sea far more ancient than human frailty.

There is something different about the cliffs behind the harbour in the middle of Sorrento. From afar they have a greyish cast, a dull uniformity, lacking all the brilliance of limestone. The streets career downwards towards the sea below the central piazza, following a steep-sided valley.

Now you can see the rock in the valley sides. It is brownish, like spiced cake, and displays little obvious structure. Look closely and you see that embedded within it, like dates in a home-bake, there are darker patches. Some are little more than wisps, others are larger—angular pieces of another rock, here nearly black, there umber brown, some including little bubbles. Then you notice that the same rock has been recruited by the local builders to construct the high walls that line the steeply sloping path, comprising blocks a few tens of centimetres across, neatly cut and used like bricks. Clearly, this rock is softer than the rough limestones that bolster the hilly vineyards and terraces. Then you notice that the same stone has been used to construct the older buildings. Down by the port there are shops and cafés painted jolly ochre and sienna, but where the stucco has peeled—or where warehouses have simply been left undecorated—the same rock is revealed as having been used for their construction. Much of the town has grown from the identical rock that forms the steep cliffs backing the harbour.

This rock is called the Campanian Ignimbrite. Its origin was a catastrophe that happened 35,000 years ago: a gigantic volcanic explosion threw out at least 100 cubic kilometres of pumice and ash. The evidence still covers an area of more than 30,000 square kilometres around the Bay of Naples, extending from Roccamonfina in the north to Salerno in the south. The violence of this eruption would make the event that buried Pompeii seem like a small afterthought. An explosion of steam and gluey lava blew out a great hole in the earth at the edge of the Tyrrhenian Sea—not so much a bite out of Italy's profile as a huge punch. A vast cloud of incandescent material buoyed up with gas flowed like a fiery tidal wave across the limestone terrain. Lumps of volcanic rock were carried along willy-nilly in the mayhem: destruction of vegetation was complete. When the cloud settled, in many places it was hot enough to fuse solid: the wispy remains of volcanic fragments testify to this welding.* There were almost certainly Palaeolithic human witnesses to this destruction, who must have thought the gods had gone berserk. The legacy of the earth's ferocity is this apparently mundane rock that looks like cake. The angular fragments of rock within can now be seen for what they are—pieces of a destroyed volcano. It

* This is what makes an ignimbrite; the general term for this kind of volcaniclastic rock is "tuff." This has nothing to do with tufa, which you can see encrusting the rocks around the limestone springs, even obliterating the feet of the saint in his niche, where lime is deposited.

is ironic that this destruction has now been reversed into constructing buildings that are "safe as houses." Naturally, nothing is safe in this uncertain world. Looking down from the limestone hills you can imagine the hot, devastating clouds settling over where limoncello is now brewed and pizzas are spun, dumping down on the low ground as a thick, lethal blanket. These kinds of rocks were deposited from pyroclastic surges. Another eruption about 23,000 years later was marginally less devastating and did not spread so widely—it produced a different deposit known as the Tufo Galliano Napoletano, the Neapolitan yellow tuff. Rather than the colour of cake, it is the colour of Dijon mustard. Once you can recognize it, you spot blocks of it in many walls and buildings around Naples itself—it is almost reminiscent of the "London stock" bricks that make the Georgian parts of the English capital so appealing. It is there in the walls of Roman remains. Most experts believe that the volcanoes that remain to this day in the Campi Flegrei are aligned around the edge of the massive hole, or caldera, left behind as the legacy of this second huge eruption. The Bay of Naples itself hides most of it. It may yet blow again.

It must be appropriate now to visit Mount Vesuvius. It is 1281 metres high—not a great peak, but grand enough. Strictly speaking, only the cone itself should be called Vesuvius: the wider region, including older volcanic remnants, should be called Somma Vesuvius. To reach the mountain, go to Erculaneo station and catch the bus waiting outside in the bustling suburb. The route climbs upwards through improbably narrow streets, and then out into fields planted with almonds and vines. The mountain produces a wine called Lachrymae Christi—the tears of Christ—as evocative a name as you could wish for an indifferent *rosso.* Now you can see close up the random assortment of houses that were visible only as flashes of light from the other side of the bay. They seem perilously close to the volcanic cone, and unplanned, as if dropped haphazardly from the sky. Higher still, there is dense scrub, and then, looming above you, the cone itself. The bus disgorges its passengers, who must then climb to the summit along a relentless upward trail. You are following in famous footsteps, for the poets Goethe and Shelley were here before you. Nor are you ever alone: the view provides an irresistible photo-opportunity for tourists of all nationalities. The conurbation of Naples stretches away below, as insubstantial as postage stamps pasted on the landscape.

The dominant colour is a warm brown-black. Nobody could pretend

Inside the cone of Vesuvius, showing the lavas produced by successive eruptions.

that a huge slope of clinker is aesthetically pleasing. The waste of the mountain has a curiously industrial feel, like slag from a smelter; it is easy to understand the classical notion of Vulcan's forges working away in the earth's interior. An excellent guidebook by Drs. Kilburn and McGuire of University College, London, reveals that these unpromising pieces of debris are scoria and lithic fragments of the March 1944 eruption. Looking down, you can make out the edges of these latest flows in Vesuvius' long history extending beyond the base of the cone itself, like chocolate sauce sliding off a steamed pudding. By the pathside, there are occasional large boulders that show black crystals the size of a fingernail; these are pyroxene minerals that had time to crystallize out deep within the chamber of liquid rock—or magma—beneath the volcano. Like everything else around, these rocks are igneous in origin: formed in fire, cooled from melt derived from the earth's depths. The crater itself is the kind of gaping hole that induces despair in those suffering from vertigo: it is 500 metres across and 300 metres deep. From the far side of the rim path you can just make out the amphitheatre at Pompeii, with the misty blue of the Bay of Naples beyond; it seems entirely plausible that destruction could be meted out at a distance by suitably violent explosions. The crater sides are sheer. Wisps of steam still arise from one side, like smoke from a poorly extinguished cigarette. Looking across the crater, you can readily see that it is made of piled layers of lava metres thick. You can also make out where flowing lava was replaced by explosive pyroclastic deposits, which are crumbly and easily eroded. Here is evidence for the origin of the kind of lethal clouds that wiped out Roman Herculaneum in the A.D. 79 eruption. The crater is not stable. Listen: there is a continuous tinkling, like ice in a glass, as pebbles fall from the sides into the interior. The volcano is temporarily quiescent, but there are many descriptions of it while awake. Here is a little of what the philosopher Bishop Berkeley had to say about it in 1717:

> I saw a vast aperture full of smoke, which hindered me from seeing its depth and figure. I heard within that horrid gulph certain extraordinary sounds, which seemed to proceed from the bowels of the mountain; a sort of murmuring, sighing, dashing sound; and between whiles, a noise like that of thunder or cannon, with a clattering like that of tiles falling from the tops of houses into the streets. Sometimes, as the wind changed, the smoke grew thinner, discovering a very ruddy flame and the circumference of the crater streaked with red and several shades of yellow . . .

Nowadays, handwritten labels creak in the wind. On the top of the cone you can make out a solar-powered Global Positioning System. So often in Italy there is this odd mixture of the improvised with the latest technology.

Vesuvius is an irritable mountain. It has been erupting for 25,000 years or so. It will do so again. It is a classic in geological literature because several of its eruptions have been accurately described over 2000 years. It also demonstrates many kinds of different volcanic activity, from sluggish flows that slowly and inexorably eat up landscape and houses, through ash-falls that suffocate, to pyroclastic flows and surges that run faster than a racing car and engulf the plains in heat and terror. It is a vulcanological case-book. After the A.D. 79 debacle, which killed 2000 people and buried Herculaneum and Pompeii, there was a great eruption that began in the morning of 16 December 1631. Within a day, ash from Naples had reached Istanbul, more than 1000 kilometres distant. This eruption killed almost twice as many people as had died at the time of the Roman Empire. The rich volcanic soil had once more encouraged the growth of a prosperous people, who had spread over vulnerable areas. Massive pyroclastic flows took their toll. So did the engulfing slurries of mud and ash that flow downhill when heavy rains blend with volcanic outpourings. Ironically, the particles thrown up by a major event serve to "seed" rainclouds in the atmosphere; thus in this tragedy the four ancient elements—earth, air, fire, and water—conspired towards a common deadly destruction. Naples itself, however, was spared. There were many subsequent minor eruptions, usually about eleven years apart, all of which avoided it. An eruption is now overdue, and might well be a big one. The latest, 1944, eruption was studied in detail: something like 37 million cubic metres of magma were expelled within a few days in March of that year. Much was released in the form of ash and "bombs," black masses often shaped like horns or twisted loaves. There were several eruptive phases separated by periods of quiescence, when the crater collapsed temporarily to block the throat of the vent. The almond blossoms everywhere shrivelled on the branch. The villages of Massa and San Sebastiano were destroyed by slow-moving lava flows. *The Times* (of London) correspondent eloquently described the process:

> The progress of destruction is almost maddeningly slow. There is nothing about it like the sudden wrath of devastation by bombing. The lava hit the first houses in San Sebastiano at about 2:30 a.m. but by dawn it still had not crossed the main street only 200 yards away, but was nosing its way through the vines and crushing down

> the small outhouses more slowly than a steam roller . . . For a while it seemed as if it would engulf the houses as they stood but then, as the weight grew, a crack would appear in the wall. As it slowly widened first one wall would fall out and then the whole house would collapse in a cloud of rubble over which the mass would gradually creep, swallowing up the debris with it . . . Masses of steam of slightly darker quality rose as cellars full of casks of wine exploded. Over all one heard a steady cracking as the monster consumed *hors d'oeuvre* of vine stalks, olive trees and piles of faggots stored in backyards . . .

If it had been a pyroclastic surge, the whole thing would have been over in a second.

A geologist amongst the crowds at Pompeii would probably be there to observe the effects of an ash-fall that destroyed a whole city as much as to wonder at the level of luxury enjoyed by the Roman elite. Nonetheless, the city's sheer size comes as a surprise. The discouragingly large crowds at the entrance soon disperse, and you are left to pick around at your leisure. Some of the roads allow a clear view of Vesuvius, and it is not so difficult to envisage the huge shower of hot ash that so effectively entombed this wealthy city for the archaeologists: the warning earthquakes, the great boom of an explosion, the dark cloud shooting into the atmosphere above the looming cone, the sky blotted out, wonderment turning instantly to horror as the fall began. The bodies of the inhabitants were moulded by the ashes that covered them. The most pathetic testimonies to the catastrophe are the casts that have been taken from these moulds, which are now stored at one end of a vineyard—in particular a figure half rising from sleep, hardly stirred from private oblivion before returning to it forever, thereby preserving something of the monumentality of a sculpture, but with a vulnerability that is wholly personal. Even the stones have their own story to tell. The streets were shod with pieces of tough basalt fitted together in a coarse jigsaw. There is something peculiarly eloquent about seeing the grooves that generations of cartwheels wore into these roads. They are imprints of time itself upon what is inevitably described in the tourist blurb as a "time capsule." The surfaces of the roads are lower than the sidewalks and the kerbs are marked by courses of large basalt blocks. The walls of the houses and shops and temples were often constructed of alternating courses of large blocks of ignimbrite and small, sienna-coloured Roman bricks. The corners were often entirely of brick, since the larger blocks chip easily. The architects knew their local geology. In the Villa di Misteri there was much use made of the Tufo Galliano Napoletano, often built into a lozenge pattern. Originally, the surfaces along the streets would have been rendered, as they are inside the villas. But since cement was originally discovered by the Romans from grinding up tuffs, so even the dressing of the buildings is geological. In the smarter buildings the walls were painted and ornamented with figures and curlicues. Mosaic floors consisted of patterned tesserae, the common black and white ones being made of marble and basalt. Discovery of pattern books has shown that these floors could be ordered "by the yard." There is a section of the hypocaust, or plumbing, on display that shows the use of lead pipes—who knows if some of that lead could not have come from as far away as Britain, where the geological circumstances are just right for lead ores? In short, Pompeii grew from the geology, just as it was eventually consumed by it.

Herculaneum was destroyed by pyroclastic flows. The inhabitants had time to attempt to flee to sea, but no time to succeed in their flight. For a small tip, a guide will let you see the skeletons of the huddled bodies down by the former harbour. The temperature must have been sufficient to kill instantly, but not to calcine the bones. Why should the outlines of the flesh at Pompeii be more affecting than these horrific accumulations of bones? It must be something to do with gesture, with individuality. All skulls smile alike.

Both Pompeii and Herculaneum have geological bones. The fire has done no more than expose them. This is not only true of famous archaeological sites. Whatever the surface decoration, geological truths determine much of the reality and character of cities; the stone that builds them, the height to which towers can rise. This is a subtler connection than that with agriculture. It is more obvious that the dark, rich soils around Somma Vesuvius are the product of the weathering of old flows, and that this explains why men and women have always repopulated the area after the latest disaster ("If San Gennaro [the patron saint of Naples] looks after us, *Deo gratias,* we will prosper"). The geology in turn is often related to tectonic plates: a deeper reality still, arbiter of the shape of the world.

It seemed appropriate to pay our respects to S. Gennaro. There is a chapel dedicated to him in the Duomo in Naples. In the third century the saint was martyred for his Christian beliefs. He was first sentenced to be torn to pieces by animals, but this was commuted to beheading. The decapitation took place at Solfatara north of Naples, where a shrine has been erected. Some of the saint's blood was scooped up after the execution. This is still preserved as a precious relic. It supposedly has the extraordinary property of turning *back* into blood when it is brought out on religious occasions three times a year. Thus is Naples' safety guaranteed, or so the Neapolitans believe. The Duomo is reached by walking through narrow streets where squalor and splendour rub shoulders. Dark alleyways draped with washing lines and oozing suspicious liquids are cheek by jowl with churches and courtyards of decayed magnificence. The silence in the Duomo is all the more impressive for the noise and chaos outside. The saint's chapel is an ebullient confection of columns and gilt, with masses of marble and serpentine, panels of onyx, busts and paintings. The saint is depicted above the right-hand altar emerging unscathed from a fiery furnace—sanctity rises above mere volcanics. The sixteenth-century crypt is an extravaganza of marble. The floor is constructed from marble lozenges and triangles of every imaginable hue: yellow and pink and all manner of mottled and blotched shades, framed in white. The walls have ranks of white marble niches capped by huge marble scallops, and flanked by urns and flowers, drapes and putti. The whole is supported by rounded marble columns—white, with gentle streaks of grey—while the ceiling sports square panels painted with the effigies of saints, angels, and bishops. A pearly marble statue of Cardinal Carata kneels on the floor facing towards a locked bay containing S. Gennaro's relic. It is like being inside a palace of spun sugar. More mundanely, the crypt serves as a reminder of the third great class of rocks besides those of sedimentary and igneous origin. These are metamorphic rocks. They may have started out as either of the other two categories, but have been altered—or even completely transformed—by heat, pressure, or by any combination of the two, when caught in the great mill of mountain-building within the depths of the earth. Marble is one kind of metamorphic rock and comes in a thousand shades, although it started out as humble limestone as plain as the cliffs of Capri. Along the Apennine "spine" of Italy there are many places where heat and pressure have altered limestones in this fashion. Renaissance architects made luscious use of these rocks. They can appear to resemble anything from blue cheese to slices of liver, or they can be pure white, like the Carrara marble favoured by great sculptors such as Michelangelo. Naples was just following the fashion. S. Gennaro's niche is clad in the products of the infernal heat to which he is now impervious.

Early visitors to the Roman antiquities of the Bay of Naples were following the tradition of the Grand Tour, a cultural perambulation around

Italy made by young gentlemen with pretensions to enlightenment. Eighteenth-century English aristocrats would have been as familiar with the classics—with the writings of Virgil and Horace—as they would have been with the salty, home-grown vigour of Chaucer, or the sonnets of Shakespeare. They would have known about the hot springs that supplied the baths in which Roman gentry lounged for the good of their health; they would have been familiar with the ambulatoria along which the pampered rich strolled with their favourite philosophers for the good of their minds. The archaeology of the area was the latest sensation. Sir William Hamilton is known now mostly for his marriage to Lord Nelson's mistress, but he was a considerable archaeologist and antiquarian and the purchaser of treasures that still grace the collections of the British Museum. He also appreciated the uniqueness of the Neapolitan landscape, and strove to make known its volcanic features. While he was British Envoy in Naples towards the end of the eighteenth century, he published a book about the Phlegraean Fields, *Campi Phlegraei,* which well described the area to which he was devoted. Perhaps it is this for which he should be remembered. Certainly, his work was known to the young Charles Lyell. Lyell had travelled to Italy in 1828 in the company of Sir Roderick Murchison, who was then the most influential British exponent of the young science of geology. If classic areas are made so by the observations of perceptive eyes, and measured by the hard currency of influence on subsequent thought, then there is a claim to be made for this modest Roman site in the middle of the town of Pozzuoli, west of Naples. But no plaque records it, no signs direct you to it. You simply come across it, as you might a standing-stone in England, or a roadside shrine in Italy.

The Phlegraean Fields! The very name sounds like some version of Arcadia (it is in fact derived from Greek for "burning"). I had seen Sir William Hamilton's portrayals of the scenery: trees giving on to a rustic scene, the odd peasant going about his business, in the distance some interesting volcanic phenomenon in progress. This was the area of Campania described by the second-century historian Florus as "the most beautiful region not only in Italy, but also in the world. Nothing is sweeter than its climate: to say everything, spring flowers twice there. Nothing is richer than its soil . . . nothing is more hospitable than the sea." Times do change, of course, and the region has been swamped by undistinguished suburbs. But the Campi Flegrei is still the same collection of craters and calderas, separated by volcanic hills as it was in Florus' time—only the flat bottoms of the former have often been filled in with army barracks or cheap industrial units. The reality of these evocatively named fields are sadly something of a disappointment.

Lake Averno was considered to be one of the entrances to the underworld in classical times. In those days, birds roosting around it were wont to fall dead from the trees. From here, Dante was led by Virgil into the *Purgatorio.* If geology is the true underworld, then Lake Averno seemed to provide a kind of metaphorical entry point to the deep regions that form the subject of this book. Can you imagine anywhere with richer associations? It is an almost perfectly circular crater lake, three-quarters surrounded by steep walls of tuff. When I visited in early spring it had a slightly sinister character: plumes of smoke arose, not from volcanic exhalations, but from burning heaps of rubbish left over from the last summer season. A few mean reeds grew from the lake bottom, and bubbles of gas arose and broke the surface—methane, I suppose, from the de-oxygenated mud. Perhaps, after all, this is a postmodern underworld, with existential gloom presiding.

There is a geological explanation for this lethal lake. After volcanic eruptions have subsided, invisible eruptions of gas may continue. Carbon dioxide is one of them. It is odourless, and heavy—much heavier than air—so it sits around invisibly, or flows into pockets and depressions. It induces suffocation. Until a century ago, a cave with these properties in another crater in the Campi Flegrei, Astroni, was shown to visitors as a tourist attraction. It was called Grotto Del Cane (cave of the dog), for reasons which will become apparent. The playwright and jobbing writer Oliver Goldsmith reported in 1774 on this grisly natural phenomenon in *An History of the Earth and Animated Nature:*

> This grotto, which has so much employed the attention of travellers, lies within four miles of Naples, and is situated near a large lake of clear and wholesome water. Nothing can exceed the beauty of the landscape which this lake affords; being surrounded by hills covered with forests of the most beautiful verdure, and the whole bearing a kind of amphitheatrical appearance. However, this region, beautiful as it appears, is almost entirely uninhabited; the few peasants that necessity compels to reside there, looking quite consumptive and ghastly, from the poisonous exhalations that arise from the earth. The famous grotto lies on the side of a hill, near which place a peasant resides, who keeps a number of dogs for the purpose of showing the experiment to the curious. These

> poor animals always seem perfectly sensible of the approach of a stranger, and endeavour to get out of the way. However, their attempts being perceived, they are taken and brought to the grotto; the noxious effects of which they have so frequently experienced. Upon entering this place, which is a little cave, or hole rather, dug into the hill, about eight feet high and twelve feet long, the observer can see no visible mark of its pestilential vapour; only to about a foot from the bottom, the wall seems to be tinged with a colour resembling that which is given by stagnant waters. When the dog, this poor philosophical martyr, as some have called him, is held above this mark, he does not seem to feel the smallest inconvenience; but when his head is thrust down lower, he struggles to get free for a little; but in the space of four or five minutes he seems to lose all sensation, and is taken out seemingly without life. Being plunged into the neighbouring lake he quickly recovers, and is permitted to run home, seemingly without the smallest injury.

It is the recovery of the dogs from their ordeal that identifies the gas as carbon dioxide. Goldsmith goes on to say that the vapour "is of the humid kind as it extinguishes a torch," an experiment still performed in school laboratories to this day. Had the gas been sulphur dioxide or hydrogen sulphide, the poor dogs would have suffered a rapid and painful death.

In 1750, a statue of the god Serapis was excavated from a Roman site at the coastal town of Pozzuoli. Serapis was based ultimately upon the Egyptian god, Osiris, but "hybridized" with Graeco-Roman gods to become the basis of a widespread cult in the Mediterranean at the time of the Emperor Hadrian. For more than a century and a half the impressive archaeological discoveries in Pozzuoli were known as the "Temple of Serapis," or the Serapion, under the belief that the statue indicated a sacred function. This is the name that appears in the caption to the illustration that forms the frontispiece of Charles Lyell's *Principles of Geology.* The image is also engraved on the Lyell Medal of the Geological Society of London, so it clearly mattered to his contemporaries. The current perception is that "the temple" is nothing of the sort, but was in truth a rather opulent market-place, or *macellum.* But it is still a geological shrine, an emblem for the coming of age of geology as a science. It is something of a disappointment that it had to be deconsecrated; after all, the "temple" is a holy place for rationalists, if that is not an oxymoron. At first sight, the Serapion is curiously unimpressive. The

The "Temple of Serapis" from the frontispiece of Lyell's Principles of Geology, *first edition (1830). Some later editions omit the philosophical figure on the left.*

old town of Pozzuoli lies on a hill, and has been out of bounds since the earthquake crisis of 1982. The site lies near the harbour beyond the hill. A few yards seawards from the ancient mart there is still a market, selling oranges and lemons and local fish. Then you see the old market-place lying beyond a well-shaded square. It lies about six metres below the level of the present piazza: it is like looking down into a sports field or a football pitch, a resemblance increased at night when the area is floodlit. The large rectangle is dominated by three huge columns, which a notice tells us are 12.5 metres high and composed of a single piece of marble. They must have been expensive items even at the time the market was built. In front of them there is a raised circle carrying a series of smaller columns that were probably reconstructed relatively recently, whereas the great columns were "in place." The circle is constructed of the familiar tuff, underlain by courses of Roman bricks. There are planted palm trees whose trunks supply a kind of biological equivalent to the marble columns. All around the edge of the market square are remains of the stalls of the traders.

The monument has none of Pompeii's breathtaking qualities, tucked away as it is so matter-of-factly in the middle of Pozzuoli. The casual observer might wonder at its importance. To discover the reason requires closer examination of the great columns: about four metres above their pedestals there are what look from a distance to be zones of blackish discoloration, where the marble appears roughened or eroded. The lower parts of the columns are, as Lyell said, "smooth and uninjured." Look more closely at the discoloured parts of the columns. You can now see that the marble has been perforated and pierced. Identical borings are caused by the marine clams that Lyell knew as *Lithodomus.* They can be found around the Bay of Naples today. The great Swedish biologist Linnaeus had originally named the destructive species *Lithophaga lithophaga,* a name that tells you all you need to know; for "lithophaga" means "rock-eater" in Greek, which is exactly what these clams do. They bore into rocks at the water-mark, eventually reducing the host limestone to a kind of framework. The implication should be clear, just as it was to Charles Lyell. The columns were—of course—constructed above the sea, but the whole market-place must then have been immersed beneath the sea sufficiently to drown the lower part of the columns. Not only that, the whole market-place must have been raised again for the visiting archaeologist to inspect the borings. It was possible to infer further details. The rock-eaters operate only in clear water, so it seems that the lower parts of the columns must have been buried beneath sediment in order to remain so clean. The borings total almost three metres of the columns; thus, the immersion, and subsequent elevation, must have been greater than six metres, and could well have been more. Lyell was tempted to associate the depression of the "temple" with an eruption at Solfatara nearby in 1198, and its subsequent elevation with the eruption of Europe's youngest mountain, the small volcano of Monte Nuovo, which appeared almost overnight on the outskirts of Pozzuoli in 1538. I went there: although so juvenile, it is already weathering away to rich soil.

Thus the interpretation of these past events was made—rationally—by reference to what could still be observed at the present time. The clams had not changed their habits. The world had gone up and down, or rather down and up, but this was not catastrophic: the street markets still plied their many trades, the same physics that had operated in past years operated again today. This statement seems almost banal now, but it provides the basis for all modern geological reasoning. There has been much puffy stuff written about whether or not Lyell's uniformitarianism permitted variations in intensity of causes, or whether he applied his logic in a consistent way, and whether he assumed indefinite stretches of geological time. And there is a good case for making the Edinburgh geologist James Hutton (1726–1797) the true father of the modern geological method. For most geologists, though, the important fact is that Lyell provided a clear way of thinking about the world. He substituted a rational system of investigation based upon present processes for a belief in a series of catastrophes, the latest of which was the biblical Flood, according to some. A marvellous cartoon by Henry de la Beche, first director of the Geological Survey of Great Britain and Ireland, shows "The Light of Science Dispelling the Darkness which Covered the World." The light in question that dispels the clouds covering the globe is carried by a female figure who also bears in her right hand a geological hammer, for this was the particular discipline that was seen as attacking reactionary thinking by main force. The Lyellian method still lies behind everything described in this book: look at volcanoes today to understand volcanoes in the past; do experiments in the laboratory to understand what goes on at depth in the earth; look at what plates do at the present time to understand how the world was made. Steven Jay Gould has pointed out that the "Temple of Serapis" was a much more appropriate symbol for the Lyellian method than temperamental Mount Vesuvius. Gould remarks in *Lyell's Pillars of Wisdom* (1986) that "Lyell presented the three pillars of Pozzuoli as a triumphant icon for both

"The Light of Science Dispelling the Darkness Which Covered the World." *Drawn by Henry de la Beche, first director of the Geological Survey of Great Britain. The cartoon portrays Lady Murchison in the fashion of the time. Notice that she brandishes a geological hammer in her right hand.*

key postulates of his uniformitarianism—the efficacy of modern causes, and the relative constancy of their magnitude through time." Scientists now tend to downplay the second part of this thesis, because they know that the earth has evolved and changed from its origins more than 4 billion years ago. Not every process has continued at exactly the same pace at all times—there have, on occasion, even been catastrophes; but the principle of reconstructing the past by using present-day analogues has remained at the centre of the process of turning geology from a pastime for dilettantes into a science. I could not have said what I have said about the ignimbrite of Sorrento or the limestones of Capri without Lyell's example. Put simply, Pozzuoli made a difference to science.

Charles Babbage should be mentioned at this point in the story. Born in 1792, Babbage was a brilliant mathematician, known today as the originator of computing and the inventor of the calculating machine. He, too, visited the Serapion in 1828, two years before the publication of the first part of Lyell's *Principles.* He presented a paper on his very detailed computations on the site at the Geological Society of London on 12 March 1834, and it is a model of Lyellian reasoning. It was not published until 1847, as a "postponed paper" in volume 3 of the *Quarterly Journal of the Geological Society of London.* The editor included the intriguing note that the paper "by the request of author was returned to him soon after it was read, and has been in his possession ever since." One wonders what sort of inhibition caused Babbage to delay nearly twenty years before printing observations that would have placed him at the start of uniformitarian reasoning. By 1847, the scientific *Zeitgeist* had already changed. However, you cannot but admire Babbage's summary:

> In reflecting . . . on the causes which produced the changes of level of the ground in the neighbourhood of Pozzuoli, I was led to consider whether they might not be extended to other instances, and whether there are not other natural causes, constantly exerting their influence, which, concurring with the known properties of matter, must necessarily produce alterations of sea and land, those elevations of continents and mountains, and those vast cycles of which geology gives such incontrovertible proof.

From Pozzuoli to the world! In this brief summary, Babbage encapsulated much of the research programme undertaken by modern geology.

This book is about explaining the character of the earth. The ultimate controls on the personality of the planet are tectonic plates—they *are* the "natural causes, constantly exerting their influence" of which Babbage wrote. So my story will inevitably lead to an outline of plate tectonics. But I shall get there through special and particular places—like the Bay of Naples—where geology and history are interwoven. I shall explore the influence of geology on the character of landscape and the character of people. Far from being the driest of the sciences, geology informs almost everything on our planet, and is rich with human entanglements. The rocks beneath us are like an unconscious mind beneath the face of the earth, determining its shifts in mood and physiognomy. The progress of our understanding is also inextricably linked with the virtues and failings of the investigators. Different eyes at different times looked at the same view across the Bay of Naples and saw different things; one perhaps saw the wrath of the gods, another relief of pressure from a magma chamber. Textbooks tend to fillet out the history and scenery in order to give the current scientific consensus, so this is a kind of anti-textbook. In truth, the same classical areas have seen a half-dozen different generations of scientists tapping the rocks or putting their instruments about. But I shall try my best to avoid what the historian James Secord terms the Whiggish view of history, by which you uncover the mistakes of your forebears in the progress towards present enlightenment. Understanding is always a journey, never a destination.

If you try to comprehend how thought has changed over a century or more, there are a few books that can act as useful benchmarks. Since Lyell, there has been a handful of authors who have comprehensively summarized the knowledge of their time—who have tried to write down the world on paper. One of those was the Austrian geologist Eduard Suess (1831–1914), whose *Das Antlitz der Erde,* published in four massive volumes between 1883 and 1904, is probably the most ambitious attempt ever to capture the entirety of everything: you feel that a fact foolish enough to escape Suess' grasp probably was not worth his notice. Suess sat in Vienna like an omnivorous spider with a web spread over the world, tugging in facts. *Das Antlitz der Erde* was translated into English, as *The Face of the Earth,* by Hertha Sollas, the wife of an eccentric Oxford Professor of Geology, and published in 1904–9. If you want a view of what geological knowledge was like at the end of the nineteenth century, when geologists using the scientific method had already scanned much of the world, then Suess is your best guide. He, too, wrote of the Bay of Naples and of the "Temple of Serapis," noting that on Capri there are *Lithophaga* borings as much as 200 metres above present sea level. He had insights that are still woven into the fabric of geological thought, and even where we now know he was mistaken it is instructive to try to understand why he thought what he did. Yesterday's cast-iron inference is often today's discarded argument. As Mott Greene has pointed out in *Geology in the Nineteenth Century,* Suess differed fundamentally from Lyell in believing that, far from operating in a kind of steady state, the history of the earth was punctuated by periods of pronounced change—revolutions during which mountain belts were thrown up, for example. We shall see that this tension between continuity and sudden change was resolved only when the plate tectonic view was established.

In 1944, forty years after the last volume of Suess' magnum opus was published, another widely influential work appeared: *Principles of Physical Geology* by Arthur Holmes. Known to generations of students simply as "Holmes," it was a tremendously successful textbook from the outset, for two reasons: it was well and clearly written, and it was illustrated throughout with photographs. Holmes was a professor at the Universities of Durham and Edinburgh, and in some ways a more radical figure than Suess (Cherry Lewis has recently written the first biography of him). Holmes did not try to write out the whole world in longhand like Suess, but sought to illustrate geological principles by selected examples, which could come from anywhere appropriate around the globe. It was a successful formula, and will do well as a record of mid-twentieth-century views of how geology makes the world what it is.

My own intention is much more modest, since I have no pretensions towards the omniscience of Suess, nor to the didactic skills of Holmes. The few places I describe in detail, many of which I have visited, show what geology *does*—to the landscape and its history, or the kinds of animals and plants that flourish there. The intention is selectively to illuminate rather than to be comprehensive.

Arthur Holmes is relevant for another reason, at the point we have reached in the journey around the Bay of Naples. He was a pioneer in methods of radiometric dating of rocks. This is vital to the narrative of the earth. Lyell may have provided the intellectual tools for understanding geological processes, but he did not provide a time scale. The span of geological time was a mystery, and contemporary estimates of the age of the earth varied wildly, from a few to many millions of years. There were even those who still adhered to the calculations of James Ussher (1581–1656), Archbishop of Armagh, which placed the date of Creation in the year 4004 B.C. However, most scientists agreed that the earth had to be old to accommodate all the rocks in their various sequences, all those Lyellian events—but how old? Leading twentieth-century physicists (notably Sir Harold Jeffreys) made estimates based upon good physics but false assumptions, such as the presumptive cooling rate of the earth. Such estimates invariably proved too young, but were vigorously defended for decades. The development of radioactive "clocks" has served to put an objective date on remote antiquity. The method is based on the natural rate of decay of the radioactive isotopes of certain elements, such as uranium, carbon, and potassium. These rates can be determined experimentally. For example, 235 Uranium decays very slowly over hundreds of millions of years to 207 Lead. Hence, if you measure the amount of decayed product, you have a measure of time elapsed: it's as simple as that, though many complexities arise along with accuracy of measurement. Holmes' working lifetime spanned an era of technical improvements—which continue today—and estimates of the age of the earth steadily increased and became more accurate as methods were refined, and older and older rocks were discovered in the field. After

Holmes' death, dating of moon rocks and solar system meteorites finally "nailed" the time of the origin of the earth at 4.5 billion years ago. These refinements of dating techniques allowed for calculation of the rates of change of what Babbage had called the "vast cycles of geology": the speed at which plates move; the time taken to erode a mountain range; the length of time of an eruptive phase; how long dinosaurs were in charge of the land. Because different radioactive elements decay at varying rates, techniques must be tailored to the problem at hand: thus, 14 Carbon decays very fast and can only be used to date comparatively recent events up to about 30,000 years ago. I have a scientific paper before me that uses radiocarbon dates extracted from the same kinds of clams that bored into the Serapion's columns to show that on the island of Ischia there has been seventy metres of uplift in 8500 years. So there are clocks for the world, completing the task that Lyell started.

It has become almost a cliché that we cannot grasp the immensity of geological time—what 5, 50, or 500 million years actually *means.* But around Pozzuoli the changes we observe have taken place over thousands, rather than millions, of years. Luciano Bagnoli has been compelled to move the mooring of his fishing boat nearly a metre lower since his father's time, such is the rate of uplift in Pozzuoli. You do not have to travel very far if you want to see evidence for the forces that are lifting the whole region from the sea. The Solfatara crater lies just on the edge of town. You smell it before you see it—a result of what Oliver Goldsmith would have called its noxious emanations and pestilential vapours. It is still hissing, furiously. Courses of angular boulders line the rim of the volcano, the remains of its last explosive phase, resulting in a volcanic *breccia.* Half of the crater floor is devoid of vegetation, and as active volcanically as you could wish. There are numerous fumaroles belching forth steam. Holmes calls these active vents *solfataras,* so evidently we are at the type locality for this kind of thing. The Bocca Grande is the largest of them all, a roaring vent that emits steam at a temperature of 160° C. It sounds like a perpetually boiling kettle of enormous size, or the gasping of some great steam engine. It has been carrying on like this for years and years. A vulcanologist called Friedlander built a hut almost adjacent to it, the better to monitor its exhalations: such is scientific obsession. The ancients called the area around the Bocca Grande the "Forum Vulcani," and you do feel that the earth is manufacturing or forging something deep beneath your feet. What actually comes out are sulphides. The stink in the air is hydrogen sulphide—the smell of rotten eggs. There are smears of reddish-yellow colour around the lips of the vents. These are condensed patches of rare minerals—compounds of arsenic and mercury with sulphur—with names like realgar, orpiment, and cinnabar that would have been familiar to alchemists five centuries ago. This is indeed where the alchemy of the earth distils precious elements upon the ground. Most of them are also very toxic—they were the preferred materials of nineteenth-century poisoners. In 1700, a high tower was built to collect condensate from the steam, and particularly for alum, which is used as a mordant in the dyeing industry. It no longer stands. This steamy natural cauldron is a place where chemistry happens in the open air. Around some of the hot holes thrive heat-loving bacteria that have probably endured on earth for as much as 4 billion years.

Elsewhere, the ground is pitted and fuming, as if a bomb had recently been dropped. In the centre of the crater, a mud volcano is still spitting. There is also the famous *stufe*—two ancient brick-lined grottoes excavated into the side of the mountain. They were once used as sweating rooms (*sudatoria*): since classical times this has been one of the alleged health benefits of the Campi Phlegrei. One room was known as Purgatory (60° C) and the other as Hell (90° C). I sat in the steam inside the entrance to Purgatory for a few seconds, but had to crouch because it was just too hot near the ceiling. Sulphur and alum were crystallizing on the brickwork. Within a few seconds the swelter became insufferable—in a word, purgatory. All these steamy phenomena are the result of a magma chamber hidden beneath the Pozzuoli area. Groundwater percolating downwards becomes superheated, and then blasts upwards, carrying with it mineral exhalations from the depths: the volcano is literally "letting off steam."

Current estimates are that the magma chamber is two kilometres below the surface, and rising. Some vulcanologists are deeply worried that the Campi Flegrei—rather than Vesuvius, where the magma is at about five kilometres deep—may be about to explode and that the Monte Nuovo eruption of 1538 might have been no more than a preamble. If so, the prospect is terrifying. If there were an eruption rapidly followed by a pyroclastic flow, the effects would be overwhelming. The chaos of the roads is bad enough on a fine spring afternoon, and speedy evacuation of thousands of people would be an impossibility. Hidden forces may yet take vengeance on lax town planners and profiteering builders in a drastic and unpredictable way. It is conceivable that the "Temple of Serapis" could be interred beneath a pyroclastic tuff, and once more consigned to archaeological oblivion. There were earthquakes in 1982–84 that were sufficiently worrying to prompt the evacuation and abandonment of part of the old town of Pozzuoli. A seismic hazard map published by the Osservatorio Vesuviano shows an area with "maximum observed values of seismic energy released and intensity" centred on Pozzuoli, and concentric zones of destruction all around. The magmatic masses are moving at shallow depth in the western part of the Gulf of Pozzuoli. The successor to the Neapolitan Yellow Tuff may yet be brewing unseen. You would never guess it from the insouciant joshing of the customers in the bars near the litter-strewn beach in Pozzuoli, even though the evidence of past volcanic fury lies in the rocks exposed all along the shore.

The problem lies in predicting when disaster might happen. It could be many years away, but it might be much sooner. This is the paradoxical nature of risk. Earthquake shocks of a particular kind should provide a signal; but you cannot live every day expecting it to be your last. By now it will be scarcely surprising to learn that attempts at scientific prediction of eruptions also had their origin around the Bay of Naples. The observatory was founded on Vesuvius in 1841, making it the oldest of its kind in the world. The scientific office has now moved to Erculaneo, where electronic monitoring can be collated at a distance, but the older, more elegant building remains on the flanks of the volcano and houses a museum, including some of the original instruments. The very first seismograph was made by Ascanio Filomarino—and there it is. It is a wonderfully simple contraption, comprising an iron pendulum 2.6 metres long with a suspended weight of 2.5 kilograms, to which is attached a pencil that draws on a scroll of paper below. When the world shook, it moved. Charmingly, there are some little bells attached to the weight, presumably so that the natural philosopher might be awakened from his slumbers if a tremor happened along. But the virtuoso had a very clear idea of what he expected of his instrument. Writing about his invention in 1797, Filomarino observed that "in the towns and villages near the volcano, use can be made of it together with an atmospheric electrometer, and observing them together with the several signs of the volcano one can sometimes, if not foresee clearly some new eruption, at least conjecture it." No modern vulcanologist could have put it better. For 200 years, the business has been trying to predict when a volcano will "blow." But even with the most sensitive seismometers, and with the help of computers of great power, it is still possible to be caught unawares by Vulcan in retributive mood.

Ascanio Filomarino would have been delighted by the measurements that are now routinely taken around the Campi Flegrei. At the edge of the Solfatara crater, I spotted another one of the solar-powered instruments that were on Vesuvius. Thanks to refinements in the Global Positioning Satellites, the smallest heaves in the surface of the ground can now be measured. These systems are now so sensitive that it is possible to detect the depression of the surface of the earth produced by heavy snowfalls, as has been demonstrated by experiments in Japan. Far from being stable, the earth seems to pulse irregularly in ways we could never have imagined. For the Bay of Naples, the Argo Project is a network for Italian geochemical and seismic observations: the communications company Telespazio has set up a network to control seismic risk for the Ministry of Civil Defence, enabling seismic and geochemical data to be transmitted to a stationary satellite. Around Solfatara there is a main outpost and two secondary outposts. The latter are powered by solar panels and have electronic sensors to measure any seismic activity, as well as changes in gaseous composition around the crater, that might indicate a change in the magma chamber beneath. The data are transmitted to the main outpost, and thence by way of a parabolic antenna to the stationary satellite. This information is then available to research stations around the world. In theory, it should be possible to recognize the potential for an eruption far sooner than has ever been feasible in the past. A sudden acceleration in the "inflation" of the ground, changes in gas emitted from fumaroles, or a juddery earthquake of the harmonic type—all might indicate an impending crisis. The most difficult problem will be making the local population take the warning seriously.

The story of geology has been one of letting go of permanence: from a world created just as it should be by God, we now have a world in flux. Volcanic eruptions may have once been considered a punishment for the sins of mankind, but they were not necessarily the symptom of a mutable world. If the "Temple of Serapis" demonstrated up and down movements beyond question, then it was only a beginning. It was evident in Pozzuoli that the land, rather than the sea, had changed level. But as the geological record was investigated elsewhere, it soon became clear that there were other times in other places when the sea was high relative to the land—as when it spilled over onto continents to leave a record of sedimentary rocks deep in their interiors. During the Cretaceous, when the white cliffs of Capri were laid down beneath the sea, there were similar sediments being laid down far and wide across the world—most familiar to the English in the white cliffs of Dover. Suess had recognized these transgressions of the sea. Fossils of a single species could be found in many localities, which enabled timelines to be cast across continents. The distributions of land and sea as they are today are just in passing, moments in the slow march of geological change. Former sediments could be raised high, to provide places where the silky-white villas of the Caprese rich enjoy a temporary vantage point, but destined to be reduced again to sea level by the slow, inexorable mills of erosion.

How much more extraordinary still to contemplate the earth *wholly* in motion. Nothing seems to be at rest. The surface of the earth dilates and collapses; the seas rise and fall; further, the very continents move. Suetonius would have dismissed me as a madman had I pointed out to him that Vesuvius is a consequence, ultimately, of Africa moving bodily northwards. But Africa really is on the move. The floor of the Mediterranean Sea is a collage of tectonic plates. Ultimately, that sea is doomed to obliteration when the main body of Africa ploughs into the European mainland in 30 million years or so. But now the sea exists in an awkward accommodation—as the saying has it, "between a rock and a hard place."

The northern edge of the African Plate is plunging downwards beneath the southern tip of Italy, which allows the infinitely slow oceanic contraction to continue. Italy is twisted and swivelled in the process; the energy released deforms rocks to marble and uplifts the limestone hills. A whole series of Italian volcanoes are engaged in this congress of plates: Stromboli, with its continuous, grumbling eruptions; Etna, on Sicily, which has erupted recently and dramatically; Vulcano, the home of Vulcan himself, and the eponymous epitome of a volcano; and more than twenty others with less familiar names spanning much of the length of Italy and scattered through the Aeolian waters. The story of how these volcanoes relate to the movement of tectonic plates is a complicated one, and geologists do not agree on the details. However, all specialists seem to concur that the region behind the African collision zone has rotated anticlockwise. Rending of the earth's crust produced the Tyrrhenian basin here 8 million years ago. The volcanoes were related to fracture zones that accommodated this convulsion in the skin of the earth; they are like sticky blood oozing from deep wounds where the earth's crust thinned. The jostling of the plates results in fractures in the crust—faults—movement along which caused earthquakes and tremors. The uplift and tilting of strata are the testimony to this activity. Ultimately, the energy needed to produce lava derives from the crunching drive of Africa northwards into Europe. As crust plunged downwards, it partly melted, creating the magma which ultimately fed volcanoes. Lavas rich in the element potassium still retain a signature that proves that a proportion of melted continental crust must have been mixed up with oceanic crust. This magma recipe also conditions the fluid properties of the lava during eruption: at some times it flows in an almost docile, if relentless, fashion; at others, mixed with gas and water, it explodes in a series of devastating pyroclastic flows. Everything around the Bay of Naples is controlled by the movements and interactions of tectonic plates kilometres below the surface: the seen is under the control of the unseen; the surface is at the behest of the underworld. Exploring such connections is the theme of this book.

Even the most apparently ephemeral of earth's memorials are a reflection of a deeper reality. What could be more superficial than man's concern with his appearance: with cosmetics—the paint on the surface—that attempt to plaster over his own mortality? What could be more trivial by comparison with geological reality? The town of Baia on the western side of the Bay of Naples was born out of hedonism and beauty treatments. The hot springs there fed Roman baths of unrivalled luxury. There are still ghosts of paintings on the ceilings of the rooms where the rich went to cool off (or heat up) and giggle at the latest gossip, or discuss the hottest scandal of the first century. What modern-day Italians call the *bella figura* was just as important nearly 2000 years ago. The baths were fed from springs that were plumbed down into a tectonic reality that cared nothing for the carnal scandals of a two-legged ape that had only moved out of Africa a few thousand years ago. These springs were the exhalations of the magmatic unconscious. Yet Baia and the Pozzuoli coast invented the concept of the holiday: time out in luxury, to tickle the palate or the fancy. According to contemporary sources, the slopes of the bay were lined with marble palaces, and walkways furnished with columns, and with cool fountains for reflection and dalliances. Eduard Suess was disapproving: "It was in the flowery fields of Puteoli, Baiae, and Misenum that under the Empire at the expense of a subjugated world the most sumptuous festivals were celebrated." This area was like Nice on the French Riviera in the days of the English aristocracy, or the Big Sur coast of California in the days of the Hollywood glamocracy—except that it was on a scale altogether more grandiose. You can still get some impression of the vanished splendour and luxury, for Baia remains, although the plaster has mostly fallen off the walls and the paintings have often faded. The hillside where the stylish Romans strolled after their bath is still there, but it now has the atmosphere of a terraced provincial park gone to seed. The view across the Bay of Naples remains splendid, and with a little effort it is possible to imagine a continuous stretch of villas lining the bay. Close up, however, all is decay. Even the diamond pattern of the yellow tuff bricks, the *opus reticulatum* that built the walls, shows mortar standing proud of the weathered bricks it supports, for the rocks produced by a violent eruption do not necessarily make strong building stones. The earth movements in this area have submerged the ancient waterfront, so that it now lies more than 400 metres from the present shoreline. Part of the drowned city is still visible in the clear water.

Plinian-style eruption of Vesuvius in 1822, drawn by George Scrope (1797–1876).

The thermal springs no longer bubble up in the ancient baths, since the tectonic wellsprings have moved on. You can still have a thermal bath on the island of Ischia, which lies off the north-western edge of the Bay of Naples and where the springs are alive and pumping. A doctor there will advise you on which particular elixir you need for your aches and pains; but fashion and geological plumbing have deserted Baia.

The western end of the Bay of Naples is the Capo Miseno. It is now a deserted spot, although it was once a great naval base for the Empire, and housed perhaps 10,000 sailors. Only scattered foundations and remains of columns attest to its former greatness. From here, you can look back across the Bay of Naples towards Vesuvius, with the Sorrento coast beyond, where this chapter began. It is a good place to reflect on historical and geological time, and the forces that have shaped everything in prospect: the bay itself, the craters around the Campi Flegrei, even the distant hills. This was the spot where geology could be said to have begun when Pliny the Younger, as a youth of eighteen in A.D. 79, observed from a safe distance the eruption of Vesuvius that obliterated Pompeii. His uncle, the elder Pliny, died in the eruption. The latter's *Natural History* was a cornerstone of natural philosophy for more than 1000 years, so it would not be too much of an exaggeration to say that the future of all science was engaged with the magma chamber of one temperamental mountain. The younger Pliny described the cloud that proved his uncle's nemesis as resembling one of the umbrella pines, or Italian stone pines, that still provide welcome shade around the villas dotted about the bay. The volcanic cloud

> shot up to a great height in the form of a very tall trunk, which spread itself out at the top into a manner of branches; occasioned either by a sudden gust of air that impelled it, the force of which decreased as it advanced upwards, or the cloud itself being pressed back again by its own weight, expanded in the manner described; it appeared sometimes bright and sometimes dark, according to whether it was more or less impregnated with earth and cinders.

This type of volcanic event is still known as a Plinian eruption. The ash column can rise to fifty kilometres into the atmosphere, and then the ash can be spread for thousands of kilometres by winds. An eruption can alter weather patterns for months. The force behind the expulsion of the column is the result of gases expanding within the magma as it rises, as the overlying pressure is reduced. A kilometre below the surface the rising mass loses all coherence. The volcano violently boils over.

As for the pyroclastic flow of A.D. 79, it, too, reached Pliny the Younger. It travelled across the Bay of Naples, shedding its lethal load and losing strength. By the time it reached the young man it was little more than a miasma. It dropped the merest layer of fine ash over his sandals.

2

Island

Hawai'i is evidently paradise. There are Paradise Restaurants, Paradise Realtors, Paradise Tours, Paradise Apartments. The *Hawaii Star Bulletin* modestly describes itself as "The Pulse of Paradise," thereby grafting paradise onto the qualities of a living being. There is a lot of business hanging on Hawai'i's paradisiac qualities.

When you first approach Waikiki, the resort town on the developed island of O'ahu, paradise is not the first comparison that comes to mind. One traffic jam is much like another, and vast hotels vaguely themed on ersatz Polynesian mementos are no vision of life in the Garden of Eden. To be sure, the lawns in front of the hotels are immaculately trimmed, palm trees lean out to greet you, there are even waterfalls tastefully constructed about the entrances, and there is a smile on the lips of every flunkey. This is paradise designed for the determined shopper: Armani and Gucci and Tommy Hilfiger outlets line up along the sidewalk (doubtless, too, in the Paradise Mall) alongside a hundred other familiar names to nudge the relaxed holidaymaker with bargains that will seduce them from their cash. The goods are mostly the same as can be bought in London or Rome or Tokyo—a kind of global cornucopia, deceptive in its promise of local idiosyncrasy. Most of the Aloha shirts are made in China. Perhaps the visitor should turn elsewhere to discover paradise.

Not far behind Honolulu, mountains rise surprisingly quickly, and they are clothed in forest. There is an easy walk through it to the Manoa Falls north of Waikiki. The little path dodges this way and that through the trees, ever upwards, often muddy. This is more like the paradise of the movies: vast trees soar away above you, and their huge trunks are decked with climbing plants, some of them with enormous leaves, or heart-shaped foliage blotched with yellow patches. These vines seem vaguely familiar from florist's shops, though here they grow on a giant scale. There is an intense and humid odour of prolific growth. You can almost hear the shoots squeezing upwards towards the sun. There are pale yellow spikes of fragrant ginger flowers on either side of the path. Within a few minutes, you are dripping with perspiration, for no breeze disturbs this extravaganza of unfettered growth. Piping noises echo somewhere in the canopy way above. Half-seen birds such as these are obligatory in Eden. When the waterfall is reached at last, some of your fellow-walkers cannot resist the urge to wade out among the rounded boulders to catch a fickle spray that has plunged 200 feet down a lava flow. Bathing in a pool in the midst of a primeval jungle is the dream of paradise on the holiday brochure.

But this, too, is a bogus paradise. Almost none of the plants that climb up the massive trees along the path are native to O'ahu or the Hawaiian Islands. Indeed, neither are the trees themselves. They are interlopers, brought to this remote place by humans. These plants settled in the tropics and thrived, displacing much of the native vegetation. The resemblance of the climbers decking the trees to pot plants is no coincidence: some of them are the same species that can be bought in a supermarket in Norfolk, England, or Plains, Iowa—as commonplace in their way as tomato ketchup. Even the sweet-smelling ginger plant that looks so at home by the pathside is a johnny-come-lately, an aggressive colonizer. This place is not so much Paradise Lost as Paradise Replaced—a paradise of aliens dressed up to look as if they belong. The massive assurance of the trees is play-acting.

The Hawaiian Islands were once pristine. As remote from any major continent as anywhere in the world, they were originally colonized by a mere handful of hardy, far-travelled plants and animals—birds, insects, reptiles—that subsequently evolved prolifically in their new Eden. The species that appeared as a consequence were endemics; that is, they were found nowhere else in the world. The islands became a natural evolutionary laboratory. There were several hundred fruit flies found nowhere else, and an array of flightless birds pecked among the underscrub. This is where a caterpillar learned to be a meat eater. The landsnail *Achatinella* evolved a thousand or more species. The familiar plant *Lobelia,* blue-flowered mainstay of a thousand hanging baskets, evolved fantastical tubular flowers that could only be pollinated by even more fantastical birds—scarlet or yellow honeycreepers with bills as narrowly curved as precision surgical

Honolulu Harbour as it was in 1854. From Manley Hopkins, Hawaii: The Past, Present, and Future of Its Island-Kingdom *(1862).*

tweezers. The Bishop Museum displays glorious, ceremonial capes of Hawaiian chiefs fabricated from the feathers of countless honeycreepers. They must once have been abundant. These wonderful birds are now, alas, mostly displaced by interlopers—mynah birds and doves—with more ordinary dietary requirements. They can feed on the crumbs and scraps left by tourists where the honeycreepers dined solely on nectar. Even the flight-

less birds had their part to play in the original ecosystem: they ate the fruits of the aboriginal trees, whose seeds could only germinate once they had passed through the avian digestive system. But then, the introduction of the mongoose saw to the extermination of the flightless birds.

So why introduce the mongoose? These aggressive little predators were brought in deliberately from India to catch rats, themselves another introduction. The rats escaped from ships, doubtless, as they have all over the world. They threatened the sugar cane crop—which, of course, was yet another introduction, and one which transformed the Polynesian mixed economy of taro cultivation supplemented by fruit and fish into the specialized cosseting of a cash crop. But then, the Polynesians themselves had already introduced the breadfruit, and several useful species of palms. You begin to wonder at what point paradise first became corrupted. Nowadays, mongooses skitter across the road in front of you when you least expect it, cheekily self-confident. They must have made short work of endemic flightless rails and other birds. But the rats are still there. Mongooses are mostly diurnal, whereas rats are nocturnal. The extermination policy that introduced them was as pointless as proposing to introduce bats in order to eliminate wasps. Both mongooses and rats prospered while the birds disappeared. The few trees that remained from archaic times have no method of reproduction. They have outlived their propagators and have become sad remnants enduring from a vanished paradise. They are not entirely doomed because attempts in the Lyon Arboretum to regenerate them from culturing slivers of their living tissue have met with some success. The Arboretum where this work goes on is where you might get a hint of what might have been in the original islands—but carefully cultivated, a kind of Garden of Eden reconstituted. You begin to wonder whether you can go anywhere to see the real thing.

Hawai'i is actually a group of islands, all of them volcanic. The island of O'ahu is the one that people tend to imagine represents the typical Hawai'i: this is where the hotels and surfing beaches of Waikiki promise counterfeit paradise. Most of the islands, however, are completely different from this urban fantasy. The volcanic activity on O'ahu is quiescent. The other main islands are: Molokai, Kaua'i, Ni'ihau, Maui, and the youngest, and by far the largest, known everywhere as the Big Island—strictly speaking, this island *is* Hawai'i, and it is still actively erupting. This is where you can go to see the world being made.

On the rainy side of the Big Island, on one flank of the Volcanoes National Park, there is a remnant of the original forest. This area is several thousand metres above sea level, and the rain is often heavy. In the mornings a drenching mist can smother the landscape, only to be fried off by the sun before midday. The dominant canopy plant is the Ohi'a, a greyish tree of medium height, often gnarled and dressed with lichens, and carrying small, oval, dark green leaves that shine attractively in a certain light. Early in the day there is a silvery brilliance in the Ohi'a forest that cannot be matched anywhere else in the world. When in flower the Ohi'a is called *Ohi'a lehua;* the flowers look like old-fashioned shaving brushes made of dense red stamens. Several of the remaining honeycreepers earn their living from sipping the nectar of their blossoms, and there always *do* seem to be flowers somewhere in the Ohi'a forest. Around Volcano Village the understory is composed prominently of tree ferns. The commonest species is the hapu'u, *Cibotium,* a majestic and graceful fern with all the elegance of a green fountain. There are so many of them it looks almost as if the earth has sprung a series of verdant, spouting leaks. Other herbs and small trees clamber among the ferns, bewildering botanists who try to place them with their more conventional relatives. In this forest there is at least a chance that the piping you hear is from a native bird. The chirring is almost certainly the native cicada. But along the pathsides there are already plantains that you recall as weeds from Europe, the scourge of every lawn; and the seductive ginger plant pokes out from the undergrowth. There is trouble already in this paradise.

This jungle is simply structured, comprising only a canopy and understory. The Ohi'a trees are not like the dominating and diverse monsters of the Amazonian rain forest, with layer upon layer of tiering beneath them. Yet within these grey woodlands there are hundreds of species unique to the Hawaiian Islands. Then we must add the fact that each island has a windward and a leeward side. Most of the rain falls prolifically on the windward side; the other side is in a "rain shadow": clouds throw down what moisture they carry as they climb the mountains, but the further side remains dry. The animals and plants are different again on the dry flanks, tougher in several ways, survivors in the face of hardship. They comprise an entirely distinct community from that of the rain forest. Growing on nearly bare lava flows, the flora includes small relatives of the familiar cranberry: the ohelo carries a bright red, succulent berry that was once a staple item in the diet of the *nene,* the Hawaiian goose. The *nene* is a handsome bird that almost became extinct in the wild but was reintroduced successfully from ones bred in captivity. Ohelo berries are pleasantly tart, like watery cranberries, but scarcely the forbidden fruit of paradise. Ritual dictates that you must offer a berry or two to Pele, the fierce goddess of fire, before you eat them yourself. Many of the drought-tolerant species like the ohelo are rarities now, too.

Because the volcanic mountains rise to great heights there are some unique species adapted to alpine conditions. The most extraordinary of these is the silversword, a huge bundle of silvery, blade-like leaves looking something like a crouching, bristling porcupine. It lives high on the bare volcanic slopes at an altitude where little else will grow. It flowers and then dies. The flowering spike arises like some huge, columnar firework from the centre of the leaf rosette. Close examination reveals that each flower head is little more than a daisy—there are hundreds of these humble purple flowers on each spike. It takes the silversword twenty years to screw up the energy for its final floral pyrotechnic. This is what Hawaiian isolation has done to one of our most familiar flowers.

The remarkable forests with their remarkable animals, the rain shadow plants, the *nene,* and the hundreds of inconspicuous fruit flies all owe their existence to the geology of the Hawaiian Islands. This is a place where the geology announces itself in everything you see. Only in Honolulu and Waikiki do the works of mankind obscure the works of nature: save for the star-fruit trees in the backyards, the neat suburban roads lined with well-tended villas surrounding these cities might be in almost any small town in Iowa or Michigan. North American domestic vernacular architecture has become the lingua franca of middle-class habitation. On the remoter shores you may see the open canopy dwellings of the Polynesians (there are very few purebred Hawaiians). On the windward shore of O'ahu these houses are often backed by steep cliffs of dark volcanic lavas. Scratch around almost anywhere and there will be similar volcanic rock beneath the dirt. The geology dictates the soil. Along the trail to Manoa Falls this rock weathers to something with the colour and consistency of gingerbread. It can be ground to crumbs in your hand. Constant tropical rain makes a mush of hard old lavas. The end product is a brick-red soil called laterite. Its colour comes from all the iron (in the oxidized, or ferric, state) it contains, its other important element being aluminium. Laterite soil is not suited to growing many crops, but sugar likes it. The great cane fields of Hawai'i used to occupy much of the lower ground, at least from the nineteenth century onwards. Feral canes are still everywhere: to a European they look like hugely overgrown leeks. The bottom fell out of the market during the Reagan era, and now the great mills stand idle, and some of the fields are reverting to scrub. Others have been taken over by pineapple plantations. In the saddle or valley between the huge extinct volcanoes on O'ahu, the Dole plantations stretch for mile after mile along Route 99. The pineapple was once considered so exotic that it became a symbol of wealth to the aristocracy who alone could afford it; pineapples sculpted in limestone often adorn the gates of stately homes in England. It comes as something of a disappointment to see huge pineapple fields so relentlessly uniform in ranks of green rosettes. You expect to see trees as exotic as the fruit itself.

Isolation created the opportunities for the species to evolve that made the original paradise. The islands all grew as volcanoes from the sea floor over the last 5 million years or so; the Big Island is still growing. Twenty miles off the south-east coast of this island there is an active volcano still hidden beneath the sea. The Lo'ihi sea mount is steadily erupting, and one day it, too, will break above the waves, and a new land will be born. Within a year or two the first pioneer plant species will arrive to colonize the bare lava, their seeds perhaps being carried on the feet of a passing booby. The process of evolution will happen all over again. This insular birth is supposed to occur in fifty thousand years or so, a geological tomorrow.

The Hawaiian Islands provide an opportunity to see the birth, maturity, and death of a landscape. Creation is here; so, too, is massive erosion: and it has all happened in a timescale of just a few million years, a span that we can begin to understand. The Polynesian peoples arrived by boat for the first time about A.D. 500, but a second wave of immigration from Tahiti was more or less contemporaneous with the Norman Conquest of England in 1066. The evolution of the web of interrelated island-endemic species took thousands of times longer—time enough for some species of birds to lose the capacity for flight and for others to develop long curved bills. Think of what has happened in the thousand years since the Norman Conquest, all the advances and reverses and complexities of history, the technological revolutions, the rise and fall of empires. It then seems less astonishing that a complete and distinct biology should arise in a span of more than fifty thousandfold again.

The Big Island is the most southerly island of the group, and near its southern side you can witness the genesis of new crust. Lava has erupted from Kilauea for hundreds of years. Occasionally it blasts out in major eruptions, hurling volcanic bombs hundreds of feet into the air. Mostly, however, the eruption is comparatively benign, by volcanic standards. The kind of basaltic lava that has built Hawai'i is highly fluid, and charged with gas. It flows rather than explodes. A new lava flow cannot be entirely stopped, but it can be outrun if your house is about to be engulfed. The flow of 1992–97 blocks the coastal road that used to run southwards from the town of Hilo. It simply truncates the paved surface at right angles, abruptly. From one side it looks as if a gigantic load of tar had been dumped by a disgruntled contractor: the road disappears beneath the flow, and sure enough, if you walk to the far side of the flow a few miles along the

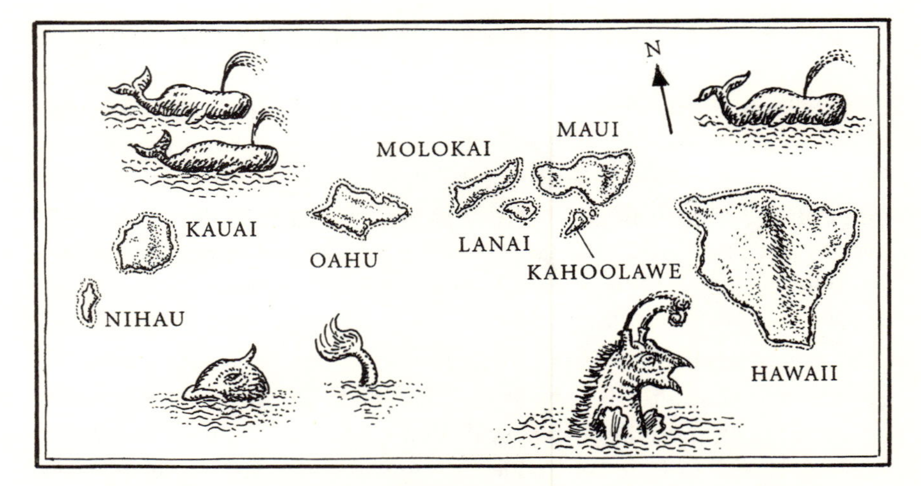

Sketch map of the Hawaiian chain, the Big Island to the right, and older islands progressively to the left.

coast, the road emerges on the other side, apparently unscathed. But the surface of the flow is hard to walk over, being a mass of black and twisted lava. Sometimes it is only too clear that it has cooled from a sticky liquid. Braided curtains or wrinkled sheets of shiny, tarry lava heave and plunge under your feet. The texture of the surface is often described as ropy, and it is an apt description to anyone who has seen tangled masses of ropes on the side of a harbour. On steeper slopes, twisted clots of entrails come to mind. It is astonishing to stand in the midst of a huge lava field with the undulating masses of volcanic material stretching away as far as you can see, for this is a fresh-made world.

You are permitted to inspect the current eruption from a safe distance. Parts of the old road have recently been revived, and you can drive very carefully and very slowly over graded parts of recent flows to get near to where the lava still debouches into the sea. Steam plumes, visible from miles away, rise continuously at the meeting between the new crust and the eternal waves, and there is a fierce hiss at the boiling junction. The stench of sulphur is everywhere, sharp in the back of your throat. If this is the breath of the deep earth, then it is poison. You must watch where you step over the recently cooled lava flows, for there are areas where the surfaces have heaved into cracked slabs, as if a vast tarmacadam airfield had been bombed almost to oblivion. It is easy to twist an ankle. On humid days water vapour combines with the sulphurous emanations to produce an acrid mist known as vog. As dusk falls, the brightness of the living lava becomes more evident. The warm red glow at the sea's edge has a brilliant orange heart. Then you witness an explosion where the combination of violent boiling, sudden cooling, and expulsion of gas result in luminous chunks—one might be the size of a pumpkin—rolling up into the miasma of steam; fireworks almost, but more inexorable. For the moment the lava wins against the ocean, but one day the balance will be reversed.

Now it is dark you can see other brilliant lights along the flanks of the landward slope. They form a line. It is easy to deduce that this line marks the current course of the flow. It is not at the surface, but flows in a conduit, a lava tube, a short distance under the inky wasteland. You can trace the lights to the vent itself, at Pu'u O'o on the flanks of the older Kilauea caldera. After the initial eruption from the new vent, the crust on top of the lava flow solidified, but the flow continued unabated beneath the skin covering over the tube. It is like an artery continuously bleeding the red blood of the earth. In places, the roof of the tube has crumbled away, and the lights on the hillside trace these windows giving into the streaming inferno beneath, glowing bright orange. There is a lucrative local trade in lava photography, and many Hawaiian photographers like to crawl close to the windows and record the streams of magma and the fountains of fire that splatter up from time to time. But there are dangers. Where the lava hits the sea it often builds out into an unstable platform. Only a few years ago two over-confident visitors were killed when the platform broke away into the violent and boiling sea. The process of creation is not merely dramatic, it can be lethal.

In the Volcanoes National Park, you can walk inside a lava tube, which has long since voided its last molten load. The Thurston Lava Tube now lies in the rain forest, and tree ferns decorate its entrance. It had a curiously familiar feel to me, and it was some minutes before I realized that this was because it had the exact dimensions of that other Tube—the Piccadilly Line tunnel in London—on which I travel virtually every day. It is indeed almost perfectly tubular. The walls are curiously smooth. There is nothing here to produce the stalagmites that adorn caves in Kentucky or Cheddar. Part of the tube has been illuminated, but you are permitted to go on into the darkness, at your own risk. It is a strange sensation to creep by flashlight into the guts of a former eruption, and just for a moment you might wonder if that rushing sound might be something other than the wind in the Ohi'a trees. But the dimensions of the tube do give you a very good idea of how much lava can be transported in an eruptive phase—by measuring the velocity of a flow, it is easy to calculate the volume of an eruption. On the present eruption at Pu'u O'o, flow velocity is measured using radar through the "skylights": since 1983, 300,000 cubic metres per day have been flowing down towards the sea. Creation can be calibrated.

The eruptions happening now are as nothing compared with those of the recent past. The Kilauea caldera remains as a testament to this activity. It is huge: 4 miles (6.4 kilometres) across. You come across it almost unawares driving through the lush Ohi'a forest. Suddenly the colossal hole of Kilauea Iki is before you, and you have glimpsed a different world. It is as if some irresistible force had pulled an enormous bung or plug out of the earth. Far below, the congealed lava in the crater still smokes from cracks in its surface. The whole pit is so clear of vegetation and so regular that for a moment you wonder whether such a structure might be man-made. It has been cauterized clean, and surely only human labour could carry out such an efficient scrubbing job? Then you can visualize that beneath the crusty surface there is still liquid rock, lying low. It somehow *looks* as if it had only recently finished cooking. Kilauea Iki erupted as recently as 1959, and there was a major flow from elsewhere on Kilauea in September 1982. The whole place is not exactly sleeping, more pausing for a moment in the middle of a bad dream. When the Volcano House Hotel was opened in the nineteenth century, the visitor could look down at boiling magma pools from the balcony while sipping a Scotch and soda. From time to time spectacular lava fountains might create a pause in the conversation, but mostly the Hawaiian lava was benignly spectacular.

Mark Twain entertained his readers in California with his graphic account of an eruption on 3 June 1886:

> A colossal column of cloud towered to a great height in the air immediately above the crater, and the outer swell of every one of its vast folds was dyed with a rich crimson luster, which was subdued to a pale rose tint in the depressions between. It glowed like a muffled torch and stretched upward to a dizzy height toward the zenith. I thought it possible that its like had not been seen since the Children of Israel wandered on their long march through the desert so many centuries ago over a path illuminated by the mysterious "pillar of fire."

On occasion, eruptions can be much more dangerous, particularly when groundwater comes into contact with hot magma. Instantly, vast quantities of steam are generated, which punch out vents with colossal force, carrying rocks, volcanic bombs, and hot steam hundreds of metres into the air. A great roaring and hissing sound accompanies the drama. Such was the origin of Halema'uma'u, now an almost perfectly circular crater within sight of the Hawaiian Volcano Observatory. Its flanks show several former lava flows quite clearly, while stains of yellow smear its sides. The crater's eruption history has been closely studied. When it contained an active lava lake in the early years of the twentieth century, the level of the lake went up and down by many metres. It overflowed onto the floor of the Kilauea caldera in 1919 and 1921. In 1924 the lava withdrew, simply draining away via subterranean fissures. The unsupported sides of Halema'uma'u collapsed and created an obstruction—until groundwater penetrated its underground plumbing. Violent explosions then showered the surrounding lava plains with rocks, which still lie scattered over the surface. During the explosive phase the dimensions of the pit increased from 800 by 500 feet to 3400 by 3000 feet (243 by 152 metres to 1036 by 914 metres). A trail takes you past a series of steam vents from which the last gasps of this volcanic fury are exhaled. Superheated steam issues from inconspicuous holes—fumaroles—which are dotted over the ground. If you get too close you get scalded very quickly. This kind of steam is colourless and undetectable, and only when it cools down does it make a wispy plume. In the early morning, when the humidity in Volcanoes National Park is high, the whole area looks as if a dozen or so kettles were boiling away, each with its own steamy signal. The yellow patches that commonly smear the ground around the fumaroles are composed of raw sulphur, and there are red patches on the flows where the steam has oxidized the iron in the basaltic rock. The bright-coloured sulphur is brimstone—the fuel of hell itself. Brought up from the depths, it crystallizes at the cooler surface. It is easy enough to scrape a few crystals onto a knife. If you examine the crystals under a lens, you will see that they form tiny but perfect prisms. The fumaroles emit a gas which is also linked with sulphur—sulphur dioxide, produced when oxygen and sulphur combine. This is the gas that once made London the perfect place for Jack the Ripper to vanish without trace after his crimes; more than half a century later the smog was abolished by banning open coal fires. You can still get the same hellish whiff in big industrial cities in China and Russia. Sulphur dioxide is distinctly bad for weak chests and is heavily implicated in acid rain, since it is readily transformed into sulphuric acid. But here, on

Hawai'i, its occurrence is wholly natural, the exhalation that accompanies the birth of the crust.

Halema'uma'u is the home of the goddess Pele. She is only one of a complex series of gods and goddesses that once influenced almost every aspect of life in Hawai'i. Everything in nature had a god responsible for it, and the gods themselves were capable of embodiment in a variety of natural things. A dolphin or a turtle may be the embodiment of Kana Loa, god of ocean and winds. All natural phenomena are extensions of the gods, who were prolific in their varied ranks and stations. In this pantheon perhaps even rocks had a kind of consciousness. The operation of taboos—or *kapu*—kept society ordered, quite cruelly on occasion. Pele, as you might expect, has a volcanic temperament in keeping with her preferred habitation. She was wont to take a fancy to handsome young men and embodied the dry and fiery leeward side of the island. Pele's lover and opposite, Kama Pua'a, was associated with rain, and with the windward side. One of her tricks was to change herself into an ugly old woman, ready to reward those who offered her charity, or launch an eruption on those who did not. After the hold of the other gods has faded, Madame Pele still seems to excite superstition. In the Volcano House Hotel there are pieces of pumice returned from all over the world. Accompanying letters claim that after taking this or that small piece of Pele's volcanic property the owners have had nothing but bad luck and so have despatched the pieces of purloined rock back to their rightful place in the hope that Pele will lift the curse. The eruption of Halema'uma'u was naturally attributed to Pele's displeasure. Perhaps she would be pleased to know that there is a volcano named in her honour on Jupiter's satellite Io.

If you rootle around among the more recent lava flows, you will find her tears and hair. Pele's tears are shiny, black, drop-shaped stones. They formed when hot lava sputtered or exploded into the air; as the tears fell to earth they cooled suddenly, frozen in shape. Now they can be found preserved in crevices in the top of lava flows. Pele's hair is spun from liquid lava: glassy filaments drawn out as the gases sped from the lava. It has something of the texture of spun sugar, but is rather brittle—the candyfloss of the earth. It has an almost golden sheen, so this conjures up a picture of Pele as flaxen-haired, but with black tears coursing down her cheeks. I also found an extraordinary light material called reticulite, a kind of froth frozen in spun glassy lava. It looks much like foam, such as might be blown up by a landward breeze on a beach, but then strangely solidified. It is very fragile, and almost as easy to damage as foam itself. Intense froth-

Frozen lava: Pele's tears (above) and Pele's hair (below) resting on pumice full of gas holes.

ing of the magma threw up this delicate confection. Eruptions can have a light touch in the right circumstances.

Lava flows can be well over forty kilometres long before they reach the sea, like the great flow of 1859, which runs north-westwards from Mauna Loa. There are two principal types of lava, readily distinguished in the field. The twisty, ropy kind that personifies flow—all congealed motion—is called *pahoehoe.* This is the easy kind to walk over: it is like traversing an irregular road that has been partially melted, heaved, and distorted. The other kind is all blocky and lumpy, a jumble of sharp chunks and clinker, rough and impassable. This is known as *a'a* lava, and it is truly dreadful stuff to walk over. I overheard a geologist say that a good pair of boots can be rasped to bits on *a'a* in a couple of months. The flows are initially as black as a Sunday hat, but as they weather a crusty brown predominates. Looking down on the landscape, from the side of Kilauea, fresh black flows overlie the older brownish ones, as if some giant had carelessly spilled a great pot of black paint down the hillside. Or it might be the shadow of a dark cloud cast upon the ground. Where the road has cut through the flows, you can easily see how one flow lies upon another. Often the surface of an individual flow is weathered, brownish, with an *a'a* topping, so that stacked flows have the appearance of a kind of volcanic club sandwich. The whole island is built of these flows, one on another. Perhaps a better image than a sandwich for the formation of the islands is a giant stack of pancakes: countless, endless sheets of congealed lava piled up, as if cooked for the fattest and most insatiable Polynesian god of gastronomy. My map tells me that the summit of Mauna Kea, the highest volcanic mountain on Hawai'i, is 13,796 feet (4205 metres) high. The other great volcano on the Big Island, Mauna Loa, is hardly less considerable at 13,679 feet (4169 metres). But this is only the part that projects above sea level. The average depth of the ocean floor in the Hawaiian region is 16,000 feet (4877 metres). A simple calculation reveals that, measured from the sea floor, their true base, these volcanoes are approaching 30,000 feet, or 9000 metres, high. Taken together, they make up one of the greatest structures on earth, higher than Mount Everest. Mauna Loa is estimated to occupy 19,000 cubic miles.

When submersibles were used to examine submarine eruptions—those unseen events that even now are building up Lo'ihi towards the sea surface—the form of the lava flows discovered was different from anything on land. Cooled rapidly as it belched out under pressure, each lava pulse congealed into gobs and lobes. The globular masses piled up. Volcanic rocks of this kind are known as pillow lavas. Ancient pillow lavas are easily recognizable. I first saw them underneath a lighthouse on the coast near Fishguard in western Wales, where they were nearly 500 million years old, but they did not look so different from some of those listed in Hazlett and Hyndman's *Roadside Geology of Hawai'i.* There are different processes again when the stack of flows rises above sea level, when sea and lava engage in violent conflict. This is the battle that I had witnessed for myself, in the voggy dusk. Explosive cooling and the sedimentation of glassy fragments results in a rock with a characteristic speckly texture known as hyaloclastite. So you can now envisage the succession of flow types that would follow one another to construct a volcanic island: pillow lavas at first, then a mixture of pillows and hyaloclastites as the island wins the battle for emergence, and then pile upon pile of flowing *pahoehoe* and its crusty companion *a'a*. All these stages are exhibited in one or another of the Hawaiian Islands.

The first western visitor to the islands was not slow to appreciate their mode of formation. Captain Cook recorded in 1778: "The coast of the Kau district presents a prospect of the most horrid and dreary kind: the whole country appearing to have undergone a dreadful convulsion. The ground is covered with cinders and intersected with black streaks which mark the course of lava that has flowed, not many ages back, from the mountain to the sea. The southern promontory looks like the dregs of a volcano." James Cook was under no illusions about this place as a kind of paradise. To him paradise would have been a much more orderly affair, a garden of organized orchards and well-tended herbs, perhaps with a passing resemblance to his home in Marton, Yorkshire. Cook was killed by the locals on his return to Hawai'i the following year, allegedly in retaliation for a flogging administered to a native thief. If there was a moment that marked an end to the islands' isolation it would be Cook's visit, for afterwards they were located accurately on a map for others to find.

The missionary William Ellis visited Kilauea in 1823, probably the first westerner to enter Pele's demesne. The perception of the place had changed utterly in forty years. In his *Narrative of a Tour through Hawai'i* (1826), he enthused that "in appearance the island of Hawai'i is grand and sublime, filling the beholder's mind with wonder and delight." The poet Samuel Taylor Coleridge read and approved. The modern version of the island as Eden was born.

The best views of the great volcanoes of Mauna Loa and Mauna Kea are from the air. From the ground, it is difficult to appreciate their colossal scale, and they are often obscured by clouds. When the sky clears, they do not tower in quite the way that mountains are supposed to. Rather, they loom. It is only when you realize that all roads directed towards them go relentlessly upwards, apparently forever, that you begin to understand their height: it takes several days to hike to the cold top of Mauna Loa. The summit of Mauna Kea is considered the ideal place to mount an astronomical observatory, for above the clouds there is only the stars and the skies are ideally transparent. From an aeroplane, you can see why these giant Hawaiian volcanoes are known as shield volcanoes: they do indeed resemble upturned Roman shields, at least until they begin to be dissected by erosion. Their comparatively gentle slopes are no less than the natural angle engendered by the solidification of countless lava flows. The shield analogy fails in detail, for at the centre of Mauna Loa (and Kilauea) lies not a boss, but rather a great depression. Such a caldera marks the collapse of the centre of the cone as the lava, which once buoyed it up, drains away to the underworld.

The ages of the volcanoes on the Big Island of Hawai'i decrease from north to south. The youngest, Kilauea, is the most active and the most southerly. Mauna Loa is the next oldest, then Mauna Kea, then Hualalai near the western Kona coast, and finally the oldest, Kohala, at the north of the island, which last erupted about 100,000 years ago and the main shield of which was probably built by 400,000 years ago. We shall see that this progression is a result of plate movements.

The first phase of igneous development is the great shield volcano. Later, volcanic vents open up wherever a tongue of magma can find an outlet to the surface. Many parts of the Big Island have experienced the sudden expression of Pele's anger, for the early phases are often the fiercest. Around the main volcanoes the eruptive centres are localized along rift zones. These zones are the traces on the ground of planes of weakness marked by faults, along which lava preferentially penetrates. The Chain of Craters road in the Volcanoes National Park follows the eastern rift zone of Kilauea. As its name implies, there are craters all along it. Mauna Ulu erupted in 1969–74, and now its legacy is a vast field of black *pahoehoe* lava glistening in the sunshine, in some places smooth and undulating like massed kidneys, in other places twisted like intestines. A little searching will turn up more of Pele's hair, so this is a truly anatomical landscape. In little crevices in the lava there are ferns growing: it does not take long for life to reclaim the barrenness. Where the lava is broken it is easy to see the bubble-shaped holes from which gas has been released during the eruption. In places, it seems as much bubble as rock—clinker, you might say, or, more correctly, scoria. Further down the Chain of Craters road there is a pagoda at Ke Ala Komo with a wonderful view of the side of Kilauea. The landscape is terraced into a series of grassy hillsides. Each scarp is thirty

A satellite view of the Bay of Naples with Vesuvius in the centre.

The view of Vesuvius from an hotel in Sorrento, and the height of twentieth-century elegance. Note the yellow cliffs to the right comprising ignimbrites produced from the explosive eruption of the volcano.

ABOVE LEFT: *Precipitous limestone cliffs of the island of Capri. Those who displeased the emperor were liable to be thrown off the top.* ABOVE RIGHT: *The pathetic figures of those overwhelmed by the eruption of Vesuvius in* A.D. *79, preserved at Pompeii.*

Mosaic floor at Pompeii showing the delicate use of geological materials, Casa del Fauno.

ABOVE: *A prospect of Lake Agnano, Phlegraean Fields, as described by Sir William Hamilton (1730–1803). From* Campi Phlegraei *(1776). The "cave of the dog" was nearby.*

LEFT: *The Lyell Medal of the Geological Society of London, with the portrait of the great geologist (1798–1875), and "Temple of Serapis" on the reverse.*

Arthur Holmes; portrait from the Geological Society of London.

RIGHT: *The earliest seismograph, designed by Ascanio Filomarino (b. 1751) and now in the old observatory on Vesuvius.*

Ropy coils of recently erupted pahoehoe lava on the Big Island.

ABOVE: *A magnificent feathered cloak kept by the Bishop Museum, Hawai'i, probably belonging to King Kalani'opu'u. Many feathers belonging to endemic species went into such artefacts.*

RIGHT: *The world's most geological restaurant: Lava Rock Café on the Big Island, Hawai'i.*

A house recently engulfed by lava flows on the Big Island, Hawai'i.

Crusty a'a lava by a roadside on the Big Island, Hawai'i.

The entrance to the Thurston Lava Tube draped with ferns, Volcanoes National Park, the Big Island, Hawai'i.

Dark, recent lava flows on the Big Island, spilling over the verdant landscape.

Beach of black sand almost entirely composed of minute fragments of volcanic rock, with Hawaiian turtles, the Big Island, Hawai'i.

A volcanic coastline fretted by erosion: the Na Pali cliffs on Kaua'i, Hawai'i.

Erosion carving away ancient lava flows; Waimea Canyon, Kaua'i.

ABOVE: *Coral reef fringing the tropical volcanic island of Fiji.* BELOW: *Svartsengi, Iceland: bathers in the Blue Lagoon, with the geothermal plant in the background.*

Undersea sculpture. A ten-metre-high carbonate chimney formed a hydrothermal vent in the Lost City on the Atlantis Fracture Zone, to the west of the Mid-Atlantic Ridge at 30 degrees north.

metres or so high, and is the graphic expression of faults that prove that Kilauea is sliding into the sea. Even as the volcanic pile heaves itself thousands of metres above the sea floor its edges begin to collapse, like the flanks of an over-ambitious children's sandcastle. The 1969–74 lava flows tip over the terraces like so much spilled treacle, and crawl all the way across the plains below to the sea. The cliffs are already more than thirty metres high, and constant erosion has forged natural arches.

Eruptions are spawned from a magma chamber—a fiery stomach—beneath the main volcano. As pressure builds up in the chamber it is released in the spewing, flaming fountains and vivid showers of lava that spill out from the crater. Tributaries of magma insinuate themselves through cracks in older flows: splits brought on by the relentless pressure from beneath. Once breached, a crack will spill forth lava until the source dries up. Occasionally, in a roadside cut, you will spot vertical feeder dykes that once fed liquid rock to the surface. They sit at odds with the general attitude of the rocks—a perpendicular injection through the layered skin of the volcano. As the magma builds up pressure within the chamber, the ground swells, as if the earth were taking a deep breath prior to strenuous activity. Sense the swelling and we may have a way of anticipating an eruption, and taking precautions against disaster.

At the Volcano Observatory on Kilauea, the eruptive activity has been monitored continuously since 1912. Batteries of instruments scattered over the ground are able to measure even the subtlest changes. Originally, the data were gathered by intrepid and persistent geologists. Nowadays, much of this information can be relayed instantly and electronically back to the Observatory, where computers do their usual job of totting it all up. Tiltmeters on Mauna Loa and Kilauea monitor the heave on the ground—they work like sophisticated spirit-levels—to an accuracy of a tenth of a microradian. That is a very small tilt indeed. Evacuation bottles take samples of the gases issuing from vents and fumaroles; levels of sulphur dioxide and carbon dioxide can be measured precisely back in the laboratory. Elevated concentrations signal hot breath issuing from rising magma. It has been estimated that between a hundred and two hundred tonnes of carbon dioxide are exhaled every day at Kilauea. This is magnified tenfold at an active eruption site: the breath of an eruption is not only choking, but stifling. A network of sixty seismometers records the pulse of the ground. Earthquakes precede eruptions as jitters do a first night. There are Global Positioning Satellites to record precise position—and any changes—by reference to satellite standards passing above in the skies. Even cleverer instruments can measure increase in pressure within the crust itself. A bore-hole dilatometer operates inside a bore-hole driven several hundred metres down into a lava pile. The instrument is essentially a bag of oil that senses changes in the "squeeze" on the sides of the bore-hole. This meter is so sensitive that one deployed on the flanks of Mauna Loa detected a rise in magma under Kilauea eighty kilometres away and about three kilometres down. The crust must have tightened under the press of the filling subterranean reservoir and squeezed the bag. Other features of the physics of the earth, such as changes in the gravity and the magnetic field, are measured periodically. Taken together, all these different measurements provide a diagnosis of Pele's temper. And when she finally gives vent to her fury, another arsenal of instruments measures the temperature of her choler. Some of them are thermocouple thermometers, which measure heat directly. Others are optical pyrometers, which convert the colour of the erupting lava to a measure of its temperature. At about 1200° C, lava is white hot; at 1100° C it is yellow; then it fades through orange (900° C) to bright red at 700° C, while the dull red reminiscent of glowing embers of the Christmas yule log is about 500° C. Any colour at all would fry you soon enough.

The Observatory was founded by Dr. T. A. Jaggar, who initiated the scrupulous series of observations that today still continue to provide the basis of our scientific understanding. He wanted to quantify the Hawaiian volcanic eruptions. At first, they had been more the subject of gasps of admiration or distressed hand-wringing. Then, through the nineteenth century, Hawaiian volcanoes came under the scrutiny of scientists who tried to deduce the role of the underworld in their formation. This was helped by the appearance of scientific journals, like the *American Journal of Science*, in which the results could be published, and the establishment of the U.S. Geological Survey, which could pay an occasional salary. C. E. Dutton published his account of Hawai'i in 1884, in which he identified the process of collapse as essential to the formation of calderas. He also recognized the rift zones radiating from the principal shield volcanoes. Of Mauna Loa he wrote: "So extensive has each and every one [eruption] of them been that the greater portions of them always reach far beyond the limits of vision, and mingling together are lost in the confusion of multitude." This paints the right picture of the bewildering profusion of vast flows and how they overlap in such a complex and braided way to build a single, great structure. Perhaps the giant stack of pancakes was not an entirely appropriate analogy after all, since the pile is built up from smeared dabs rather than sheets. For many years, the definitive account of Hawai'i was based on field studies that mapped out individual flows and that was based on an understanding of the relative ages of the main volcanoes. It was written by the great American geologist James Dwight Dana, who was seventy-seven when his work was published in 1890. He himself had the stature of a Mauna Loa compared with most of his contemporaries. But the eruptions of 1912 and onwards allowed many more physical measurements around the craters themselves. Curiously, Dr. Jaggar, for all his pioneering objectivity, was convinced that changes in volcanic behaviour were related to an 11.1-year sunspot cycle. He looked to the sun rather than to igneous plumbing here on earth. His statistics are not now taken seriously, but the observations on which they were based are still used. There is a pattern, but it has nothing to do with the sun.

Most important, it is clear that the plumbing systems of the great volcanoes are all separate. They do not originate from a single magma chamber. If they had been intimately connected, movements in the lava fields of one should be obviously connected with events in another, just as a domestic heating system requires central equilibrium. This patently was not the case: a drain-out from Mauna Loa was not necessarily accompanied by a reaction from Halema'uma'u. With modern techniques it is possible to distinguish the source magma for each of the volcanoes; each bears a distinctive signature in the elements it contains, and these elements, especially the very rare ones known as the Rare Earth Elements, can be measured with precision of parts per *billion,* thanks to modern technology. The main volcanoes, then, are separately connected to the deep underworld.

Yet to the casual eye the rock type making up the lavas looks extremely similar almost wherever you are on the Big Island: when fresh, it is black and fine-grained, occasionally glinting slightly if you turn it in the light. It frequently contains holes, or vesicles, especially nearer the surface of a flow where gas has escaped. Ultimately, it weathers to the gingerbread rock we saw near the waterfall on O'ahu. It is called basalt,* and is one of the commonest rock types on earth. There is nothing with which you can exactly compare its texture: heavy German black bread comes closest, perhaps. The pottery manufacturer Wedgwood went to a lot of trouble to produce a black porcelain called "Basalte," which is now much sought after—it is cer-

* Correctly speaking, Hawaiian basalt is not just basalt, but is a special variety enriched in silica known as "tholeiite." This variety is typical of basalts extruded in the "hot spot" situation.

tainly easier to collect examples of its natural namesake. You cannot often make much of basalt in the field. However, if you cut a slice from a fresh basalt deposit, and then grind it thin enough to see through, you can examine it under a petrological microscope. You will then see that it is a mass of tiny crystals, many scarcely larger than sandgrains. Basalt is evidently a composite rock. If you turn up the magnification further you will soon recognize that the most obvious of these minerals is transparent, forming tiny, elongate crystals, shaped like the fine chaff you can make if you rub brittle straw stems between your hands. These crystals are of one of the most abundant minerals in nature—feldspar; furthermore, they are of a particular kind of feldspar rich in the element calcium, termed plagioclase. A mineral displaying a pale green colour makes up much of the space between these prismatic crystals. This second important component of basalt is the dark mineral pyroxene, a compound of predominantly silica, iron, and magnesium. Then there is a green mineral called olivine. The olivine crystals in most basalts are imperfect, lacking clear crystal faces. Gem-quality olivine is known as peridot, which has a subtle green light all its own. Chemists will recognize that the green is the colour of the element iron in its ferrous state. When you have studied a few rocks in thin section, these minerals become old friends, and much of their chemistry can be identified just by studying their optical properties, such as their refractive index. But some little black grains between the crystals never become transparent, no matter how thin you cut the section: these are metallic sulphides, usually iron pyrites. In many basalts, the memory of flow is retained in the orientation of the longer feldspars: they seem to blow in one direction, pointing all one way, like chaff in the wind. It is easy, then, to imagine this broth of crystals forming as the liquid magma cooled downwards progressively from white-hot, separating out into different minerals, which were then entrained in further flow. Cooling was rapid enough to ensure that the individual crystals had no time to grow large: all was flux. There is an urgency about this rock. When the flow cools even more quickly, there is no time to form visible crystals. A *glass* is the result: Pele's tears and hair are glassy, the product of the most ebullient phases of eruption, when Pele's irate lava was thrown skywards and froze in mid-air.

Hawaiian natives had an eye for the subtlest differences in the rocks. The lavas on the summit of Mauna Kea are glassy and flinty. They make very good stone tools, since their edges can be "knapped" to keenness. All around the world, native wit has appreciated the technology implied by such stone. One of the craters near Kilauea is called Keanakako'i—literally, the "cave of the adze," which was another source of superior stone. For pestles and mortars, more massive basalt, *pohaku,* was required. There were also game stones for rolling, a kind of Pacific bowls—*'ulu maika.* And stone has always been the recipient of carvings, or petroglyphs—in Hawai'i's case, often pleasantly rudimentary figures with arms and legs akimbo. The basalt was also piled to outline large squares in the ceremonial centres known as *heiaus.* They are forgotten now, but in the eighteenth century they would have been heavily populated, and cultivated all around. Somewhat forlorn in decline, they are slowly becoming engulfed with scrub and secondary forest. Even the red pigment that results from prolonged weathering of basalt had its use as a dye, a feature that some imaginative entrepreneur has revived, so that you can purchase Red Dirt Shirts from many outlets. The Hawaiian Islands were Stone Age homelands at one time, and their cultural possibilities were circumscribed by the potential of basalt. Rock ultimately rules.

The story of the volcano is not over when the great shields are complete and the calderas collapse around their summits. The magma reservoirs still have life in them. Late-stage volcanic eruptions are different. First, they are more violent, and they build steep-sided cinder cones, which resemble everyone's first notion of a volcano. Second, they are smaller by far. From high in the air they look like boils or carbuncles on the flanks of their parent volcano. They are built from a different kind of lava, much gummier than that of the free-flowing tholeiitic basalt. This is the result of changes in the magma chamber, where the liquid mixture has had time to evolve and so produce a more viscous alkali-rich lava. Magma evolution proceeds little by little over several thousand years, as heavier minerals crystallize out and drop slowly to the bottom of the chamber, enriching and altering the composition of the magma that remains. It's a kind of underworld fractionation. The type of rock that results is often described as trachyte and contains abundant crystals of a different kind of alkali-rich feldspar, such as sanidine, rather than plagioclase. The spattered cones can maintain a steeper slope than the fluid flows. There are several cinder cones near Honolulu on the island of O'ahu. The Punchbowl Crater standing high above the town is home to a tasteful monument to the war dead of the Pacific war. The crater floor is now all lawns and plantings, and you would scarcely guess that it was born of a violence commensurate with the conflict it memorializes. Diamond Head, which bounds Waikiki's western edge, is wilder, making a feature against the sky, its steep sides supporting rather miserable scrub (this is the drier side of the island), and displaying ranks of rough, brown volcanic rock-beds plunging obliquely towards the sea. Some spectacular surf gets up off Black Point. A few kilometres to the west, Hanauma Bay is a former cone breached by the Pacific. Shaped like a deep bite into the coastline, it is one of the few places where the novice can swim over corals by a close and protected shoreline. It has become a favourite tourist stop as a result, and the corals have not benefited from the abrasion of a thousand flip-flops. But you may still get a glimpse of Hawai'i's famous incarnation of Lono, god of rain, peace, and agriculture, a little bluish fish called Lumuhumunuknukuapua'a. On the way down to the bay you pass much clinker-brown volcanic rock in the cliffs. I thought my eyes were deceiving me when I examined this rock closely because it quite obviously contained chunks of white coral amongst its volcanic matrix. Bleached by time, the coral lumps seemed suspended like figs in a cake. Evidently, the explosive violence had been sufficient to punch a hole through a fossil reef and the smashed coral got caught willy-nilly in the blast.

There are three kinds of sand in Hawai'i: black, white, and green. The commonest kind is white sand, the staple feature of those brochures promising paradise. It creates the beaches on the older islands, always behind the fringing coral reefs. It constitutes the sea-front at Waikiki, and at Kahana on Maui. Coconut palms really do tower over these ivory sands, making little puffs of shade. Visitors like myself are accustomed to white sand being, well, sand—that is, finely rounded quartz such as you find on the beaches of New England, New South Wales, or Newquay. Hawaiian sand is utterly different: it is comminuted coral and algae, the broken detritus of a hundred thousand storms washed up against the land. It has been churned to fineness in the massive winter breakers upon which surfers perform their improbable feats. It sticks to you. Quartz sand brushes off as you dry, but coral sand seems determined to add itself to your skin. It is light stuff, startlingly white, not rock at all. It is curious to experience its opposite just a day's journey away. On the southern shore of the Big Island, Punalu'u Beach is backed entirely by black sand. The beach seems to suck light in rather than reflect it. Take a handful of this sand and smear it out so that you can examine it under a hand lens, and it is easy to see that the "sand" is composed of tiny, smoothed-out fragments of glassy basalt, which are quite shiny when they are moist. The grains have something of the lustre of black pearls. This sand is from a geologically young beach, freshly ground from the newest part of the island. There are no coral fragments offshore to dilute its dark purity. On this particular beach, huge Hawaiian green turtles bask. Perhaps they appreciate the warmth that the black sand quickly picks up from the morning sun. The third sand colour is a pale olive green, and it is much rarer. It occurs near Ka Lae on the Big Island, a promontory that has the distinction of being the most southerly point in the United States. This bleak shore is where older flows originating from Mauna Loa confront the full force of the Pacific Ocean. It is a slightly spooky place. An oddly uninterested man in an information centre (which contains very little information) offers to guard your vehicle for a small fee. From whom? There is hardly anyone around. The only trees are stunted and pressed downwards, reaching away from the sea. Near the sea's edge little pockets of green sand are tucked among the rocks. If you scoop up some of this material, a hand lens will reveal that each grain is a transparent crystal, rounded, green like bottle glass. These are olivine crystals. They have been concentrated from the eroding lavas in this special place by the particular combination of sea and wind winnowing away all other minerals.

This variety of sand is the most superficial aspect of the aging of the islands. The greatest and newest island still lives and grows, with the red ichor of lava streaming through its arteries. The other islands are all about erosion and decay, after the death of the volcanoes that gave them substance. Maui lies northwest of the Big Island and is the youngest of the other islands. Maui may—perhaps—be dead, but it feels as if the corpse is hardly cold. Erosion is still in its early stages. Charcoal recovered from soil beds buried by late lavas can be radiocarbon-dated to show that there were large eruptions not much more than 1000 years ago. In 1786 a Frenchman called La Pélouse mapped a bay near the south-western tip of the island, which had already been obscured by fresh lava when the navigator George Vancouver passed the same spot seven years later. This eruption originated from the end of a rift zone, and could conceivably revive once more. Maui is an amalgamation of two volcanoes, West Maui to the north-west, and Haleakala to the south-east, connected by a lowland known as The Isthmus, an area scattered with ancient dunes. The shield of West Maui was complete 1.3 million years ago, while the main shield of Haleakala was built up layer by layer about 700,000 years ago. Later, explosive eruptive phases followed, and pock-marked the landscape with cinder cones. Some of the original form of the volcanoes is preserved. On the wetter eastern side of Haleakala there are thousands of cascades that are eroding deep valleys. But on the summit basin it can seem as if the acrid smoke faded away only yesterday. The scale is vast: the floor of the basin is more than 610 metres (2000 feet) below Kalahaku Overlook, and the highest cinder cone within it is 180 metres (600 feet) high, but by a natural trompe-l'oeil contrives to look much smaller. The cinder cones are arrayed in a range of reds—

some raw and startling, others like old blood—that were painted on the rocks as the result of alteration of the lavas by steam and volcanic gases after the main eruption. The cone is only the accumulation of the coarsest scraps of lava coughed up during the eruption, built up from tossed volcanic bombs and gobs of lava tumbling down the growing slopes. Larger bombs still lie around on the bluffs: they are often rounded by their passage through the air and, where they are broken, their glassy skins prove how quickly they cooled. Much of the finer stuff would have been carried away by the wind as ash. The last blasts left craters in the top of the cones, like raw carbuncles.

O'ahu is older again: it grew above the sea nearly 4 million years ago. It, too, is an amalgamation of two volcanoes, but the dual mountainous spines of the island are the wreckage of the shields that once made it. The mountains are only minutes away from the urban anonymity of Honolulu. You can drive into them along the interstate Highway 3, a road that Senator Inouye managed to get built to connect the two strategic bases of Pearl Harbor on the leeward and Kanehoe on the windward shore of the island: allegedly it was the most expensive road in the world. After emerging from a tunnel you seem to fly alongside the most precipitous cliffs; yet these mountainsides are carved and grooved with valleys, which are so steep they seem to be draped over the scenery. No matter how sheer the slope, everything is covered with a green blanket of ferns and trees. Within minutes the view is wrapped up in clouds and it begins to pour with rain. Mark Twain perfectly described this eroded volcanic panorama in April 1866: "It was a novel sort of scenery, those mountain walls . . . Ahead the mountains looked portly—swollen if you please—and were marked all over, up and down, diagonally and crosswise, by sharp ribs that reminded one of the fantastic ridges which the wind builds of the drifting snow on a plain . . . the whole upper part of the mountain looking something like a vast green veil thrown over some object that had a good many edges and corners to it." Twain was free with references to Eden and to paradise in his letters. But this island, like all the others, is doomed to return to the sea above which it once had the temerity to rise. Is paradise necessarily eternal?

Twain was right to use the analogy of hidden struts propping up the improbably steep mountain slopes. They are actually partly bolstered by the vertical supports of basalt feeder dykes. These once supplied the great lava flows that built a shield volcano, long since ruined and dismantled by time. Just as reinforced concrete pillars sustain a soaring tower, dykes act as struts supporting the beetling precipices. The steepest drop of all is at the Pali Lookout. You can drive there easily enough through forested slopes, but when you reach the top, the prospect westwards into the sheer drop of the Ko'olau caldera is enough to make your legs go peculiar. A thousand tourists try, and fail, to capture the view on film. It was here in 1781 that King Kamehameha I (known as "the Great") drove 300 warriors of the King of O'ahu over the cliff to their deaths, thereby uniting most of the islands under his rule. A highway named after him runs around the island.

If you follow King Kamehameha's road close along the windward shore of O'ahu, you see something of the island as it was before the appearance of the first tower block. In spite of the dominance of secondary vegetation, paradisical thoughts come easily. It is a lazy-feeling place, green and lush, with open verandahs dotted along an almost continuous coral-sand beach. There are delicious bananas, and papaya, and fresh coconut milk to drink. I thought of the medieval Earthly Paradise—an alleged island of perfection and immortality located far out to sea. (It is actually marked on the *Mappa Mundi* in Hereford Cathedral, located near India in that eccentric geography.) Perhaps the memory of this ideal place does linger somewhere in our psyche. A remote island with coconut palms fringing it, like hair round a tonsure, and coral reefs' surf delineating a line no more than a swim away from shore, seems to chime with an idea of escape from the worries of life into primordial calm. Hawai'i was made this way, and the causes are wholly geological. Along the shore, cliffs of bedded lava of the Ko'olau caldera sometimes squeeze the road into a narrow belt beside the sea, without room even for a dwelling. At Kualoa a marvellous vertical-sided valley almost reaches the sea; offshore, there is a stack known locally, and appropriately, as the Chinaman's Hat. This side of O'ahu, far from being the product of slow and peaceful weathering of the crust, was the result of a catastrophe. The steep topography was the consequence of an enormous collapse of the eastern half of the original Ko'olau Volcano. It slid off the edge and into the Hawaiian deep as if it had been amputated, creating a tidal wave of almost inconceivable power. This event is known as the Nu'uanu Slide. Nor is it unique. On the island of Molokai the eastern volcano was virtually bisected when its northern flank collapsed into the Pacific Ocean. The great slide left a scar that now makes the sea cliffs along the northern shore; and they are possibly the most spectacular cliffs in the world: 24 kilometres long and rising vertically to 11,000 metres. Molokai is but a sliver now. When their volcanoes die, islands perched in the midst of an ocean begin to founder.

The most north-westerly, most eroded, and oldest island is Kaua'i. This is where you must go to see the maturity of a volcanic island. It is the gutted remnant of a single, huge volcano, Wai'ale'ale, which still sits at its centre, 1576 metres (5170 feet) high, and almost always veiled in cloud. Although much of the island has been weathered down to soil, and is heavily cultivated, the mountains remain extraordinarily rugged and remote. The Alaka'i Swamp has the reputation of being the wettest place on earth, and it is so isolated that relatively few interloper plants have yet established themselves there. It is one of the last havens for native birds: there are still twittering honeycreepers among the lichen-clad branches of the Ohi'a trees. There were massive slides on Kaua'i, too, the largest of which left as its legacy the green-clad, inaccessible, and mighty cliffs of the Na Pali coast (the movie *Jurassic Park* set its dinosaurs down here). The forbidding cliffs mean that no road can circumscribe the island, and part of it remains inviolate. From the viewpoint at the top of Na Pali you can look down on helicopters taking tourists along the cliffs. From this height these machines look like mosquitoes skimming alongside the hide of a great, wrinkled elephant. It is hard to believe that, in former times, even the most remote valleys on this coast supported native Hawaiian villages.

Kaua'i is dissected by huge, eroded valleys. They are ripping into the volcano's heart and will one day strip out its lava flows and dykes. The island will not be ground down so much as wasted away. You can get some idea of the rate at which this happens at Wailua Falls, where twin, 25-metre waterfalls plunge over a massive lava flow. This resistant flow is underlain by a rubbly looking flow that is clearly more susceptible to erosion than its overlying neighbour, thus becoming continually undermined; pieces of it eventually break off and fall into the waters below, ensuring that the precipice remains sharp. Hence the falls fitfully retreat, and a gorge grows seaward of them. This gorge was excavated through some of the later, stickier lavas that are a couple of million years old. Since the gorge is between 3 and 5 kilometres long, very approximately, the lavas retreat in the gorge at 1.6 kilometres per million years. There are all kinds of complications possible—for example, the relative sea level has not been stable—but this is probably about the right order of magnitude. We can imagine that 20 million years or so would have to elapse for similar erosional processes to traverse the island. When the island finally reaches close to sea level there will no longer be sufficient elevation to produce the rainfall that drives the erosional machine. The island will be ready to sink slowly beneath the waves.

The most spectacular valley is Waimea Canyon, which has been described as "The Grand Canyon of the Pacific," a description that is woefully inaccurate as far as the geology is concerned. However, the epithet "grand" is unquestionably apt. For the Waimea gash has opened up into lavas that reflect only a geological instant compared with the billion years or more of strata that are on display along the Colorado River: but what a display this brief moment of geological history has given us! The Waimea River cuts into two series of lavas and their related volcanic rocks. The canyon is at least 22 kilometres long, 1.5 kilometres across, and up to 228 metres deep. A road runs up its western side, so that if you stop regularly on the way upwards you are vouchsafed intermittent views across this vast dissection through a volcano; and every view you have is a surprise. The sides of the valley are striped with the flows that give them a spurious resemblance to the strata of Colorado. There are pinnacles and ridges, often wisped in cloud. You wonder that there could be so many shades of red, which change and change again with the angle and intensity of the sun. The slopes are dappled with reflections of the clouds, which enhances the drama of it all. As you get higher, native vegetation takes over, so a prospect might be framed by the dangling leaves of the beautiful *koa* tree, each leaf a gently curved, trembling sickle. The course of the river is shown only by a thin line of green impossibly far away in the base of the valley—how could such an insignificant trickle grind away such a mountainside? Then you notice isolated buttresses of weathered lavas, some beginning to totter. Where there have been cliff falls and landslides—the evidence of debris lies before you—in your mind's eye the whole process can be imagined. As in a cartoon where the speed absurdly redoubles, you envisage the rise of the shield volcano, catastrophic collapse of its sides, and initiation of the rift zones, then the blasting of late lava cones and the inroads of streams into the caldera, and finally the rotting of its lava by the elements and vegetation and its eventual erosion to a stump of its former mightiness. It is more challenging to imagine the sheer weight of these vast volcanic constructions on the ocean floor. Once they have ceased being built, they start to sink. This is because of the kind of isostatic adjustment that was explained above. Everything is against the survival of the islands: once born, they struggle above the waves, only to be assaulted by the elements, but they are already sinking under their own mass of basalt. If this is paradise, it is a kamikaze paradise.

There are messengers from the underworld preserved among the lavas. Occasionally, geologists have discovered places where the contents of the magma chamber itself survive. Because they cooled more slowly, these rocks are composed of large crystals, which catch the light on a fresh example, but they are similar minerals to those in the basalts: plagioclase feldspar and pyroxene and some olivine. These lend the rock a comparable dark colour: it's heavy and oppressive in a hand-specimen, but lovely and lustrous black-and-grey when polished. It is known as gabbro and is thought to form the lower layer of the oceanic crust at about four kilometres depth and to comprise a layer up to six kilometres thick, making it one of the most abundant, if least recognizable, materials on earth. Then there are small boulders of other kinds of coarse-grained rocks, much rarer than gabbro, brought up from the innards of the earth in the explosive events that accompanied later phases of eruptions. These are envoys from the most alien parts of the underworld. They are collectively known as xenoliths, Greek for "foreign rocks," and they are indeed out of place on our green planet's surface. Their natural habitat is in the nethermost kingdom of Pluto, from which they have been abducted by the pull of eruption. They look like rocky chips, or triangular to spherical bodies dotted within the main mass of basalt. They often weather to a different colour, a leprous brown, as if distressed by their journey to an alien sphere. They rot readily. Most are varieties of peridotite, and it will be recalled that peridot is the more glamorous name for the green mineral olivine, so coarsely crystalline olivine is an important constituent of this rock type. Varieties of peridotite are distinguished by their mineral composition; for example, dunite is a peridotite which is virtually all olivine, and a dense, dark green in colour. The most important variety is called lherzolite: its list of constituent minerals includes a lot of olivine but also feldspar and pyroxene, and often small quantities of two other minerals: garnet and spinel. These last give a clue to the conditions from which these strange fragments were recruited, for they form at high temperatures and pressures. These superficially unimpressive fragments of rock are incredibly important, for they provide us with just about the only direct evidence of what rocks are like, deep, very deep, *down there.* They have been hijacked from the mantle. The inner earth is green, too, but a rich green such as is seen on the leaves of climbing plants in the rain forest, the green of the forest floor at Manoa Falls. Incredible though it may seem, these little chips are what most of the earth is made of.

The basalt lavas of Hawai'i are derived from the selective melting of such mantle rocks. As a whole, rocks of this type are referred to as mafic, because of the importance of magnesium and iron in their composition. Chemists appreciate the subtleties of differences between their various species, but to us the extraordinary thing is the differences that appear once the breaking and dissolving powers of rain and sun and plants have wrought their metamorphosis. So much variety wrung from a seemingly uniform raw material spewed from the depths is a wonder worthy of an Earthly Paradise.

Putting the facts of the Hawaiian chain of islands together, what can be deduced? First, think about the age of the islands, and the state of their decay. The youngest island is Hawai'i itself, the Big Island, erupting steadily, still building itself from its deep magmatic wells, shields still turned to the sky. Except to the south-east of it, a new island is growing beneath the sea, and will one day take its place. Even *within* the Big Island the younger volcanoes are to the south-east. Alas, the great shields afford no protection from geological inevitabilities. Maui to the north-west of the Big Island is extinct, but hardly cold (it may still have life in it). Its recent history is written all over the landscape. The small island of Lanai was once part of it. Following the same line to the north-west, Molokai and the Oah'u are still older, more worn down, with major slices from their sides collapsed into the deeps, presaging their ultimate ruin. The edges of the old calderas are now fantastically fluted by erosion; a lot of the volcanic rock that was once black and tough is weathered soft as cake, and is much the same colour. Thence north-west again to verdant and rich Kaua'i, the oldest island of all, gouged by chasms of erosion digging towards its very heart. Geologists can see into the deeper layers of volcanoes thanks to this progressive erosion.

As an aside, I should not forget to mention Ni'ihau, west of Kaua'i. This is the island that nobody can visit; it is known locally as "The Forbidden Island." Can this be where Forbidden Fruit still grows? Is this the real Garden of Eden? It is easy to see it from the Na Pali heights on Kaua'i. It is a small island, representing just one flank of an ancient and decayed volcano, the rest of which foundered into the ocean. It is privately owned by a family called Robinson, who forbid access to outsiders. Of course, there is nothing like the word "forbidden" to make you itch to go to it. It is now the last bastion of the Hawaiian language; McDonald's has no franchise there, and it is beyond the dominion of Colonel Sanders. Fish is still farmed in the traditional way. Its total privacy has guaranteed the survival of a culture which has blended with the American way elsewhere to the point where the enquiring visitor begins to wonder what, if anything, is authentic. There is such a thing as an authentic hula dance, but it is not the one you see in the Hilton Hotel; and even grass skirts did not originate on Hawai'i. The Robinsons have taken on the role of the traditional Lord, and thereby kept this island race alive: the honeycreepers of *Homo sapiens.*

The islands age on a line from south-east to north-west, progressively and regularly. Each island was created by lava, but then the source died. One way to encompass these facts might be to propose that the heat source of the volcanoes migrated south-eastwards, one by one. Such was an early explanation. But what if the tables were completely turned, so that the fiery source of magma was fixed, and instead it was the volcanoes that moved? Or, to be more accurate, crust passed over a fixed heat source, so that its volcanic progeny erupted in sequence in harmony with this movement. As long as a favoured volcano lurked above the source of magma, it would grow and grow above the sea, flow by flow, eventually forming a shield, like Mauna Loa. This would be how islands were created. Life would colonize the new land, seeded from the next oldest island. Isolation from the neighbours might then engender another endemic species. Thus, even biological richness might be accounted a geological phenomenon. The effect can be simulated, more in spirit than in truth, by moving a sheet of thickish polystyrene foam a few inches over a lighted candle—carefully, I should add, and outdoors to avoid any toxic fumes. When I tried this experiment at the right speed, the sheet did not simply melt in two; rather, it became corrupted with blobby, burned holes in an approximate line—some large, some small. Islands in the negative, you might say.

Then, as the tectonic conveyor moved the plate onwards, away from the heat, the island would begin to founder slowly under the intolerable load of its massed lava layers. The crust would adjust to the weight loaded upon it; unstable slopes would break along faults, and, once in a few dozen millenniums, huge slides would collapse and tumble to the same sea floor above which the islands so impertinently rose. Another of these catastrophes may happen even now, carrying a slice of paradise to oblivion.

The moving crust is part of one the great tectonic plates of which the earth is composed: the Pacific Plate. Hawai'i is almost in the middle of it. The fixity of the magmatic heat source supplies a wonderful test of the reality of plate movements, because it provides a fixed marker on this peripatetic earth of ours. In a world that shifts, here is an igneous reference whose fidelity has been tested over millions of years. If the theory is right, it will be relatively easy to calculate the rate of movement of the Pacific

Plate over the volcanic source. The first direct measurements using modern instruments were made on Maui, at Science City on Haleakala. It seems that the Pacific Plate moves north-westwards at about ten centimetres, or four inches, a year. Hair grows much faster. Cumulatively, this is all that is needed to throw up new crust, make an island, and move it onwards to begin its destruction.

What of the heat source that has been so steadfast while invading species, and belatedly humankind, have settled upon its productions? The lumps of peridotite preserved as xenoliths—those strange rocks so far from home—give evidence that there is more or less direct plumbing down into the depths, way below the crust itself. The persistence of the site suggests a permanent flow of energy coming from far inside the underworld, from which strange emissaries with strange names like lherzolite bring news of alien spheres. They come from a high-pressure world, brought upwards by what has been called a mantle plume. Despite the name, a plume is not liquid; at great depths solid rocks behave in a curious way. A plume is a high energy, hot "jet" of solid rock that rises through the mantle. Many geologists believe that plumes originate at a depth of nearly 3000 kilometres. Where the plume hits the crust, it is called a hot spot. The paradise of the Hawaiian Islands is thus the product of the most direct route to Hades we have on earth. Heat is lost from the interior of our planet by means of these ducts, which maintain the same position for millions and millions of years. Not all the magma that erupts as lava on the surface, however, has streamed directly from unfathomable depths. Much of it is generated by heat melting of peridotite—strictly speaking, its variety lherzolite—at about a hundred kilometres below the surface. The kind of magma that is produced is dependent on the exact conditions of temperature and pressure, which change with depth. This can be simulated by "melting" experiments carried out inside special pressurized apparatus in the laboratory—a kind of pressure-cooking. The peridotite is partially melted to generate the most stable magma under the ambient conditions of pressure and temperature: basalt magma is the result. The magma then seeks out weaknesses above and, once the ocean floor is breached, the growth of the shield is relentless. A magma reservoir forms within the growing volcano at two kilometres or so below the surface, which is where the density of the magma matches that of the surrounding rock. Magma is delivered into the chamber in "pulses." Its upward passage has been monitored by measuring the small earthquakes generated during its movement: a newly cooked stew of magma can erupt just a month or so after it starts to rise in the mantle. When the new batch arrives in the magma chamber, the ground draws its breath, and the tiltmeters register; then comes the exhalation of lava. Thus goes the respiration of the living earth as it generates new crust.

The detailed structure of Mauna Kea on the Big Island is currently being investigated by means of a deep bore-hole in the old airfield at Hilo, which is scheduled to reach a depth of 6000 metres (the lava at this level should be 700,000 years old). It is hoped that this deep probe will reveal some of the secrets of the mysterious mantle plumes. The time represented is long enough to record the shift of the volcano from off the centre of the plume. There is a lot of clever science that can be carried out on the fresh, unweathered rocks recovered from the bore-hole. Very precise measurements of abundances of elements like barium show when the later, alkali phases of eruption are reached. Other rare elements, like neodymium, exist in more than one isotopic form, the ratios of which can provide a measure of the depth from which the flow material may ultimately have been derived. These are like xenoliths, but at a molecular level: hidden messages from the underworld that require the most sophisticated tools of modern science for their translation.

So the increasing age of the islands towards the north-west is now readily explicable as the ancient track of the moving Pacific Plate. Modern studies suggest that the Hawaiian hot spots are probably paired: there may well be two of them, side by side. In spite of their common cause, rare earth elements tell us that each volcano is in detail an individual: every version of magmatic plumbing is unique. The logic of the plate tectonic explanation for islands demands that north-west again of Kaua'i there must be further islands, still older expressions of the hot spot(s). So it proves to be. There is a whole chain of islands, their rock outcrops now effectively submerged, aligned along the Hawaiian Ridge; northwards again there is a train of submerged volcanic cones known as the Emperor Sea mounts that extend all the way to the Aleutian Trench at the edge of Asia. Once deep beneath the waves, sea mounts are protected from further major erosion. In total, this chain of volcanoes extends across more than 30 degrees of latitude. Drawn upon the sea floor is a line of geological time, burned into its very fabric. There were still dinosaurs alive at the time the first islands over the hot spot were erupting. It has taken more than 70 million years to build this memorial recording the slow march of plates across the face of the earth. For once the cliché is true: this trace is literally a graphic illustration of the span of geological time. Recall that everything I have described in the Hawaiian chain—from the birth of an island to its erosion—has taken less than 10 percent of the time recorded by the complete chain of islands related to the same hot spot. Now, multiply the time taken for the formation of the *entire* chain of islands by ten, and this will still be only about one-sixth of the age of the oldest Archaean rocks on earth. To put it another way, imagine a plate spinning about six times around the whole world with a speed not much exceeding the rate at which our fingernails grow. This is a more realistic—more graphic—way of signifying the sheer immensity of time involved in earth history than the twenty-four-hour clocks, or running tracks, or the other commonly used analogies that attempt to bring geological time down to a domestic scale. To collapse time by whatever analogy is to misunderstand it. You have to think hard in order to appreciate the geological scale. As John Keats said:

> *Time, that agèd nurse,*
> *Rocked me to patience.**

The islands of the Hawaiian chain to the north-west of Kaua'i are mostly atolls—circular reefs surrounding a warm lagoon. The volcanic islands themselves may have sunk below the waves, but the corals that are ultimately founded upon them continue to live close to sea level. Charles Darwin himself, when he was naturalist on the voyage of HMS *Beagle* in 1837, made observations that allowed him brilliantly to deduce an explanation for atolls. As in the reefs that form a clear white line along the shore of Oah'u, coral growth begins close to land in the early stages of the life of a volcano. Most reef-forming coral can only grow in shallow water, since the algae that live inside the tissues of massive corals need light—without it they fade and eventually die. And, of course, the shield volcanoes are surrounded by deep water, so corals may only perch on their margins. As the islands sink, as sink they must, the corals grow upwards. They are as vigorous as forest trees. Built on the firm base of their own dead skeletons, the coral animals comprise no more than a skin of living tissue. When the island eventually subsides below the waves the corals keep abreast of sea level, and they continue to fringe their vanished host. Their waste fills the lagoon inside the reef, white sand that reflects the sky so perfectly as to produce the most aquamarine water in the world. It is astonishing how much debris a healthy coral reef can yield, for as much or more as ends up in the lagoon is tipped seawards to buttress the whole structure. Darwin's sce-

* *Endymion: A Poetic Romance* (1818), bk i.

nario remains definitive. In 1910, R. A. Daly added further observations about changing sea levels that would have inevitably been associated with the glacial epoch. During the maximum extent of the glaciers in the northern hemisphere, so much water was locked up in ice that global sea level was greatly lowered. Then surf would have renewed its onslaught on some of the volcanic islands. So the actual history of the coral reefs on any given island or atoll might prove to be rather complicated, depending on the relative trade-off between sea level and subsidence rate. There will be places where the sea deepened so rapidly when ice sheets melted between glacial pulses that reef construction was perforce abandoned, as in the battle that happens on every sandy beach in summer as the tide comes in, when small children vainly pile sand against implacable waves to protect their carefully constructed castles. In other sites, a coral reef may have grown when sea level was high, only to be stranded as a bleached fossil when the waters once more withdrew. These reefs are like ghosts of the real thing: wan, skeletal, and retaining no trace of the brilliance of marine life, but the intricate lattices of the corals are still unmistakable, even in petrifaction. There are several of these bleached reefs on the island of Molokai, which has probably stopped sinking. Eventually, however, steady and unstoppable plate motions carry any island beyond the tropics to a place where corals cannot grow. In higher latitudes they are preserved only as fossils on the sea mounts where they once prospered. Such is the end of paradise.

Plate motions provide a marvellous, yet simple explanation for a range of apparently disparate facts about the earth, several of which have been the subject of speculation for more than 150 years. The march of erosion north-westwards through the Hawaiian Islands, and the ever-southerly birth of new volcanoes; the prolific endemic species on those islands; the line of sea mounts and atolls and volcanoes, of which the Big Island forms the largest and the latest; Pele's fury and her subsequent quiescence; even the colours of the beach sands: they are all explicable by a combination of hot spots* and the movement of the Pacific Plate. Can there be a more gratifying example of how plate tectonics has transformed the way we look at the world?

* Science moves on. As this edition goes to press, a "counter culture" questioning the reality of mantle plumes has started to surface in the scientific literature. It is early days, but this may prove to be the basis of another paradigm shift in the coming decades.

3

Oceans and Continents

The Hawaiian Islands are remote. It is hard to appreciate quite how far away they are from any continent, when an aeroplane can whisk you there in a few hours from San Francisco. As you look through the cabin windows at endless blue water, time seems to collapse, hours blur . . . yet all the time you pass over the vast and apparently featureless Pacific you are also traversing an incalculable stretch of basalt. Down, down, far below the waves, into the dark abyss where no light reaches except the luminous flashes of deep-sea organisms, the sea floor is dressed in sediment—but this is only a thin blanket atop five kilometres or so of basalt. If some terrible explosion in the sun were to vaporize the seas and blast clear its sedimentary top-dressing, the Pacific hemisphere would be black—the dark, matt signature of basalt. The area floored by varieties of this igneous rock covers about two-thirds of the earth's surface. It is invisible to a passing satellite because this is an area that coincides exactly with the major oceans. Basalt provides the lining of the basins that cradle the seas. There are 363 million square kilometres of sea floor; there is a lot of basalt in the world.

The depth of much of the ocean floor lies at about 10,000–16,000 feet (approximately 3000–5000 metres). It has been observed that we know more about the topography of the moon than we do about some parts of the ocean floor. It is an extraordinarily difficult place to reach. Bathyscaphes are designed to withstand the enormous pressures that come with depth. They are like small, riveted coffins of steel, lowered for hours into the dark. Modern successors include the submersible vessel *Alvin* operating off the research ship *Atlantis.* Few people have been privileged to make the journey to the Deeps, and until a few years ago our knowledge of the ocean floor was negligible. Anything could hide there. One of my early cinematic memories is going to see Jules Verne's *20,000 Leagues Under the Sea* with James Mason playing an obviously deranged Captain Nemo (Latin for "Nobody"). What stays in the mind about this film is not so much the jerkily animated giant squid as the sheer mystery of the depths. When Eduard Suess was writing—not long after Verne, and a little more than a century ago—it was still possible to imagine whole continents foundered upon the ocean floor. The lost land of Atlantis—paradise, even. The dredging of samples was technically difficult. Maps made by soundings were hardly possible. The greatest single advance in the nineteenth century was the voyage of HMS *Challenger,* the first ship equipped for, and dedicated completely to, oceanographic research, which steamed out of Portsmouth in December 1872. *Challenger* made thousands of measurements, and collected an equal number of samples. Most of what was brought up from the ocean floor was just mud. Soft sediment is the easiest thing to collect from a simple dredge. The chances of picking up a good rock sample are about the same as catching a fish with a pair of tweezers at the end of a long pole. The ubiquity of the basaltic underlay was not at first apparent; but the *Challenger* did discover something of the ocean's extremes. On 23 March 1875, near the island of Guam, more than five miles of sounding line were

HMS Challenger *at St. Thomas in the West Indies during the* Challenger *Expedition of 1872–76.*

run out before the bottom was reached. The crew did not know what they had found: we now know that it was the first sounding of an ocean trench, part of the Mariana Trench that forms part of the marginal zone of the Pacific. Not far away is the deepest spot in the ocean, 11,034 metres (36,200 feet) deep—6.85 miles. In January 1960 the famous Swiss physicist and derring-doer August Piccard descended in a bathyscaphe of his own design, *Trieste*, to the bottom of the appropriately named Challenger Deep. It's worth remembering that it was only nine years later that footprints were first left on the moon. Robert Kunzig in his book *Mapping the Deep* points out that Mount Everest tipped upside down would tuck comfortably into the Challenger Deep, leaving room for a minor alp on top. Nonetheless, it is interesting that the tallest mountain and deepest depth are of the same order of magnitude, with mean sea level providing the zero calibration. This may not be a coincidence.

Challenger also made the first reliable measurements on one of the other great features of the ocean. Near the middle of the Atlantic the soundings unaccountably diminished. Depths of considerably less than 2000 fathoms (3660 metres) were commonly encountered. The conclusion that there was a mid-oceanic shallowing seemed inescapable, if unexpected. To visualize this as a mountain range was more imaginative. We would now know these measurements were a consequence of peaks aligned along the mid-ocean ridges. But, at first, images of foundered lands came to mind—could these represent land bridges that had once connected the continents? In *Das Antlitz der Erde* Eduard Suess had written:

> This contrast between the outlines of the Ocean basins and the structure of the continents shows in the clearest manner that *these Ocean basins are areas of subsidence* reproducing, but on a far larger scale, the subsidence with which we have become familiar in the interior of the continents . . . [his italics]

For Suess, the ocean floors were but foundered continents. He reasoned that the processes that generations of geologists had so meticulously investigated on land should still apply even in the abyss. The realms of Neptune and Jupiter observed the same rules, at least according to one of the greatest geologists at the end of the nineteenth century. At the time that Suess wrote, this uniformity of concept must have seemed what is now termed "good science." The fundamentally different geological nature of the ocean basins was only slowly accepted, but it is always easy to be wise after the event. It does Suess and his contemporaries more justice to realize that they were seeking to apply general principles in a general way—a reliable scientific practice in most circumstances. They would no doubt have answered any potential critics that they were observing good Lyellian principles—to infer from the secure basis of processes you understand to ones you wish to investigate. If we can observe subsidence in the Bay of Naples even in historic times, then is it not reasonable to anticipate the same processes elsewhere, even if it is difficult to make the necessary observations to confirm what you surmise? Lost continents were *expected*—and here they were.

As for the composition of the ocean floor beneath its sedimentary covering, direct evidence from dredging was elusive. It was only samples from mid-ocean ridges that were relatively easy to gather, for the simple reason that sedimentary cover is thinner there. Currents sweep away a thin dusting of sediment in many places. What was retrieved, of course, was basalt. On occasion, other rock types were recovered from different ocean-floor sites that served to confuse the issue. More than twenty years ago, when I was a new boy at the Natural History Museum in London, I received a sample dredged, I believe, from the deep Atlantic. The samples were black shale, and on the shales were some ancient fossils called graptolites. They were instantly identifiable as of Ordovician age. Had they been retrieved from the crust of the ocean, they might have caused a major problem for plate tectonic theory (since all the crust should have been much younger than Ordovician). Under the lens, however, the pieces of rock showed the remains of a crust of marine creatures—the blocks must have been lying around on the sea floor just waiting, as it were, to be dredged up. A little more research work revealed the likely source of the shale as a smallish area of New York State. It is likely that the fragments were ballast, taken on board to weight an empty hold, a common practice in the days of sail. Jettisoning the ballast for whatever reason showered the sea floor with an alien load. There are other rocky strangers out there: granite boulders rafted by ice floes from the far north, foundered blocks slumped from the edge of the continental slope. It is not surprising that the basalt underlay was not immediately obvious.

The development and application of seismic methods, especially in the 1950s, had an effect upon our knowledge of the earth's anatomy comparable with that of the discovery of X-rays on the diagnosis of disease. Seismics provided a way of peering into regions where eyes were useless. Nature provides seismic information in the form of earthquakes, but these are not focused necessarily in the areas you might want to explore. For the investigation of the structure of the ocean floor, timed artificial explosions (often dynamite) were set off from a "shooting" ship, and the pulses so released probed the deep layers of the ocean floor. It is necessary to know the water depth fairly precisely for the method to work, but that can be obtained by sonar. Where there are changes in the composition of the rock layers, these seismic waves should be reflected in different ways. A receiving ship some kilometres away from the "shooting" ship would then pick up all the consequent signals; the timing of returned signals can reveal the thickness and characteristics of the rock layers traversed. Different kinds of rocks are able to transmit waves at diagnostic speeds. The calculations involved are not complicated as calculations go. However, it is important to know what *type* of waves are being recorded. The several kinds of waves are conveniently distinguished by mnemonic capital letters. P-waves are the first to arrive; that is, they travel fastest. These are "push-pull" waves analogous to sound waves, in which each particle moves back and forth as the waves propagate. Then come S-waves—shear waves. In this kind of wave, every particle moves up and down. If you lay out a longish rope along the ground, holding one end, and then shake that end vigorously up and down, waves of this kind travel along the rope, like shivers. P- and S-waves do different things: P-waves can travel through liquids, for example, which is something that S-waves cannot do. So this is like having more than one kind of X-ray at your disposal, capable of seeing more than one kind of internal organ. Later still, long-period or L-waves arrive at the receiving station. These are surface waves, confined in their movement to the crust. They have different characteristics again, notably that they travel further more strongly than either P- or S-waves.*

The instrument that measures the arriving effects of an earthquake or explosion—natural or artificial—is a seismometer. The measurements are recorded on a seismograph, usually a continuous paper log of the intensity of seismic activity, scratched by a needle (or a light beam) onto a rotating drum. Early instruments often had to be tuned to receive a particular frequency, but modern broadband instruments can measure and record the whole gamut of amplitudes and frequencies within the same black box:

* This is because their energy dispersal is inversely proportional to the distance they are from the source; whereas for P- and S-waves this figure is the square of the distance. Hence over a distance of three kilometres an L-wave will diminish to one-third, but a P-wave to a mere one-ninth. L-waves also come in more than one variety, named after prominent scientists in wave theory—[John Strutt, 3rd Baron] Rayleigh waves and [Professor Augustus] Love waves.

they listen to the music of the earth. Amplification of the faintest signals is now routine, including those that have travelled through the innermost core of our planet.

The statement that the main refracting layer beneath the sedimentary cover has the capacity to transmit P-waves at 5 km/sec will now, I hope, make sense. This is the velocity figure appropriate to a dense, iron-rich igneous rock like basalt, and it is almost ubiquitous in the ocean basins. It is underlain by another layer with a velocity of 6.7 km/sec—appropriate for gabbro. The seismic reflection profiles tell us also that the oceanic crust is thin: no more than ten kilometres thick, and in places half that thickness. By contrast, the crust under the continents is up to forty kilometres thick (around twenty-four miles), and always much thicker than under the ocean basins. This is, quite simply, the most important distinction on the face of the earth. Eduard Suess did not know it, but Arthur Holmes did. The fundamental and invariable distinction between oceans and continents was essential to seeing the physiognomy of the world drawn in plates. The earth can be divided into basalt and the rest. Not that a ubiquitous basalt underlay is the only interesting thing about the ocean floors; as we shall see, its sedimentary cover has provided evidence crucial to several other important scientific advances.

At the time of writing, the ocean floor has been core-sampled at more than a thousand sites by the Ocean Drilling Program. The technical problems and expense of this kind of research mean that it has to be funded internationally. "Sea time" is always one of the most expensive items on any project budget. Since January 1985, the Ocean Drilling Program has recovered almost 160,000 metres of drill core, much of it sedimentary, from the deep oceans. It is sobering to compare this with the scattered recovery of odd rocks from dredges at the time of the first *Challenger.* The Deep Sea Drilling Program ship was christened, appropriately enough, the *Glomar Challenger* in deference to its pioneering predecessor. The current "state-of-the-art" ship, the *JOIDES Resolution,* is able to collect samples from below a water depth of more than 8.2 kilometres, and carries an on-board team of about thirty scientists. The technical problems of drilling literally miles away from the mother ship are now almost routine. The most challenging problem is that of maintaining position above the hole while being buffeted by waves and winds, and the ship is generously provided with stabilizers. Acoustic beacons near the drill-site on the ocean floor ensure that the *Resolution* does not drift from where it should be. As for the drill itself, its housing is constructed in sections and extruded through the bottom of the ship. There is hardly any part of the ocean that can escape its probe. Even this paragon of a ship will be updated. A new vessel is under construction in Okayama, Japan, that will be able to drill holes in the ocean floor as much as seven kilometres deep—three times more than the current ODP fleet. This will penetrate into the gabbro layer that underlies the basalt. There is still much to discover "down there."

It is one thing to take samples; it is quite another to make a map. Yet understanding is often rooted in a map. Problems often need to be anatomized first, before they can be tamed by explanation. The circulation of the blood was inferred in part from anatomical charts of veins and arteries. Elucidation of the principles of stratigraphy accompanied the publication of the first geological maps. So it was with the oceans. The map of the ocean floor that often adorns the walls of geography classrooms around the world was prepared by Bruce Heezen and Marie Tharp and first published in *National Geographic* in 1967. It almost glows with reality: here is the ocean drained. The sea floor is a mass of fissures and mountain ranges, and everything seems to be arranged in lines: the mid-ocean ridges are the most linear features on the earth, one of them bisecting the entire length of the Atlantic Ocean. At its apex is a neatly marked rift. There is the Hawaiian chain and its extension northwards to the Aleutians. Sea mounts teeter. In places the ridges are displaced by craggy lines, as for example opposite the Guinea coast. In other places the tip of sediments from the great rivers—Amazon and Mississippi—is patent alongside the steeply sloping continental edge. It's all so exact. I remember wondering when I saw this map for the first time: how do they *know* that? It was clearly prepared by an artist well trained in the skills of suggesting three dimensions. In fact, much guesswork was involved. Since the days of HMS *Challenger* depth soundings had been developed using sonar equipment—more sophisticated versions of the kind of equipment that is now standard in modern ocean-going yachts. Even so, there were parts of the ocean across which there were remarkably few surveys: there was much need for imaginative extrapolation. Tharp constructed a long stretch of mid-ocean ridge, around 6000 kilometres, between Australia and Antarctica on the basis of one line of soundings; this portrayal, as she said later, "was of necessity sketched in a very stylised manner." So it was, but the guesses were masterly. This was the map that made clear to the interested observer that there were natural seams in the oceanic crust. The world began to look like a kind of irregular sports ball, sewn together from disparate patches. The joins were apparently located at the mid-ocean ridges, those most persistent lines at once elevated above and graven into the surface of the ocean floor. Those who knew that the Great Wall of China could be seen from a space satellite orbiting the earth would instinctively know that these basalt walls were more fundamental to the planet than anything humans could construct. The spot depths that the *Challenger* had discovered were now rationalized into ocean trenches—mysterious and profound dark blue arcs skirting the edge of Asia. For all its imperfections, the Heezen/Tharp map made the imaginative leap for the workaday scientist. However, it did exaggerate the vertical scale twentyfold. This had the effect of making the mountains and slopes look imposingly steep. This is a distortion of real nature; for example, the Hawaiian volcanoes do not rise so precipitously: if they had the proportions shown on the map they would founder immediately. It is important to remember that the oceans are full of gentler slopes, generally of a magnitude that would not trouble the submarine version of an all-terrain vehicle.

Modern multi-beam sonar can enormously increase the accuracy of portraying the ocean floor. Beam technology discriminated the giant landslips that had left Molokai looking a mere sliver of an island, whilst some of the "hills" on the sea floor proved to be enormous, foundered blocks over 300 metres (1000 feet) high. Much of the island must have collapsed. Modern mapping techniques have discovered entire submarine mountain ranges that were not portrayed on the Heezen/Tharp map. The Foundation Sea mounts in the South Pacific form a chain of submerged volcanoes 1600 kilometres long and rising 3000 metres above the sea floor. This is like omitting from the picture of the world a Great Wall of China, but fiftyfold higher. Nor has discovery stopped. In 2001 a paper in the magazine *Nature* reported the occurrence of volcanoes possibly nearly 900 metres high on the Gakkel Ridge under the Arctic Ocean. You can fly over this region a few hours from Heathrow Airport in London. Who would imagine that there were discoveries of such moment still to be made on one of the standard Great Circle routes? The ocean floor has certainly not yet been mapped in enough detail to be certain that there are no other surprises waiting down there. Perhaps it is as well to retain something of the awe once felt for the biblical monster Leviathan; if not giant serpents, there could still be realms in the deep sea of which we have never dreamed.

The standard basalt is known as MORB—Mid-Ocean Ridge Basalt. Its chemistry is the bottom line for oceanic crust. Other kinds of rocks are often compared with MORB as being enriched in this particular element or depleted in another. The mid-ocean ridges are where new crust is added to the earth, so MORB is the basic building material of our world. The extrusion of submarine basalt flows supplied by feeder dykes at the mid-ocean ridges may be less spectacular than the foaming lava fountains of Hawai'i, but it is more important to the world. For this is how the ocean floor grows. It grows by stealth in the dark. Hidden in the gloom there are pillow lavas, such as those that make up the lower flows on the Hawaiian chain. There are tremors accompanying eruptions, although they mostly pass unobserved except by the special instruments trained to spot them. World seismic maps show a thin line of weak earthquakes closely following the ridges. The ridge as a whole is buoyed up by the heat that comes from below. The rift at its apex is the seam at the suture of creation. This is where plates are born and where they part company forever, despite being joined at birth. The new volcanic material added at the apex becomes part of one plate or another. In the simplest system to read, the Mid-Atlantic Ridge, the new crust is destined to move in the direction either of the Americas or of Europe and Africa. One analogy might be with paired conveyor belts, both tuned to run at the same speed and running back to back. The proper term for the process is "sea floor spreading." Even so, the image is too active: the eruptions do not push the system, they are a consequence of it. The lavas insinuate themselves into the cracks along the rift system as the oceanic plates on either side spread apart. This is nature abhoring a vacuum, rather than nature punching a hole in the crust by main force. The high heat flow is a measure of an upward movement of the convection cell below the ocean ridge, a movement that both drives the system and engenders the magma that in its turn gives rise to new ocean floor. Spreading rates vary. At some mid-ocean ridges they are slow, including the North Atlantic at 2.5 centimetres a year. This has been compared with the speed at which fingernails grow. The fastest rates are 15 centimetres a year, as measured on the East Pacific Rise. The total length of all the ocean ridges added together is something like 60,000 kilometres.

Water from the sea can seep into the spreading system. In some of the more active spreading ridges it becomes superheated and charged with minerals in solution. This water then discharges through hydrothermal vents, the name being no more than classical sciencespeak for "hot water." Sulphur and iron we have met before on Maui and the Big Island, painting the rock rusty or bright yellow. But at depth, iron sulphide—or pyrites—builds fantastical chimneys. There are encrusted dark towers, and crazily teetering tubes. The surrealist painter Max Ernst drew fantasy landscapes uncannily resembling this scenery, as if his untrammelled unconscious mind had had prescient access to another immersed store of weirdness. Or the ridges may remind you of the delirious castles designed by Mad King Ludwig of Bavaria, piled masses of turrets and pinnacles, daring gravity to do its worst. In the depths of the ocean they can grow undisturbed, for no wind will topple their extravagance. Vents discharge iron-rich, superhot water, and the pyrite is deposited where this water meets the sea water; the chimneys grow progressively around the hot streams. Water does not boil under the pressures at these depths, so temperatures of 300° C are common. The vents belch forth their sulphurous exhalations as dark miasmas. They have come to be called "black smokers." This may seem slightly curious: in a world of blackness, how do you recognize a black fog? It was not until they were seen by the lights of the submersible *Alvin* that these factory chimneys of the deeps could be recognized and christened. Among biologists they soon became famous for supporting a whole ecosystem not previously known on earth. The sulphurous tubes supported sulphur bacteria in profusion, and they in turn gave birth to a food chain comprising crustaceans and worms, and then predators upon these animals—virtually all of these species, in the language of the taxonomist, "new to science." Some of the tube worms that lived near the sulphurous springs were almost Leviathans.

That there is always something to be discovered in the deeps is proved by the recent recognition of a new kind of hydrothermal vent. In 2001 Deborah Kelley of the School of Oceanography at the University of Washington in Seattle reported masses of white pinnacles growing from the sea floor, building what was romantically called the Lost City. The chimneys can reach sixty metres in height. Fluted and branched, they look like the façade of Gaudi's famous cathedral of La Sagrada Familia in Barcelona, or like vast quantities of candle wax partly melted and piled into columns by some megalomaniac sculptor. They are largely composed of the common mineral calcite and its relative aragonite, both plain old calcium carbonate—the stuff of chalk. Apparently, there are microbes aplenty there, too. Water emanating from the vents from which the columns sprout is much cooler than that associated with the black smokers—75° C or less. The site, too, is different. The Lost City lies on that part of the Mid-Atlantic Ridge opposite north-west Africa, and positioned a little further from the centre of the ridge than in the case of its boiling-hot relatives. This is a slow-spreading ridge, while most black smokers have been found on "fast" ones. The deposition of calcite is the result of warm vent waters high in calcium coming into contact with cold sea water. The ultimate origin of the fluids may relate to processes deep in the underworld, where the seepage of sea water reaches the top of the mantle. This results in alteration of peridotite rocks to serpentine, and the residual calcium-rich fluids migrate away towards the sea floor—ultimately to lay down their chemical burden 800 metres down from the surface of the Atlantic Ocean. The research vessel *Atlantis* carries the submersible *Alvin* for this kind of exploration. Photographs of the Lost City were taken by digital camera mounted in a remotely operated vehicle called *Argoll.* The scientists aboard the *Challenger* certainly set something in motion when they pioneered the technology of undersea exploration. Will the sea floor ever become the ultimate tourist destination? There seems to be as much interest to go there as to the moon, and there is always a chance of coming face-to-face with some new new species of deep-sea creature. For myself, I would prefer this last, inaccessible, wild place to remain in its dark security. The human touch has been so devastating elsewhere. This may be one place where we should satisfy our curiosity and then move on.

The continents comprise only about one-third of the earth, yet they also compose the greater fraction of all dry land. This is a self-evident fact, yet stating the obvious has one useful function: it makes one wonder why the obvious is also true. They make dry land because continents stand up. They extend outwards under the sea, as continental shelves, which may be up to 100 kilometres wide but are frequently narrower. In geological terms, shelves are no more than drowned portions of the continents. The real division between the continents and the realm of basalt is the continental slope, the steep edge that bounds the shelf so as to appear almost sheer on the Heezen and Tharp maps (but remember that vertical exaggeration). The edge of the shelf frequently parallels the coastlines of continents with which we are familiar, as in Africa; but in places, such as to the west of the British Isles, there is no relation between the shape of adjacent shores and the outline of the European continental margin. The continents are geologically diverse: being stitched-together masses of a myriad rock types, including granites, gneisses, sandstones, and shales. If this book were to list them all it would be a litany of incomparable dullness. In this fundamental respect continental crust contrasts with oceanic crust, which is a hundred variations on a theme of basalt.* This is the result of the manufacture of

* Basalt rocks also occur on the continents. Basalt is just one of the many ingredients in their complex makeup.

oceanic crust at the mid-ocean ridges: there is a hot umbilical cord connecting its birth with the pervasive mantle rocks beneath. Melting inexorably leads to basalt magma. By contrast, each continent has its own biography, its own genesis and evolution. Continents are *complicated.* But on the simplest level, the difference between oceanic crust and continental is one of weight, or rather density.

Continental crust on average is less dense (upper layer 2.7 gm/cc) than oceanic crust (greater than 3 gm/cc). You can just about feel the difference if you hold a piece of black gabbro in one hand and a comparably sized piece of gneiss in the other. Continental crust is also much thicker than oceanic crust, generally something like thirty to forty kilometres, but may be double again underneath the great mountain ranges like the Himalayas. Oceanic crust, by contrast, does not exceed much more than ten kilometres in thickness. Continental crust rides up as a consequence of its thickness of less dense rock. So where the light continents are thickest (e.g., Himalayas, Alps), there also is the greatest tendency for them to rise: prod a chunk of floating wood down with your finger and see how it bobs up. But this is a misleading analogy, for the wood is too buoyant and the motion too fast. The asthenosphere that underlies the surface plates is not liquid in any vernacular sense, although it *is* relatively plastic. This means it deforms by flow rather than by fracture. Nevertheless, the flow is sufficient to accommodate the Himalayas rising, as erosion by ice and torrent continually wears down and removes some of the load from the summit of the range. The mountains are perpetually delivered into the teeth of storms and the vice of ice, and presented again continuously by uplift to the worst that the elements can do. There is an upper limit to such an uplift process, dictated by physical limits related to the thickness of the crust, so that no mountain can rise much higher than Mount Everest. To put it as teleologically as possible, every mountain will eventually be ground down to the point where it wishes to rise no more. The stable, average continental thickness eventually attained in this way can endure for thousands of millions of years above mean sea level, or at least above the level as it stands today. We will return to mountains in more detail later.

Within the oceans, there are few places where the crust breaks the sea's surface. The abyssal deep is the stable position for cold oceanic crust. Even on the mid-ocean ridges the volcanoes rise towards the surface but rarely disturb it as islands. Had there been no continental crust, there would have been no evolution of land animals, no trees, and assuredly no thinking bipeds. Intelligence may have become the property of the squid, who communicate through colour changes on their skin. In another world this book might have been written in shimmering, swirling modifications of tens of thousands of chromatic cells. Hence the most important, but often unacknowledged, distinction on the face of the earth—between continent and ocean—is a consequence of the properties of basalt. In order to give habitation to penguins and Polynesians, and bring about land, dense basaltic crust needs an extra "push" of heat. This brings us back to paradise. The "hot spots" provide such an additional boost. Hawai'i is an exceptional place, as were its geological ancestors that bulged upwards over the same deep feature, and as its descendants will be in their turn. Pushed upwards into the atmosphere beyond their natural height, when they lose touch with their private artery back into the middle of the earth, these islands will sink below the waves to remain as sea mounts—unless a persistent coral reef grants them a stay of banishment from Neptune's kingdom.

There are oceanic hot spots elsewhere. Most geologists consider that Iceland is one: it sits atop the Mid-Atlantic Ridge. We have seen that the ocean ridges are prominent features, but submarine ones. Iceland alone forms a substantial island, 480 kilometres long, with enough shallow waters around it to ensure that the sensible, conservationist Icelanders still have a deep-sea fishing industry. Like the Hawaiian islanders, they, too, have a respectful, rather than a fearful, relationship with their volcanoes. The eruptions are mostly of the comparatively benign, basaltic kind. Like a northern hemisphere Hawai'i, the island has been built up from countless flows piled on top of one another. However, Iceland is not part of a chain. Since it is situated so closely on top of the Mid-Atlantic Ridge, plates do not drift over the hot spot; instead, Iceland sits almost still at the centre of creation. The apical rift zone is produced by extension where the two plates are prised apart. It is normally accessible only to probing echo sounders far beneath the sea's surface, but in Iceland it runs right across the centre of the island. The American Plate lies on one side of the island, the Eurasian Plate on the other. Uniquely, on Iceland the plates spread apart on land; so there may be a place on Iceland where you can stand with one foot on America and the other in Europe. Iceland is where geologists can go to study the mechanisms of plate birth. No paradisiac luxuriance decorates the lavas here: mosses and grasses and lichens are what ultimately soften the blackness. The volcanic centres are very much alive, and usually clearly related to the fracture zone. Hekla in the south of the island erupted vigorously in

1947–48. Surtsey was a new island born in the sea off the southern coast in 1963. As I write, the Grimsvotn volcano is erupting underneath an ice sheet: this combination of ice and fire is especially elemental. On 29 September 1996, there was a magnitude 5 earthquake associated with the eruption at Vatnajokull: thus the earth records its birth pangs.

In Iceland you can wash in water heated by the process of creation, making the island a world centre for the use of geothermal energy. Hot water is piped into offices and private houses for central heating. It is used to heat greenhouses, too, so that you can eat fresh tomatoes and smell fresh roses even as glacier ice grows not so many kilometres away. Water percolates towards the hot plume all the time, and there is an endless supply of heat. Iceland is the home of the original geyser, at a place called Geysir. The spout lies at the centre of a bowl of geyserite some twenty-five metres across. Geysers "blow" when the water that fountains up so spectacularly is released from overlying pressure, often rising in a single gush up to sixty metres in height. The water spouts out at 75° to 90° C. At that temperature it can retain silica in solution, but as it cools the silica is deposited as a white lacquer which builds the geyserite bowl, layer by layer. This siliceous sinter will outlast the geyser itself, and "fossil" geyser sites can be recognized by the stony remains of the warm ponds to which they once gave birth.

Not far from Reykjavik there is a geothermal lake where you can sit in hot water for the good of your health, or maybe because you like being simmered. The Blue Lagoon at Svartsengi is backed surrealistically by the silvered pipes of a hydrothermal station. When the sun is upon it, the lagoon is genuinely caerulean blue, coloured with something like the intensity of the utterly different lagoons in Hawai'i. The Icelandic bathers dip in the hot draught that rises from the underworld. You cannot help remembering the spring of eternal youth in Rider Haggard's *She.* If anything is going to have the properties of an elixir surely it should be these special waters. Unfortunately, there is no sempiternal paradise here, either—just as there was none in Hawai'i. Islands grow, islands fade. But there is something miraculous about such pinnacles of life in the vastness of the basaltic oceans. Miraculous, and fragile, too. We shall discover that they are all—islands large and small—doomed to obliteration.

4

Alps

Every river in Switzerland has a companion highway, perched precariously above the white water below as it seethes and rushes downstream between giant boulders. In the canton of Glarus the little road that runs to the village of Elm is no exception, closely dogging the course of the River Sernft, and dotting back and forth from it like a persistent beggar. This is comparatively low alpine country, and the steep valley sides are densely clothed with trees. Winter drapes their crowns with brilliant clots of snow, imparting airiness to the landscape; but in summer, when the geologist does his fieldwork, the trees are altogether denser and more serious. Glades of beech and conifers provide deep cover. When I visited the Sernft Valley, wisps of low cloud clung to its sides, as if smoke from some bonfire burning slowly deep within the woods had drifted out heavily upon the humid air. There are cleared pastures among the woods, where the ground slopes less steeply; these are richly green and splashed with yellow daisies. Discreet wooden hay stores poke out of the hillsides as if they were entrances to underground tunnels. Larger chalets are scattered apparently almost randomly over the slopes, with white-rendered foundations and dark brown planking above, each with shutters as neat as can be, and no casement without a window box planted out with bright-scarlet geraniums or extravagant petunias. In the summer there are hardly any tourists, so the inhabitants appear to deck out their houses for their own pleasure. It is hard to believe that these primly picturesque dwellings not so long ago housed peasant farmers whose lives were relentlessly hard, irrevocably linked to the rocky terrain and the imperatives of transhumance. You do not have to look hard for evidence of the geology that controlled their lives, for crags break out from verdant surfaces of the fields and, higher up, on the peaks, there is little else but rock all the way up to the glacier's edge.

The Alps—particularly that part that lies within Switzerland—are the type example of a mountain range. Many great geologists have spent their lives trying to unravel the mysteries of Alpine structure. For this region of the earth's crust is where everything is topsy-turvy, where great slabs of rock may be flipped over like badly tossed pancakes, and where a mountain of a height to challenge the most experienced alpinist may be no more than the tip of a vast geological fold. It is a place where nature has apparently relished stirring up the strata on such a scale as to make wriggling rock conundra to torment the minds of scientists. The Alpine orogeny (Greek *oros,* a mountain, + genesis) took place over more than 60 million years, continuing into the Pliocene period. These highest parts of Europe are also among the most disturbed—as complex in their way as the Hawaiian chain is simple. Hence you cannot just describe the Alps in lifelike detail, for there is no end to their complications. Instead, I shall take one small piece to represent the whole: a selected microcosm to illuminate the macrocosm. I shall start in the Glarus area of the eastern Alps, where some of the secrets of the range were first decoded from the rocks. It is classical terrain. This simplification is rather like taking one of Shakespeare's sonnets to represent his whole oeuvre—it will not do justice to it, of course, but it is enough to suggest the outline of greater things.

Lochseiten in the Sernft Valley is one of geology's holy places. It does not look much when you arrive: just a wood by a minor road. To find rock, you have to scrabble up a steep bank covered in leaves derived from the dense canopy above. It is rather dim under the trees on an overcast, damp day, and hard to keep your foothold on the slippery bank. The path is underlain by a rock termed the Flysch, meaning something like "slippery slope" in the local patois. After a few minutes of scrabbling along a path through the wood you reach an overhanging cliff a few metres high. A projecting rock ledge provides shelter from the rain. This is where a famous contact between rock formations is exposed. The lower part of the rock section over which you have just climbed is the Flysch: dark grey slaty rocks that come out in small shards if you pick at them with your fingers. These rocks were once soft muds beneath the sea, hardened now by time and pressure. The cliff above is made of a completely different type of rock. It is a massive stone, and, if you scrape off bits of moss and lichen, a rich red colour is exposed: the same ripe berry tones can be seen in the Old Red sandstone of the Welsh Borderland. Look closer and you notice that the red rock is a mass of pebbles of various sizes and colours, some as large as your fist—the red is mostly concentrated in the matrix that encloses them. It is a conglomerate—nature's approximation to concrete—such as might be formed from the cobbles on an ancient beach, or from the pebbles accumulated on a lake bed after storms. The pebbles themselves were the product of weathering of still more ancient rocks: limestones here, a lump of sandstone there, perhaps some resistant volcanic rocks—in fact, something of a potpourri, all cemented together by the red matrix. We have all had a comparable experience when picking out the differently coloured cobbles on a beach. This distinctive rock is known as the Verrucano. Look closer again, and you notice something else: the pebbles in the conglomerate are all elongated. They have been turned into parallel sausages, telling you that the whole outcrop of rock must have been extended—stretched plastically on some kind of tectonic rack. This is something that only happens at depth in the earth's crust—where pressure makes rocks "flow" rather than fracture. So now you can tell that there was once a great thickness of rock above you—where only brooding clouds now float thick and grey—which erosion has removed to expose the rocks seen today. You are looking into the interior of the earth, and back in time. So far, so much deduction.

But there is more. At the contact between the Verrucano above and the Flysch below there is a thin interval of limestone—you might call it a "bed," but it undulates in thickness. Limestone, of course, is calcium carbonate, and thus chemically utterly different in composition from the clayey rocks below and the hard, pebbly ones above. It is known as the Lochseiten limestone—named after the type locality. At this spot under the trees it is about thirty centimetres thick (at other localities it can be two metres thick, but never much more); it forms a recessive notch in the cliff's profile—the eaten-away softness of this stratum explains why the overlying, and much tougher, Verrucano forms the overhanging cliff. The limestone is a creamy colour. In several places under the overhang it projects downwards into the Flysch below as a kind of lobe. A closer look reveals that the Lochseiten limestone is all ground up, and contorted; it has been pulverized, mangled, and mashed. In the middle of the limestone there is what appears to be a horizontal crack that extends straight as a die along the length of the outcrop.

At first glance, you might assume that what you were looking at is a normal sedimentary rock succession: oldest rocks below and progressively

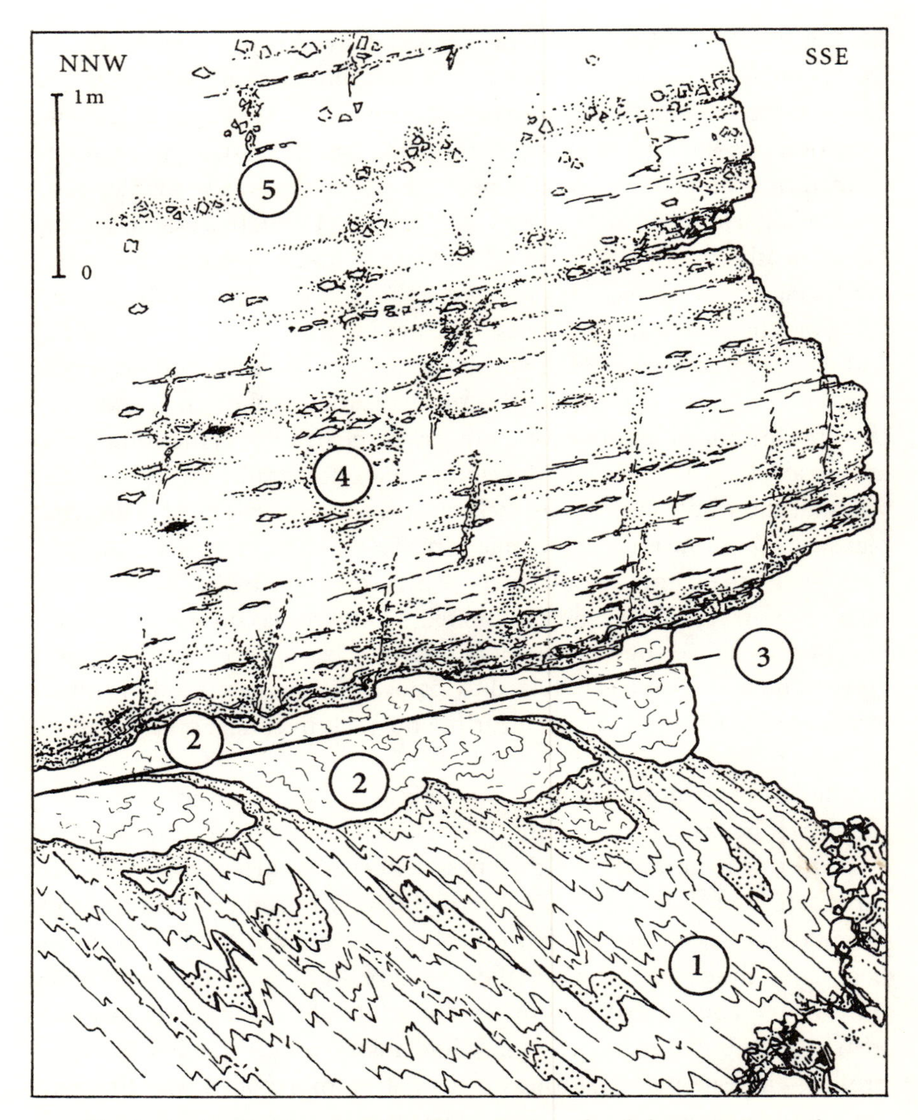

The Lochseiten section showing the succession of rock formations. 1—the Tertiary Flysch; 2—the Lochseiten limestone (mylonite); 3—the "fault gouge"; 4—deformed Verrucano flattened constituents; 5—purplish Verrucano (Permian) with undeformed pebbles.

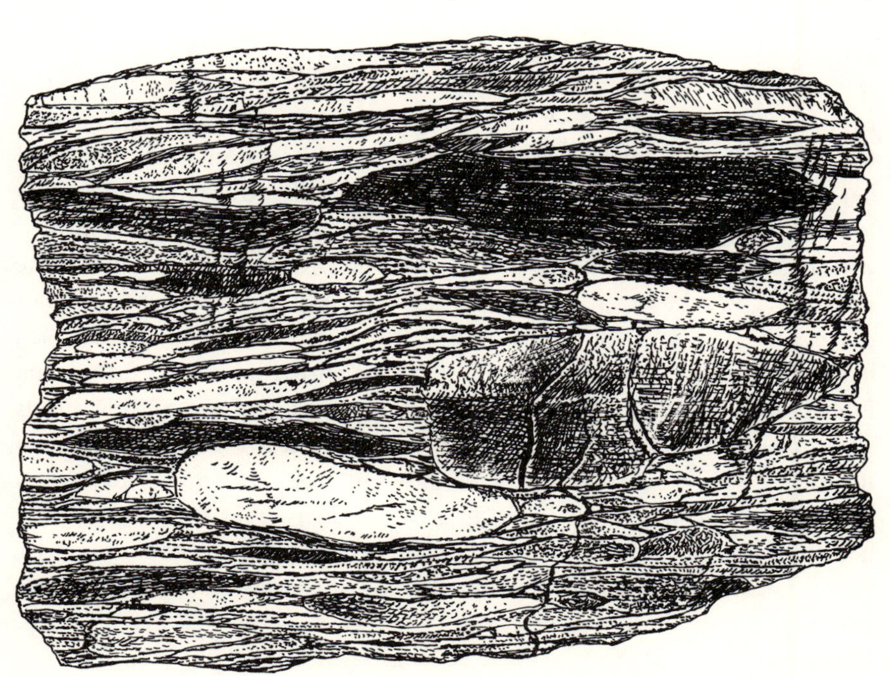

Pebbles elongated by Alpine deformation. Under extreme conditions such pebbles deform plastically.

younger ones above—Flysch, limestone, and Verrucano respectively. This is the kind of sequence you can see in cliffs of undisturbed strata by the seaside all over the world. Rocks that were originally laid down under the sea as sediments and then elevated above sea level stack up this way: it is as easy to read, as a diary of geological time, as the sequence of basalt lava flows on O'ahu. After all, the succession of names for the intervals of geological time shown on the endpapers of this book was worked out by placing one set of strata on top of another in their proper sequence, Jurassic before Cretaceous, Cambrian before Ordovician, and so on, eventually building up the narrative of geological time. Clearly we need to ask what evidence of age is there for the rock sequence in the Sernft Valley.

A few kilometres up the river there is an old slate quarry in the Flysch at the hamlet of Engi, today little more than a scattering of houses and a sawmill. The valley is wider here, floored by a few luscious fields in which pretty grey-and-tan Swiss cows munch contentedly. In the nineteenth century it was a bustling place, employing 150 people in extracting and dressing the roofing slates. Attractive dark roofs with slates arranged in geometrical patterns can still be seen all up the valley. The inhabitants once built a special barge annually to move the slates downstream to the Rhine, and ultimately to Rotterdam for sale. Today you can walk a steep trail up to the abandoned workings: slabs of steel-grey slate lie about all over the hillside among the wildflowers, and a tap with a hammer will neatly split them into smaller plates. Like the rocks on the cliff face further down the hill, the slaty Flysch has been squeezed in a tectonic vice, which is why it splits so cleanly along its "cleavage." The commercial seam is just one part of the hillside and is now marked by a ferny cave where miners once crawled into the belly of the earth. The modern observer cannot but feel a frisson of awe when imagining what it was like wriggling into a deep seam to bring out the heavy rock, and all done by hand. But the rocks gave up other treasures, too: fossil fish. The slates yielded exquisite skeletons laid out like bony ghosts on their dark surfaces, some as large as salmon. Their gasping, toothy jaws splayed out flat have something of the grisly intensity of a Francis Bacon painting. They were described in a famous monograph by the great nineteenth-century palaeontologist Louis Agassiz. His work was lavishly illustrated by 400 plates (it is said that he kept the lithographic company of Nicolet in Neuchâtel in business and that once the monograph was completed the company went bust). Agassiz named fifty-three species from Engi, but he had reckoned without the distortion that the slates had endured in the grip of the earth. He sometimes described a long and short species, which, in fact, were the same species that had simply been stretched within the rock to differing degrees; after correcting for this, the modern tally of species is twenty-nine. Irrespective of their true number, the fossils told the age of the rocks—for a suite of fossils is as diagnostic of a particular prehistoric time period as the face of a particular Caesar on a coin is of an historical epoch. Those spiny-mouthed terrors could not lie: the Flysch rocks were Oligocene in age, laid down on the sea floor about 28 million years ago.

This is where things get really interesting. For the Verrucano is well known from occurrences outside the Glarus district as being of Permian age—that is, more than 250 million years. Hence in the roadside section, in the cliff under the trees, much older rocks lay *above* Flysch of Oligocene age, with the Lochseiten limestone forming a thin layer in between. Rocks from the age of mammals were overlain by those dating from before the

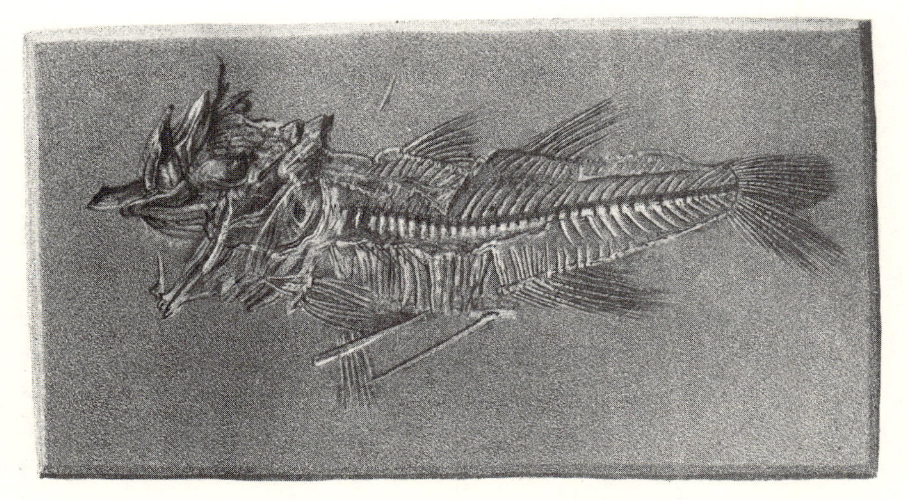

Evidence from the bones: one of the beautiful fossil fish from the Flysch at Engi described by Louis Agassiz (1807–1873).

rise of the dinosaurs: the reverse of the normal sedimentary succession. For a while this caused consternation. Could all the edifice of geological time built up so carefully by dozens of palaeontologists around the world be in error? Could geological time itself be as insubstantially founded as a house of cards? Or, if the ages were correct, what convulsions of the earth must have happened to stack the strata in their current order?

Closer inspection of the Lochseiten limestone proved crucial. The contortions within it made sense only if it marked the location of enormous earth movements. Could it have acted as a kind of lubricant as the vast mass of the Verrucano slid over the Flysch beneath? Towards the end of the nineteenth century the crush rock known as mylonite had been recognized, a rocky paste that was ground when one mighty rock mass was pushed bodily—or *thrust*—over another. Very often, younger rock was forced over older. The older Verrucano must have travelled to its present position as it slid and ground its way on the back of the oppressed Lochseiten limestone, which buckled and churned under the superincumbent load. This explained why the limestone fluctuated in thickness according to the local pressure, and, on occasion, pushed its way down into the softer Flysch, but never upwards into the harder Verrucano. If this scenario were true, another inference is now obvious: this mighty earth movement had to have occurred after the *last* sediments of the Flysch were accumulated—

that is, in the Oligocene or later. The sliding motion cut short the history of the Flysch like an assassination, providing one way to date the formation of a mountain range. The next question is: how far did this massive horizontal movement extend? And what was the great "push" that caused the movement?

To answer the first question, one very basic piece of research is required: a geological map. Strata can be plotted out upon the ground—plain or mountain alike—then coloured in on the map to make their distribution clear, so that the extent of the movements that transported Verrucano over Flysch can be traced. Mapping geological strata, or formations as they are usually known, is a skilled business. A good appreciation of subtle changes in rocks, and an accurate grasp of three-dimensional space, are de rigueur. In Switzerland, it also requires stamina and persistence. My guide over this ground was Geoff Milnes, formerly at the renowned Technical University in Zürich, who had tramped many kilometres in the area with "topo map" and notebook in his rucksack, hammer in hand, marking out this boundary or that in order to try and understand how the Alps were put together.

There was an heroic age of rock mapping in Switzerland. In the earlier part of the nineteenth century there were pioneers like Arnold Escher von der Linth. Escher had acquired the last part of his name from his father, an engineer who had built the Linth canal, and drained malarial marshes in the process, so his son had reason to feel proud. Eduard Suess was an admirer of Escher the younger, whom he had met in 1854. In the preface to his own great work, *Das Antlitz der Erde,* he remarked: "Escher with all his simplicity was a remarkable man. He was one of those possessed of the penetrating eye, which is able to distinguish with precision, amidst all the variety of a mountain landscape, the main lines of its structure." He went on: "Escher and Studer's map of Switzerland . . . may serve as monuments of the subject." Geological insight, thought Suess, was all about *seeing* a deeper geological reality beneath the complex, folded rocks that presented themselves to the observer: a special eye. This remains as true today as it ever was. Albert Heim was Escher's student and intellectual successor, and his *Geologie der Schweiz* in three volumes remains the biggest single-handed contribution to understanding the Alps. Professor Heim did not just make maps—he made mountains. In his old department in Zürich, they sit in glass cases: models to the life of the peaks he had studied, with the strata painted beautifully and accurately, passing over *arrête* and valley alike. They induce the same feeling of wonder that is experienced on seeing one of those model towns in which a sparrow hardly bigger than a pin-head perches upon the town hall clock. Almost every

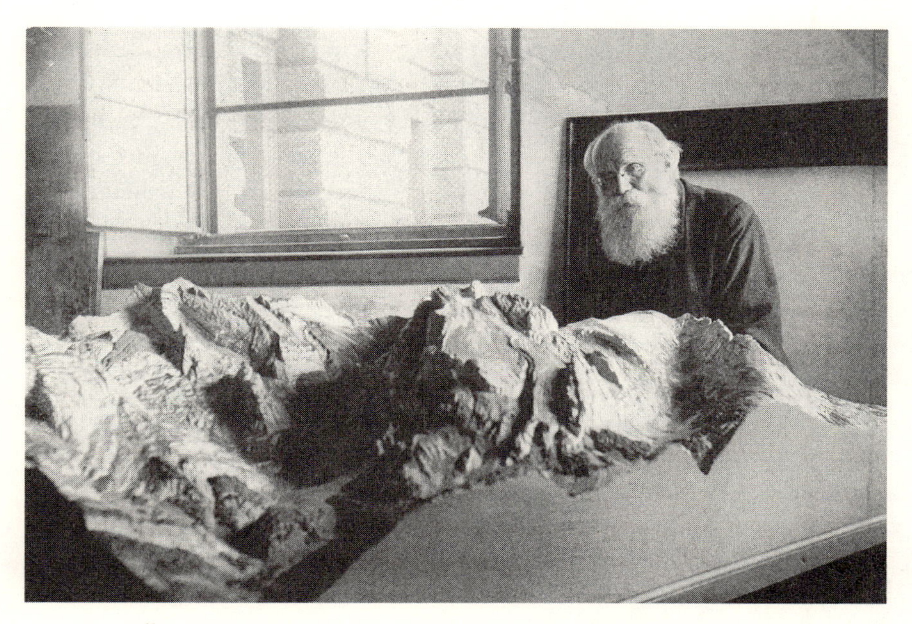

Albert Heim (1849–1937) with one of his superb models of Swiss mountains with their geology laid bare, 1905.

piece of snow—almost every tree—was modelled on Heim's replication of Saentis on the northern edge of the Alps. No detail was too small to be beneath recording. Heim's modeller, Carl Meili, deserves much of the credit for these marvellous miniaturizations.

Heim used his meticulous scale models as teaching aids for his students—to tame the mountains, to cut them down to size. There was nothing so vast that it could not be conquered by the geologist: see, a whole wild landscape has been caged for your instruction! You could regard Heim's models as embodying in concrete fashion a century's change in attitudes towards great mountains—a change in aesthetics as much as science. Mountainous terrain had once been thought of as terrifying and untamable, a place belonging to wolves and unquiet spirits. An eighteenth-century observer wrote of "those places where mountains deform the face of nature, where they pour down cataracts, or give fury to tempests." Rather than exemplifying nature at her most untrammelled, the mountains were a deformity, having little to do with true beauty, which might rather be found in farmed and orderly lowlands, or perhaps in a productive garden. A wise man might well prefer to keep away from them. A grand switch in perception had already happened by the time Ralph Waldo Emerson could write (in *The Conduct of Life*, 1860): "The influence of fine scenery, the presence of mountains, appeases our irritations and elevates our friendships." Mountains had become the object of Romantic sensibility, where emotion seemed to be concentrated in direct proportion to the thinning of the air. They became sublime. Mountains were now places for the sensitive soul to visit in order to experience exaltation and grandeur: wild places that revealed deeper truths. Geological elevation was first cousin to spiritual. The hikers that stride off along mountain paths today, equipped with the latest boots and binoculars, probably subscribe to a less passionate version of the same attitudes; but, whether they realize it or not, these modern explorers are following the footsteps of Johann Wolfgang von Goethe and Lord Byron:

I live not in myself, but I become
Portion of that around me; and to me
High mountains are a feeling, but the hum
*Of human cities torture . . .**

Understanding followed upon the removal of terror from the heights. On a mundane level, the heights simply became more accessible as better roads made the mountains negotiable. Spectacular railways followed within a few decades, and are still the arteries of Switzerland. Paradoxically, better engineering helped foster an aesthetic of wildness. Geologists turned their attention to scientific explanations of the action of glaciers and the genesis of high-altitude landforms, the great Louis Agassiz first among them. For other pioneers, the carefully plotted geological map was the primary tool in understanding Alpine structure; Heim's models succeeded the early maps; and then the consideration of deep causes followed in due course, right down to the plate movements we know today. Filleted of fear, even mountains could be encompassed by the powers of deductive reasoning.

We can now return to the structure underlying the Verrucano, mapped by Albert Heim. Geoff Milnes showed me that the notch produced by the erosion of the Lochseiten limestone could be traced quite readily, rising gradually from the top of one peak to the next above the quiet little village of Elm. The "package" of thin, pale limestone and overlying Verrucano could be seen through good binoculars. The line of contact of the limestone with what lay below seemed to be as straight as a die for several kilometres.

* *Childe Harold's Pilgrimage,* III, lxxii.

Above Elm, far above the tree line, there is a natural hole in that outcrop of Verrucano that forms the jagged crest at the culmination of the mountain. This is called St. Martin's Hole (Martinsloch), and it is said that the sun shines through it directly onto the village church every St. Martin's Day, 11 November. To get a closer look at this part of the "Glarus thrust" we need to make a detour, moving slowly southwards, towards Italy.

Just to get to the other side of the mountain you have to drive for several hours round three sides of a very large rectangle and cross over into another canton. There are no shortcuts over the Hausstock (3158 metres) without a helicopter. The town of Flims is much more the classic ski resort town than little Elm, retaining a well-groomed charm in the summer that must be the legacy of the many euros, dollars, and pounds spent when the snow lies crisp upon the piste. Even out of season, ski lifts are still available to carry you up the mountain: two chair lifts in succession and a cable car to get you to the very top. The lifts soar above the summer chalets until there is a fine view across the valley of the western branch of the upper Rhine (Vorderrhein) to the mountains to the south. The car continues higher, above the last, stunted conifers, past a magnificent cliff exposure of the Malm, of Jurassic age: a massive, thick-bedded, yellow-creamy limestone that gives rise to features in many parts of the Alps, and which was once laid down under a shallow, warm sea. You feel insignificant measured against the great crags. Upwards and upwards you climb aboard your dangling craft until you debark at the summit of Flimserstein (Cassons). There is a trail leading out over the gentle slope. The close Alpine sward has splashes of the most intense blue in nature: the flowers of the gentian, deeper than cornflower, brighter than human artifice can achieve—except for Raphael's drapery perhaps. D. H. Lawrence described them "darkening the day-time torch-like with the smoking blueness of Pluto's gloom." On either side of the path there are other plants to get the botanist excited: creamy milk vetch, rough-leaved lungwort, humble creeping polygonum, delightful little pinks. Here, saxifrages grow so close within cracks that they really do seem to live up to their Latin name for "rock breaker." It is a special community of flowers in thrall both to altitude and geology.

A little further along the path, the profile of the reverse side of the mountaintops that towered above Elm is outlined against the sky. There is a line

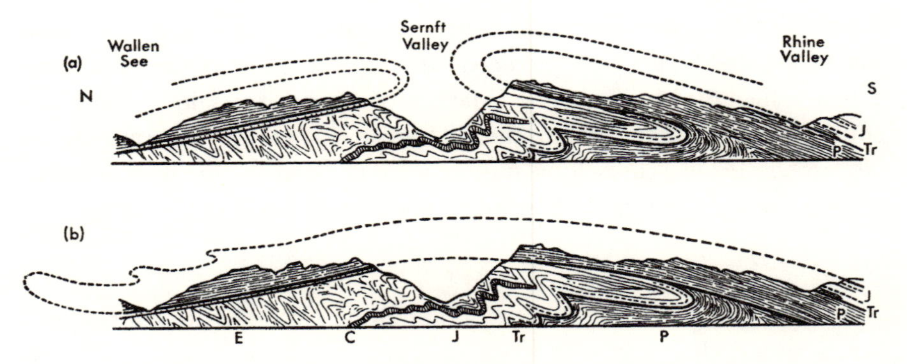

Arthur Holmes' portrayal of the competing hypotheses to explain the Glarus structures. (a) above: the double fold of Escher von der Linth wherein the Verrucano is pushed over the Flysch from opposite sides (b) below: a single nappe pushes the entire Verrucano over the younger Flysch in one mighty slice.

of jagged peaklets, like an array of dog's teeth, called the Tschingelhorner, and beyond them, St. Martin's Hole. On a closer, dumpy peak named Ofen you can see the famous contact very clearly: it is so straight that it looks as if the top of the mountain has been sliced clean through just off the horizontal and then placed back again along the line of severance. The Lochseiten limestone is a pale band, and it undulates in thickness along the length of the outcrop. There is even the same sharp line, or cut, *within* the limestone that was observed in the roadside section. Below the contact, a Flysch (here called Sardona Flysch, South Helvetic Flysch) forms soft-looking, greyish scree slopes. A small glacier creeps down from the heights. It is a wonderful view, and one that attracted attention from geologists from the early days. Sir Roderick Murchison's party sketched the same scene in the 1840s. Murchison was an English aristocrat, imperious and brilliant, author of *The Silurian System* (1839), the great attempt to systematize rocks belonging to what we would now call the Lower Palaeozoic era (545–417 million years ago). Thus the pragmatic geologist followed hard on the heels of the Romantic poets, and the Flimserstein became an obligatory stop for Alpine geotourists. Arnold Escher von der Linth made a painting of St. Martin's Hole from near the same spot on which we stood. Famous eyes had taken it all in before us. Looking further afield, the same geological contact could be clearly seen towards the top of several other adjacent peaks: on Piz Dolf (3028 metres) and Piz da Sterls (3114 metres). The temp-

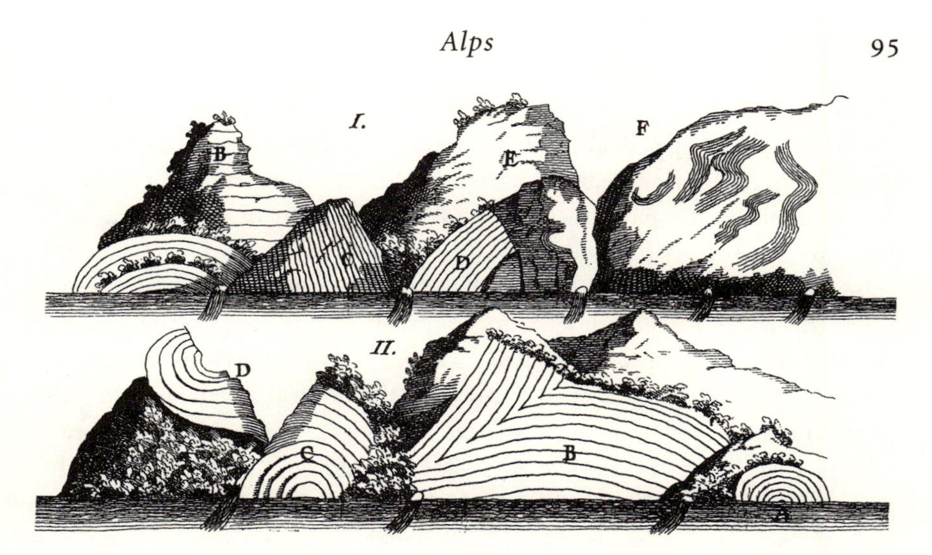

Early views on strata. Drawing of limestones on the shore of the Urnersee by J. J. Scheuchzer (1672–1733) showing little attempt to join up the limbs of the folded strata.

tation to draw a kind of dotted line through the air to connect the separate elements of the contact together was overwhelming: it must, surely, be a single, vast plane.

The story of the development of understanding of the Glarus structure encapsulates that of the Alps. We have traced it from the lower part of the Sernft Valley all the way to the Flimserstein and beyond: older rocks above younger. Clearly, there is a vast perturbation of the earth to be considered. There are several ways to bring older strata on top of younger: thrusting has been mentioned previously. The other way is to fold and overturn the strata, much as one turns back the underlying sheets and blankets at the head of a bed. It was clear from the early days of geology that parts of the Alps were intensely folded. Whole mountainsides could evidently be twisted as the heights were built up. There is a drawing dating from the earlier part of the eighteenth century showing one shore of the Urnersee, the easterly lobe of Lake Luzern. It was reproduced in Vallisnieri's *Lezione all'origine delle fontane* in 1715, one of the earliest systematic accounts of mountains. I stood on the very spot from which the drawing had been composed by J. J. Scheuchzer. The same lakeside had other resonances, too, for Tellskapelle was where Wilhelm Tell escaped to freedom from the boat that was to take him to Kussnacht castle and a grim death. Goethe's friend Schiller retold the story in incomparable fashion in 1804; it seems that geological and Romantic locations have a tendency to get tangled together. On the far side of the lake, disturbed Cretaceous limestone strata are splendidly displayed. Scheuchzer clearly portrayed folds in several of the mountains and, overall, made a passably accurate sketch. What he did *not* do was connect any one series of strata from one mountain to the next; each outcrop was its own self-sufficient story. Scheuchzer is one of those unfortunate men who is chiefly remembered for an error. In 1776 he published, in the *Transactions of the Royal Society of London,* a fossil find from the famous Miocene deposits of Oeningen as proof of the biblical flood. He called the fossil *Homo diluvii testis,* which speaks for itself. It is a kind of spread-eagled object, with obvious legs. Sadly for Scheuchzer, a hundred years later the fossil was shown, beyond dispute, to be that of a large salamander. Diluvial man was but an amphibian, and Scheuchzer joined a select few—like King Alfred—unfairly immortalized for a gaffe. That this was unjust is shown by his competent drawings of strata.

A hundred years later and the strata were joined up. Geological maps showed that one limb of a massive fold might connect the base of one mountain with the top of the next: interconnecting strata had been eroded away. In cross sections, the "air" became full of dotted lines showing where strata had once passed, only to be removed grain by grain as a result of the operation over millennia of frost and water and wind. Ghosts of vanished mountains could be inferred, faint spectres of topography past imagined. Scheuchzer's outcrops were linked together by the outline of folds. Writing of the same area, Eduard Suess said: "Those who have seen the cliffs of the Axenberg on the Lake of Lucerne will have viewed with astonishment the inextricable entanglement of the limestone beds." It became a familiar observation that massive rock beds might be turned on their sides and crumpled like the bellows of a concertina. Overfolding did indeed turn whole piles of strata upside down. Mapping distinctive rock beds helped prove this structural suffering of strata beyond doubt. The thin, iron-rich Jurassic rock bed known as the Dogger is but one example. Near the little village of Bristen, Geoff Milnes showed me how its distinctive signature could be traced from one hillside to another, bucking and twisting like some demented fairground roller-coaster. You could see how the earth buckled, and how such complex pleating shortened its crust by folding strata onto themselves. I am reminded of unpicking the bud of a poppy: so much gaudy petal crimped and folded into so small a volume. For many years the careful mappers mapped, and folding became quite the thing to explain virtually all geological problems in the Alps.

Following the *Zeitgeist,* Arnold Escher interpreted the Glarus structure entirely in terms of folding; and Albert Heim, model maker *extraordinaire,* apparently followed suit. Heim was a patriarchal figure: authoritarian and sure of himself. He would not be an easy man to cross. Escher and Heim explained the Glarus structure as a "double fold" (see p. 94): two mighty overturned and recumbent folds that faced one another across the Sernft Valley. The Verrucano was brought back over the younger Flysch in two folds: imagine turning back *both* the head and foot of the bed, so folding back two swathes of the underlying sheets to oppose one another. One fold came from the north, the other from the south. It was certainly a striking concept, which implied a lateral movement of the Verrucano of at least fifteen kilometres. When Escher first proposed the double fold it seemed a radical, even revolutionary, idea to have so much crustal movement. In his *Tectonic Essays,* Sir Edward Bailey quoted Escher's caution at the time: "No one would believe me if I published my sections; they would put me in an asylum." Eduard Suess later summarized the Glarus story as a "magnificent conception, unheard of in the views of that time, of a double folding of certain parts of the Alps." Albert Heim defended the idea of the *Dopplefalte* for many years—and who would have dared to disagree to his face? It would be wrong to pooh-pooh these great geologists with the wisdom of hindsight, for what they saw was an honest interpretation of the high ground they knew so well.

The only alternative scheme seemed even more radical—to "join the dots," as it were: to unite the sole of the structure marked by the thin sliver of Lochseiten limestone as one stupendous line of movement, skipping from one summit to the next like a tectonic ibex. I had experienced the temptation to draw such a line atop the Flimserstein. Everything from the Vorderrhein to Glarus would belong to a single structure: you would have to more than double the distance travelled by the great mass of the Verrucano to some thirty-five kilometres. A mighty heave, indeed. Perhaps Escher shrunk from the sheer magnitude of the crustal movements implied. Could the earth shrug its skin so momentously? We know that rocks could crumple and fold, but could huge chunks of crust slither laterally for such great distances? And if they did, what forces could propel them? The proposer of the single Glarus Thrust, the man who attempted to rub out the double fold in 1884 (in a seminal paper published in the *Bulletin de la Societé géologique de France*), was Marcel Bertrand, Professor of Geology at the University of Neuchâtel, in the French-speaking part of Switzerland. What may have galled the meticulous mapper Albert Heim is

Albert Heim contemplating spectacular folding in the Verrucano at Mels in St. Gallen, Switzerland.

that fact that Bertrand never actually visited Glarus. Bertrand admitted he had "utilisé les descriptions de Heim du 'double pli' " to draw his new lines and his new conclusions. How dare he be so presumptuous? Even today, there are hardened "field men"—working geologists on the rough end of a hammer—who feel that the laboratory- or computer-based scientist is something of a softie, a spinner of hypotheses away from the coalface of real observations, and who might have more than a sneaking sympathy for Albert Heim, worsted though he was in the end.

Further geological mapping in the greater Alps—much of it magnificent work by Heim and his colleagues—revealed that nearly all the structures in the rocks faced northwards. Thrusts were directed there; major overfolds bulged outwards towards the north. This was familiar to Eduard Suess, for whom the Alps were home ground. He pointed out that the whole mountain system was bowed outwards in a convex arc to the north. Here the mountains encountered what he termed a rigid "foreland." Whatever it was that caused all the rucks and dislocations, it seemed to be "pushing" from the south. Yet the double fold of Glarus required quite as strong a shove from the north to produce its massive, south-facing overfold, and this force apparently must have operated at exactly the same time as that on the north-facing fold, and exacted the same geological effects; after all, the rock sequences seen in the field in Glarus were similar wherever they occurred. This was all too much to ask; besides, if Bertrand's single thrust were accepted, it was accountable—once again—to movement from the south towards the north, and no need to make an exception for a small area in eastern Switzerland. The double fold, it seemed, was doomed. Albert Heim, with some generosity of spirit, finally acknowledged as much in 1901 in an open letter published in a memoir by Maurice Lugeon (1870–1953) on the structure of the western Alps.

By now, it was apparent that in mountain belts great masses of the earth could move long distances at low angles. This was the key that unlocked the mystery of the Alps: from Glarus to the world. By a repetition of the same process, mountain chains could be understood as collections of slices piled high on one another; this was how the earth's crust both shortened and thickened. The name given to a slice was *nappe* in French—the word for a tablecloth. It is not clear who first used this word, which has now become the accepted term. Using an analogy appropriate to the name, the historian Mott Greene described the Alpine strata before their elevation as a richly patterned tablecloth on a polished table: "If you should place your hand flat on the table and push forward, the cloth will begin to rise into folds. Push more and the folds will flop over (forward) and the rearmost fold will progressively override those before it, producing a stack of folds." The generation of an Alpine mountain range is a matter of piling on the nappes. The early nappes pile up, and are then carried piggyback as a new active zone develops: pile is heaped upon pile. In the Alps, three great "packages" of nappes moved in succession. Of course, today you very rarely see the whole pile, for erosion has deeply dissected away the landscape, removing some of the uppermost series of nappes. Even at the top of the Flimserstein, we have to imagine several kilometres of rock above our heads. Time has sent it packing. Continuing with Greene's tablecloth analogy: "Take a pair of scissors and cut away at the stack from various angles, removing whole sections of the pile of folds. Having done so, push the pile again so that the segments become jumbled against each other." In other words, you add chaos to confusion by adding erosion and later earth movements to the recipe, and that is why you stare at a cliff face in Switzerland and wonder at the spun complexity of its strata. But the recognition of nappes set out the research project: you need not simply throw up your hands in despair at perverse plications and ceaseless crenulations in the Alps. The idea now was to map the nappes, work out their sequence, and identify the rocks of which they were composed. There were many feats of heroic fieldwork: Pierre Termier, Rudolf Staub, Rudolf Trümpy; these geologists will be forever associated with the mountain ranges they unscrambled, as surely as if their names had been etched upon some mighty cliff face. Nappes were named, so that hitherto intractable mountains might now be recognized as comprising several slices stacked one above the other. In science, naming is often taming. The Glarus Nappe it became, and remains, to say nothing of the Saentis Nappe, the Murtschen Nappe, and all the rest—thousands by now—each named for a locality where it was well developed. The deformation of the rocks was often ductile—they were stretched, not broken. Albert Heim himself had collected distorted ammonite fossils that had been deformed along with the rocks that contained them. I examined a few of them in Zürich. They looked somehow hunch-backed, squeezed out of shape, lopsided, their original regular spirals twisted and wonky. In some places, fossils were stretched up to ten times their original length. Mathematical calculations can be made to compute the distortion these rocks have endured. They tell us how the earth behaves in extremis. Forget tablecloths for a moment: an image of pasta dough stretched and spun back and forth seems more appropriate. Alpine mountains might be seen as badly made lasagne, crudely layered and buckled in the cooking.

Perhaps the last of the Alpine heroes is Rudolf Trümpy, scion of the line of "field men" going back to Arnold Escher von der Linth. I saw him receive the Wollaston Medal (p. 220) of the Geological Society of London in 2002. Wilhelm Tell himself could scarcely be more legendary. In the urbane meeting rooms on Piccadilly, the grand old man seemed somehow as craggy as one of the mountains he had unscrambled. You just knew that his natural habitat was with a party of geologists, bellowing at them through a megaphone on a small cruise ship in the middle of Lake Lucerne. For another important advance, to which Trümpy contributed, was the naming and mapping of the three main "packages" of nappes. They were emplaced in bundles: tectonic messages on sheets delivered one after another. Each of these packages was the record of a separate mountain-building event. The chronology of the Alps became calibrated by very careful geological mapping, later augmented by radiometric dating. The main events saw the arrival of what were termed Helvetic, Pennine, and Austro–Alpine nappes, respectively; these packages would eventually make sense in the context of plate tectonics. Since nappes preserve chunks of the scrunched-up earth's crust, all out of place, you could find different bits of ancient geography as you moved across Switzerland—you have to imagine taking a map of the past, cutting it up, and then moving chunks around, hither and yon. It was all most confusing. You begin to understand how a scientist might spend a lifetime working on no more than a few mountains. Sometimes the effort entailed in reconstructing the original disposition of the rocks seems like that involved in unscrambling an omelette.

Further south again, we follow the valley of the other branch of the Rhine—the Hinterrhein. All the routes dog the rushing river: the old road alongside the impeccable new autoroute, and the railway to Thusis, all dodging in and out of sight of one another. The modern road misses no excuse to shoot into a tunnel. The Swiss are masters of such engineering, and now the new roads enter and leave tunnels with abandon, apparently ignoring folds and thrusts and nappes in their eagerness to arrive. Older roads tend to wind up precipitous slopes in a series of hairpin bends. Five kilometres south of Thusis the old road takes you into a dreadful chasm—the Via Mala—literally, the Bad Way. It appears to be as much an irregular crack in the earth as a gorge; in places it is only a few metres wide, but as much as 300 metres high. The sun does not reach into it so it is always gloomy, like a grotesquely narrow, unlit passageway. It provokes awe; a small party of schoolgirls we were with even stopped giggling and jostling as they craned their necks to get a view of the torrent. The river has gathered urgency in its confinement, the white water swishing past and over

The dismal Via Mala, with the bridge of 1739 viewed from below.

rounded boulders far below. A few spindly trees cling to the sides of the gorge, but their roothold on the tiny patches of soil gathered in cracks is precarious. It is a place to make you shiver, and pull up your collar even on a warm day. A bad way indeed it must have been in the Middle Ages, when a precarious path allowed for the passage of just one traveller at a time. Traces of the old path can still be made out; in places, chains hammered into the rock helped the traveller retain his foothold. Seventy metres above the river bed there is a remnant of one of the subsequent roads, a single-span stone bridge, built in 1739, crossing from one side to the other. From below, it looks, absurdly, as if it were actually propping the two walls of the chasm apart, like a buttress in a cathedral. The Via Mala owes its eldritch, claustrophobic atmosphere to the geology. The Hinterrhein has cut through the limy schists known as the Bündnerschiefer, which are part of the Pennine nappes. Their remarkably uniform character has ensured the narrowness of the gorge. As the mountains have been uplifted again and again, it is as if the river acted in the manner of a circular saw cutting into a particularly dense log of teak: the gorge maintained its vertiginous integrity as the mountains were delivered into the jaws of erosion. So the character of the rocks determines the character of the land. The Bündnerschiefer are known to be the time equivalents of the massive limestone cliffs of Malm that we passed in the cable car above Flims. They are an utterly different rock type, originally deposited, some geologists believe, as lime-rich muds in a deep-water basin rather like the Gulf of California. Sorted, squeezed, and reshuffled by tectonics, the nappes dictate the lie of the land.

It is something of a relief to emerge on the other side of the Via Mala into the Rheinwald, where the well-wooded valley broadens and comfortable-looking farmhouses reappear. Even so, perched on almost every rocky promontory along the valley there is a castle, or at least the ruins of one. In the latter case you have to look hard to recognize them for what they are, because they are roughly built of local stone and merge naturally into the crags. In the past, they must have protected the important Rhine trade route. The old towns along the route are as solidly built from the gneiss of the region as you might expect. The roofing "slates" are actually thin slices of gneiss: they must weigh heavily on the massive beams beneath. As so often when natural stones are used for this purpose, these "slates" have an attractive irregularity, a dappled richness, especially when lichens paint them. The Posthotel Bodenhaus in Splügen, where I stayed overnight, has stone walls as thick as those of a fortress. It was somehow comforting.

When pile upon pile of nappes are superimposed, the earth's crust thickens. Not just the Alps, but virtually all mountain belts are zones of shortening and thickening: the opposite, if you will, of the thinning of the crust over the mid-ocean ridges mentioned previously. When the crust clots together sufficiently, something else happens: the rocks themselves transform. This is because dramatic thickening also entails burial, and burial means that the ambient pressure on the rock increases—as does the temperature. This encourages metamorphism. Minerals change from one to another; the very fabric of the rock is altered. The original rock becomes disguised, and eventually unrecognizable. If the rocks were once sedimentary, any fossils they may have formerly contained become obliterated. Schists and gneisses are the most abundant products of metamorphism in mountain belts. Nature's metamorphism is not like the magical change that happens to Gregor in Franz Kafka's famous tale *Metamorphosis,* who wakes up transformed wholesale into a monstrous insect. Natural metamorphism can only work with the chemistry of the original rock: all the mineral changes that occur are expressible in equations and can be experimented upon in laboratories. These rocks are, to return to the pasta analogy, the result of a kind of cookery—or more correctly, pressure-cookery. The dish is utterly transformed in the oven, which, however, cannot produce anything that is not in the ingredients. This topic will appear several times in this book: for now, I have said sufficient to take us further southwards to the Passo del San Bernardino.

The old pass of San Bernardino (not to be confused with the Great St. Bernard Pass), at 2065 metres (6775 feet), forms the watershed between the Italian part of Switzerland and the German part to the north. Nowadays, the main road passes scornfully beneath it through one of those tunnels, but not so long ago all the heavy traffic had to grind up and then down its relentless hairpin bends. Today you can have the summit to yourself. There are wispy clouds below you, and rank upon rank of mountains in the distance; just the place for Emerson's irritations to be appeased and his friendships elevated. The grey rocks emerge from the ground in low hummocks, like a school of whales breaking the water. Some are obviously *roches moutonées,* polished and shaped by the ice sheet that covered this whole landscape during the last million years of the Pleistocene ice age—a geological yesterday compared with the formation of the Alpine chain itself. Major remnants of this great ice cover remain on parts of the Alps, although the recent retreat of the glaciers is one of the causes célèbres in the debate over the effects of global warming. Elevation of the mountain chain to the altitudes at which ice sheets can grow is itself a consequence of the tectonic thickening of the light crust—as nappe piles on nappe—relative to the dense mantle layer below. As the folding and thrusting doubles the thickness of the light crust, compensation is inevitable, and the mountain chain rises. It is bounce-back. If you look for ultimate causes, then very deep forces sculpted the features of the face of the earth.

The rocks of the San Bernardino are part of the Adula Nappe; to use the geological language unadorned, they comprise coarse gneisses, which have been deeply buried and metamorphosed within the Alpine pile and subsequently exhumed by erosion. Crudely shaped blocks of similar rocks were employed in the thick walls at the Posthotel at Splügen. They are pallid, banded, with greyish to creamy layers of the minerals feldspar and quartz alternating with dark wisps and stripes composed of several different minerals. Shiny plates that catch the light are black biotite mica. In some parts of the outcrop the layers are distinctly striped, like the fabric in old-fashioned pyjamas. In other places, the layers undulate and curl in folds that often just peter out: you can almost visualize the squeezing that produced these writhing graffiti. Or there are dark curlicues, resembling Arabic writing. It is all massive, uncompromising, and highly impermeable. As a result, hollows are occupied by mossy little bogs, fringed with cotton grass, looking like so many miniature white sticks of candy floss. Tough, coarse grass and the tiny, creeping dwarf willow grow on the slopes and in wandering cracks and fissures. Animals tend to scream at you up here: alpine swifts screech shrilly; dark crows croak cries of "quark," and Geoff Milnes assured me that a curious, persistent, piping angry squeak was the call of the marmot, a comical-looking mammal that gets as fat as it can during the short summer season before hunkering down for the long winter. There is a thick vein of milky quartz cutting through the gneiss; it looks as if it might be a precious seam, but it is actually the commonest of minerals. Nonetheless, I could not resist picking up a white morsel of it to take home. The vein must have been a response to some later event when gashes opened up in the obdurate mass; quartz is always ready to fill up cracks created by tension. What struck me was how similar this place was to other gneiss terrains I had visited, no matter what their geological age: bare hills in the Highlands of Scotland, barren boggy wastes in the middle of Newfoundland, ancient areas of Sweden, all obeying the bidding of their geological foundations.

Assiduous fieldworkers have discovered little pods of a rock called eclogite in the rocks around San Bernardino. This rock has nothing at all to do with Virgil's pastoral verses, the *Eclogues,* good though it would be to imagine an earth scientist taking inspiration from a Roman poet. But it *is* a stone that makes geologists as lyrical as they ever get, because it is composed of special minerals, including one type of garnet, that only form very far down in the crust. Rather like the xenoliths we met in Chapter 2, eclogites tell of deep things. The area now populated by crows and marmots was once buried deeper than we can imagine—"to exceed 70 miles," thought Arthur Holmes, and modern estimates are not so different. What revolutions of the earth have been played out on this bleak mountain pass, and what work time still has to put in to wear these peaks low again! When I asked Geoff Milnes what age the rocks were *before* they were turned into uncompromising gneisses by the Alpine orogeny he made a rueful face. As with the landscapes they produce, gneisses are much of a muchness and have lost their original personality. Gneisses can start out as shales, or granites, or even another, older gneiss. Heat and pressure is the great leveller; radioactive clocks can be reset. In the case of the San Bernardino rocks, the last option is probably the right one. In the profound tectonic scraping through the crust, older rocks, dating from still earlier revolutions of the earth,* were caught up in the new vice of Alpine mountain-building. Having already endured what the earth could inflict upon them once, they were subjected to it all over again.

But where did all these nappes come from? They have moved northwards, in pick-a-back piles, and, as Eduard Suess noticed even before nappe theory had been fully worked out, "as a rule the highest lying sheet has been carried furthest." We have already worked our way steadily southwards into the heated heart of the pile. When the mappers had completed their monumental surveys it became clear that the "sheets" were tucked downwards towards the south (Maurice Lugeon called such regions root zones). If the nappes were sheets, or recumbent masses, they lie at a low angle, but in the root zones everything is tipped to the vertical. It was as if the mountain belt had been squeezed out progressively from these narrow belts. Cross-sections prepared by modern geologists show the geology arising from these zones like huge cartoon genii whipped up out of the spout of an Aladdin's lamp. A fat series of nappes will taper down to little more than a will-o'-the-wisp at its root zone. It must be time to move further southwards to look at this place where mountains apparently arise in the fashion of gigantic fungi from a small crack in a log. "Off to Africa!" my guide exclaims. It is impossible not to respond enthusiastically.

The road winds and rewinds upon itself downwards, apparently forever. At some point trees make their reappearance, and there is the plangent dong-dong of cow bells, so the zigzag route must really be shedding altitude. When buildings appear by the roadside you realize that you have crossed into another country. As the bad-tempered crow flies, it is less than twenty kilometres from Splügen, but you have effectively entered Italy (though you're actually in Italian Switzerland). All the signs are in Italian, and the ristorante offers superlative coffee at last. It is easy to understand

* These Hercynian, or Variscan (to use Eduard Suess' term), rocks are the product of tectonics some 300 million years older. Rocks of this aspect surface in the Vosges, the beautiful mountains to the north-west of the Alps and part of Suess' "foreland."

how effective a cultural barrrier the mountains must have been in the past: geology separating languages. It is harder for a stranger to comprehend the glue that has retained this canton for so long as part of Switzerland. The road follows the Val Mesolcina towards Bellinzona, downwards all the way, but gently now. All the gneiss roofing slates have vanished, to be replaced by pantiles painting patchworks of all possible orange hues. Broad-leaved trees—chestnuts, limes, beeches—have replaced the conifers. Neat vineyards make green stripes on the south-facing slopes. We are heading for the main root zone of the Alps, which is marked by a major fault, the Insubric Line. It is one of the most important geological lines in Europe. Somehow, you expect to see it drawn out in black on the ground, as it is on geological maps; but of course it is more subtle than that. The San Andreas Fault in California has the same modesty, as we shall see.

We stop in the village of Pianazzo in the Val Morobbia: the river valley follows the famous line here. Streams have a way of seeking out major junctions in the earth's surface, picking out old weaknesses. An ancient and somewhat decayed path leads down to the river. Its walls have been carefully constructed from the local geology, so it must have been important once. It was probably the path that was used by the villagers to conduct their flocks to the summer pastures on the high ground beyond the river, but there is little sign of that ancient peasant life today. It is a pleasant walk downwards underneath the sweet chestnut trees. The ground is moist enough for apricot-coloured chanterelle mushrooms to sprout among the leaf litter, and there is a smell of damp in the air—almost like walking through jungle. When you see the rocks poking out at the sides of the path you are within earshot of the tumbling waters below. You notice at once that the "grain" of the rock is vertical. Ribs of rock flank the hillside like dense palings. There is a lot of the mineral mica, which glints in the sunlight; every flat little plate of this mineral is also orientated vertically. The "grain" of the rock is its foliation—the way it splits in response to the massive tectonic forces that have squeezed its every particle. Here, you realize, the rocks have been compressed and mangled in a way that has oppressed their very atoms. Seeking to accommodate unbearable pressure they have realigned themselves at right angles to the lateral forces generated within the earth. A strange-looking lumpy rock is all that remains of a granite, its feldspar crystals reduced to little knobs. Evidently, even the most obdurate of stones has succumbed to relentless pressure. There are creamy-looking rocks, too, which are reminiscent of the Lochseiten limestone: these are mylonites, the paste that results when rock grinds against rock in tectonic torture.

All these highly metamorphosed rocks are the roots of the Austro–Alpine nappes, those that overlie everything else: the roof of the Alps. This is where great sheets of strata come to ground. Here is the aperture that spun out those gobs of tectonic pasta that extend 100 kilometres to the north. It is a wonder how so much can arise from so little. It is known exactly when it happened, too. The dark, lustrous mica called biotite grew in response to the mighty squeeze; it contains a radioactive "clock" that was set at the time of the tectonic events, so we know it all happened 20 million years ago. The mountains were made high, and since then the action of eons of erosion has allowed us to see into their guts, as we did at the San Bernardino Pass. This story cannot but seem dramatic to the imaginative observer, but the drama, of course, is in our minds: we inevitably anthropomorphize. Even the words we are obliged to use come with our own human baggage: *squeeze, pressure, grind.* I have tried to avoid using the word *titanic* throughout this book, although it could be strictly applied since the ancient Greek explanation was that mountains were thrown up by the Titans. As science has advanced, the Titans have been sent packing. The rocks are indifferent to humankind; they move to the slow beat of geological time. But our language inevitably still entrains them with our lives. The Scottish poet Hugh McDiarmid put it well (in "On a Raised Beach"):

We must be humble. We are so easily baffled by appearances
And do not realise that these stones are one with the stars.
It makes no difference to them whether they are high or low,
Mountain peak or ocean floor, palace, or pigsty.
There are plenty of ruined buildings in the world but no ruined
stones.

Onwards—to Africa! Close to the swirling river there is a little rustic bridge that carries the old track across to the other side of the valley, before it snakes upwards again through yet more woods to reach the pastures above. When you lean over the bridge you can see rocks in the river bed. What is immediately obvious is that they are directed at right angles to those we examined by the descending path. They are something quite different. They look completely distinct from the crushed and shiny rocks, too—greenish schists for the most part. On the mountain slopes beyond they are in turn overlain by Triassic age limestones, which have no counterpart in the Alps to the north at all. The conclusion is obvious, yet it is hard to believe: the little bridge spans two worlds. South of the bridge there is a strange continent: it meets the main body of the Alps along the Insubric Line. The fact that the rock sequence to the south is neither squeezed nor deeply metamorphosed like those north of the bridge tells us something else: there must have been something like twenty kilometres of vertical movement along the line. Squeezed and heated rocks that have been deep inside the earth now lie alongside others that have not. Looking over the bridge at the stream making its way along the ancient junction, you expect to observe something more profound: but no, all that movement has long since sunk into quiescence, and only the waters recognize convulsions that once shook the earth. For this is the place where Europe meets Africa to the south. The bridge spans the two continents. The collision between the masses pushed up the Alps—which is grossly to oversimplify, as you must at first. When we had a closer look we found that Africa actually extended to just the other side of the bridge, so it was possible to take a photograph of my daughter with one leg in Europe and the other in Africa. I can imagine the photograph, by then fading slightly, puzzling one of our descendants—why the legs akimbo on a leafy path somewhere that seems without particular distinction? What would they say if they realized that the curious posture straddled a significant fraction of the world?

Now we come to what, to use a happy pun, is the crunch. For now the confrontation of tectonic plates comes into the argument. In our mind's eye, we have to draw backwards from the particular spot where my daughter stood on the surface of our kaleidoscopically complex world; or rather upwards, as if we were carried on a rocket into the ionosphere. Distance lends its own clarity. We see that our footpath terminates at one point in the south-eastern part of the Alps. Further upwards and we can see the Mediterranean Sea with its pattern of islands, and the great boot of Italy hanging downwards. We might even make out a plume of smoke from one of the classic volcanoes with which this book began, marking a current engagement of the plates. We can now appreciate the whole sweep of the Alps, and its continuation eastwards into the great arc of the Carpathian Mountains, and beyond that into Turkey, where we know the ground still shakes disastrously when the earth shrugs. We can comprehend that this is all one interconnected system, and that our path through Switzerland was no more than an amble through one tiny piece of a great mountain chain, which in turn links other ranges all the way eastwards to the Himalayas. Then we see Africa: a vast continent, and an ancient one. We can appreciate its presence, its mass, its unyielding solidity. Africa's sheer extent dwarfs that of the Mediterranean Sea. From this height, mountain ranges begin to seem more like wrinkles; nappes might be no more than tics on a spasm of the crust. If Africa is moving northwards—and it is—it suddenly seems plausible that all those monstrous rucks and buckled limestones and squeezed gneisses, over which our insignificant bodies clambered in the Alps, might be no more than the desperate scrambling of the earth away from the encroachment of the inescapable giant. The nappes fled northwards; the whole Carpathian chain curves away from the oppressor. The ancient mass of northern Europe—Eduard Suess' "foreland"—braced itself against the onslaught, and the effects are felt far beyond the Alps themselves. The basement rocks of Europe shivered, and the rocks that covered them were flexed. Even the shape of south-east England is influenced today by the approach of the distant giant. Everything scales to the great event: the intimate history links with the grander design. You cannot look on too large a scale to help you understand a solitary footstep through the landscape, a single crystal on a cliff.

The mountains grew as ancient oceans vanished. The oceanic lithosphere was consumed until continent collided with continent, Europe with Africa; this whole process was the "motor" of mountain-building. There were other "Mediterraneans" before the present one, seas that came and went in the area between Africa and the European foreland. Sediments that accumulated around their continental shelves became the limestones, sandstones, and shales that subsequently fed the nappes. The greatest of these seas was what Eduard Suess called the Tethys (named for a Greek sea goddess), which largely predated the growth of the mountains. Much of the Tethys was shallow and warm during the Mesozoic, and full of life, a marine paradise of sorts: ammonites and sharks thrived, and are now preserved as fossils that date the rocks. Suess had inferred greater depths in some parts of the Tethys because calcium carbonate shells had apparently dissolved (something which only happens at depth): "the fact that the calcareous base of the shark's teeth has been dissolved and removed [proves that] the Tethys, where it extended over many places now occupied by the Alps . . . can hardly have been less than 4000 metres depth." Some scientists may dispute Suess' exact figure now, but nobody doubts that the Tethys spanned everything from the shallows to the deeps. But what of the dark Flysch, over which we had slithered by the roadside where this chapter began? As the mountains rose above the sea in their early stages they were immediately set about by the forces of erosion: the waste derived from the growing chains fed into adjacent marine basins—especially sags lying to the north of the growing chain—as masses of dark sediment. These sediments, the debris of tectonics, comprised the Flysch. As the mountain belt continued to evolve, and the crust shortened further between the opposing jaws of the tectonic vice, the Flysch itself became entrained with the folding and thrusting. In places it became deformed: hence the squashed Flysch fish. The Glarus Nappe then slid over it, trapping it forever. By bringing in the movement of plates, the explanation of mountain chains becomes simpler than the recondite geometry of folds and nappes, and all the twisted extravagance of strata, might have suggested to those patient mappers.

Except it isn't so simple: nothing in nature is so obligingly straightforward. The *principle* of the Alps being squeezed out inexorably between the moving mass of Africa and obdurate Europe is correct enough. However, the margins of both "sides" were not one single piece of continent—like the jaws of a woodworker's vice. Rather, both the African and European edges comprised several smaller pieces that behaved with considerable independence. A better analogy than a vice is to think instead of two huge ice sheets jostling in the sea, with pieces fragmented from either edge grinding together or free-floating by turn as the gulf between the larger sheets waxes and wanes. The plate that lay on the far side of the bridge in the Val Morobbia should correctly be called Adria—of African origin, surely, but with a tectonic history only loosely tied to its great parent. So it was the collision of Adria—not Africa in the strict sense—with the adjacent part of Europe that was responsible for the heave of the Alpine nappes. By the same token, the margin of "Europe" actually comprised several blocks—massifs—of old rocks. The field mappers had in fact recognized these masses of ancient (Hercynian) rocks underlying nappe complexes within the Alps. This was a triumph of deduction considering that this "basement" could look much like some of the baked younger rocks. Most of the massive igneous rocks, like granite, within the Alps actually belong to these ancient massifs, including mighty Mont Blanc itself (15,782 feet, 4810 metres). As we have seen in the San Bernardino Pass, the older rocks could also be remobilized in the Alpine movements, just to add perplexity to complexity. These ancient blocks provide an explanation of the separate "packages" of nappes. A single crunch—Africa (even Adria) on Europe—could not by itself explain the complexity of tectonic events that built the great piles. Instead, there was a long evolution of the mountain chain, event piled on event, which threw one series of nappes above its predecessor. There must have been several closures, block to block, not just the single great one when the opposing continents came into direct confrontation. When one massif collided with its neighbour this produced earlier phases of deformation, the earlier packages of nappes. There was even a pattern to it, a southerly migration of collision events as the African blocks progressively approached the European. There must therefore be more than one root zone—for each of the major events should connect to the line of its own genesis. There was, in short, squeeze following upon squeeze. The last, and arguably the greatest, was the one that left behind the tortured rock along the Insubric Line. So great, in fact, that morsels of "Africa" were carried bodily northwards over Europe. Each of these events was profoundly important in shaping mountains, and yet, viewed from the high vantage point of a passing comet, none was more than another pulse in the slow and inexorable progress of the plates. Lubrication was important: several of the nappes glided on salt, deposited after one of the ancient seaways dried out during the Triassic. This was the oil that smoothed the passage of mighty segments of the earth.

Such is the moulding of mountains: each one explicable in terms of tectonic principles, yet each stamped with its own individuality. To this extent, mountains are like people, cast from common genetic rules, yet every one having a personality of its own. Mountains are often the personification of strata, and indeed they are often named for their fanciful resemblance to white teeth or a virgin's breasts. Alpinists struggling with difficult peaks always seem to address their adversaries in human terms rather than as examples of tectonics in action. The Matterhorn, on the border between Valais (Switzerland) and Piedmont (Italy), is celebrated for its precipitate sides and for the fact that Colonel Whymper lost four men in its conquest in 1865. Less well known is the—far more astonishing—fact that its upper portion is actually a part of "Africa" thrust bodily northwards over Europe. The character of the mountain, and the face it presents to the climber, is as surely a product of its geology—and the subsequent effects of weathering—as our own character is a consequence of our genetic makeup acted upon and modified by the vicissitudes of our upbringing.

It is time to leave "Africa" and head northwards again, back into the Alps to look at one of the ancient massifs. From Bellinzona, this means following the Ticino Valley upwards, to the west of our path down from the San Bernadino Pass. The climb on the old road to the St. Gotthard Pass proved to be a reverse replay of the descent. My visit there was obscured by low cloud: what should have been a splendid panorama was reminiscent of a late Turner painting, a kind of dramatic but milky obscurity. But it was still possible to inspect—up close—the rock of the Gotthard Massif. This is the heaviest kind of grey granitic gneiss, a rock of ancient and implacable solidity, grim and giving no quarter. It is easy to imagine it as one side of a tectonic vice, or as some kind of bulwark against which nappes are no more than pliable mash. The sparse vegetation clearly has a tough time extracting the most meagre rations from its impoverished soil; it is the Scrooge of rocks. Time to move on.

The next massif northwards is the Aar Massif—more grim gneisses forming intimidating cliffs. Sandwiched between the Gotthard and the Aar, like a thin slice of Parma ham between two halves of a loaf, is the root zone of another of the great nappe packages—the Helvetic nappes. At the foot of the St. Gotthard Pass, the town of Andermatt is nearby, one of those places with a kind of storybook Swissness. On the edge of town, near the railway, and in a much less picturesque quarter, is a large, dull grey barracks, behind which the root zone crops out on the hillside. The cliff displays a series of yellowish crushed limestones and—as along the Isubric Line—they are vertical, with a squeezed mien. We have come full circle, for this is also the root zone of the Glarus Nappe, where this journey began. This is the time to envisage the mighty piles above you; how erosion of the higher nappes has opened "windows" giving on to the earlier tectonic events; how the mountains are still rising, while frost and rivers incessantly seek to render them low; and how all this is accountable to the movement of plates. Modern research is focused particularly on deep processes and structures, in the lower crust and upper mantle, as revealed by seismic reflections. Here you can see how the opposing continental masses really meet. For example, it is likely that a "wedge" of Adria impinged into the guts of Europe, such was the northward heave of the greater continent to the south. This, the most studied range in the world, still has secrets to reveal.

There is one rock type to mention as a postscript: molasse. The typical locality is the mountain known as the Rigi, west of Lucerne, but the whole area on this shore of the Vierwald is made of it, an old-fashioned plum-pudding kind of rock, a pebbly potpourri, including rounded cobbles of all manner of other rock types. I saw many pale limestone pebbles, some sandstones, even some pieces of what I took to be gneiss, and all dusted with a kind of pinkish skin. It is a massive rock formation, underlying whole mountainsides. The molasse provides a record of the wearing down of the Alpine range at a late stage; unlike the Flysch, it is largely non-

The waste of a mountain range: the Rigi by Lake Lucerne is composed of molasse derived from erosion of the Alpine chain.

marine. The cobbles and pebbles had their edges knocked off in raging torrents, and they were rounded further in their inexorable progress downhill. Every cobble in the molasse represents a bit broken off a mountain: they were collected together indiscriminately in the deposit. A mountainside of this conglomerate is a mountain destroyed elsewhere. The rivers of 10 million years or so ago drained northwards into a basin that now occupies the Swiss Plain, and so molasse underlies the serious, business side of the country: even commerce bows to geology. By the time they had reached this point, much of the energy of the waters had abated and was insufficient to move pebbles, so sandy and silty rocks predominate under the plain. Around Lake Lucerne, though, there are coarse rocks laid down in ancient channels or spread out by floods. Down by the lake the molasse slopes are gentle, and the soil is a reddish colour that I do not recall seeing elsewhere on our journey. There are old orchards everywhere, but the most profitable crop these days is tourism. Down at Weggis, the Seehof Hotel du Lac is the very picture of luxury, with a waterfront having Venetian pretensions, and wrought-iron balconies and blue shutters. It was fashionable when built a century ago and is now quietly dignified and expensive in maturity; the foreigner's vision of Swiss elegance, perhaps. But I like to think first of the apple orchards on the hillside above, growing on red soil that has been derived from the weathering of peaks long since vanished. The revolution wrought by tectonics has been finally converted into fruit.

We have been on a journey through a small part of a great range, and that range is itself connected to others that extend as far as the Himalayas—and further, through the Malayan Peninsula to Indonesia. Although it is a classic area, it would be foolish to expect the Alps to be a surrogate for all mountain belts everywhere: the world is too rich and complex for that. The Alps are, for example, poor in granites, those most implacable of mountain makers, and most of those that do occur are ancient rocks caught up in the later earth movements.* No matter: we shall meet them elsewhere. Nor are nappes as important in some other mountain belts—the Alps were the product of a particularly hard "docking" of one continental mass against another. Push invariably came to shove: but it does not have to be so violent. And we will encounter in other places fragments of ocean floor caught up in tectonic mayhem. Actually, these *do* occur in the Alps, for the mappers had accurately recorded "serpentines" in their accounts; it just took a very long time for their significance to be appreciated. According to Rudolf Trümpy, if they had been recognized for what they were, the whole theoretical edifice of plate tectonics might have been deduced from the Alps: but this is to anticipate things to come. What can be said without blushing is that the elements that we met in our journey round Glarus are common to mountain belts through much of geological time and in most orogenic belts. There is "Flysch" and "molasse" associated with many folded ranges. I have laboured long hours trying to find fossils in Flysch nearly 500 million years old: no fishes to gape at me then, just the little "hack-saw blades" of graptolites. Then, too, the earth's crust has buckled and shortened in every mountain chain, and where it has thickened, metamorphism has produced gneisses not so different from those we saw on the San Bernardino Pass. Strata from ancient seas have been thrust up to cap high peaks (there are fossil shells at the top of Mount Everest). Virtually all ranges, too, are composite like the Alps, with one event following another as the mountains evolved over several tens of millions of years. Convolution and concatena-

* There is one important granite associated with the Alpine orogeny. It cuts through the Pennine nappes and therefore must postdate them. It is thus important for calibrating the timing of the events in the whole range.

tion, folding and refolding, buckling and cracking or heating and bending, squeezing followed by collapse when the pressure is off: it is about as complicated as geology gets. Sometimes rocks are so perturbed and disguised that it is hard to say *what* they once were.

Mountains demand an explanation beyond the fury of the Titans. After all, the thoughtful traveller soon realized that many mountains came in linear ranges: there had to be a *design* to the earth beyond the pettish whims of the gods. The seventeenth-century Jesuit Athanasius Kirchner saw mountains as an internal skeleton, the bones of the earth. When you view the Alps from a distance, perhaps this does not seem quite so absurd, especially since hard and resistant rocks often take the high ground: rocky bones indeed. Kirchner's *Mundus Subterraneus* of 1664 continued to be quoted until well into the eighteenth century and was, at the least, a stimulant for thinking about mountains in the context of a global explanation. The search for a theory of mountains was to occupy many more able geological minds for the next 300 years. The Alps were always central to the arguments. With their gift for abstraction, it is perhaps not surprising that the French played a prominent part in the process. Since Descartes had freed the wondering mind from theological baggage, the world had become their philosophical oyster. Léonce Elie de Beaumont (1798–1874) was the director of the *Carte Géologique*, a mighty undertaking to map the rocks of France completely for the first time. He was by training a mathematician, and inclined to view the world as deducable from mathematical principles. He was gifted enough to become the top graduate from the École Polytechnique in Paris, forcing ground for the *élite*, and became a distinguished field geologist. His achievements probably induced in him a certain arrogance; and Elie de Beaumont did indeed become a monolithic authority. For thirty years, between the 1820s and 1850s, he published influential works on the causes of mountain-building. Recall that this period coincided with the acquisition of basic data from the rocks themselves—the heroic efforts of those who climbed the Alps with pencil and paper in their knapsacks as well as a hunk of bread and cheese for lunch. So Elie de Beaumont was theorizing in tandem with new and hard-won discoveries in the field.

The same period also saw the rise to prominence of the scientific methods advocated by Charles Lyell with which this book began. Lyell himself had considered mountain formation in some detail. In line with his uniformitarian notions, he had conceived of grand elevations by the continuous accretion of existing effects, particularly the uplifts in topogra-

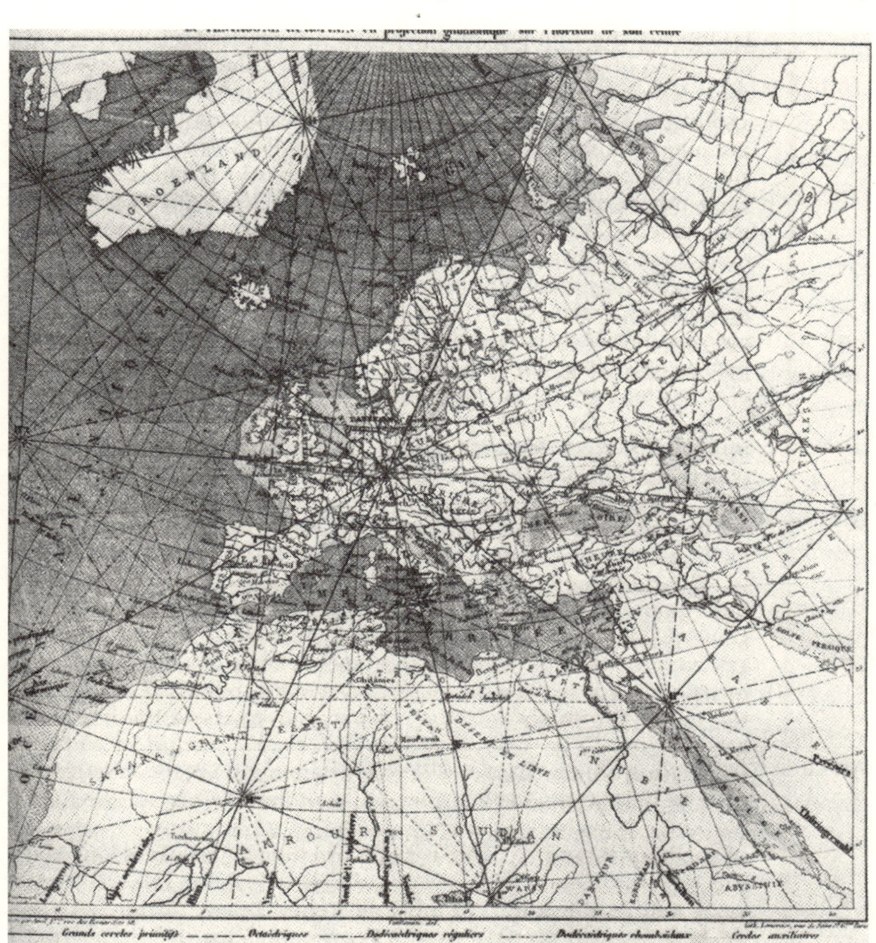

An early attempt at global synthesis of tectonics: Elie de Beaumont's (1798–1874) attempt to line up mountain chains into his geometrical system of the réseau pentagonal.

phy that were known to be produced by earthquakes and volcanoes. It was a case of excelsior little by little. By contrast, Elie de Beaumont saw violent revolutions, narrowly confined to zones of weakness in the earth, and limited to special periods of global unease. He was aware of the growing evidence for powerful folding in the Alps: no mere nudging could produce such contortions. He was also attracted to measuring the lie of the mountains, and plotted their linear trends with an eye to their systematization. Elie de Beaumont was in little doubt that the folding and thrusting he and others had observed in the Alps resulted in an overall piling up of the strata—what we would now recognize as crustal shortening. Some of his results were published in the *Philosophical Magazine* in 1831 as "Researches on Some of the Revolutions of the Surface of the Globe: Presenting Various Examples of the Coincidence between the Elevations of Beds in Certain Systems of Mountains, and the Sudden Changes Which Have Produced the Lines of Demaracation Observable in Certain Stages of the Sedimentary Deposits" (*Philosophical Magazine,* volume 10). If an author were to submit such a title today the editor would faint with shock . . . but it does save me the trouble of explaining what it is about. It is tempting to seize upon the "Revolution" in the title. After all, revolution was likely to figure high in the consciousness of a Frenchman, and Lyell's modest changes, which effect much in small steps, provide the model for accomplishing reform through continuous parliamentary democracy. Lyell's ideas were also crucial to the young Charles Darwin: evolution, not revolution, has been claimed as the mantra of democracy.

The clash of opinion on the origin of the Alps is often presented as a test case for the great nineteenth-century debate on "uniformitarianism versus catastrophism," but it is not as simple as that. Many geologists of fundamentally Lyellian persuasion were impressed by Elie de Beaumont's results. His political equivalent as the head of the British Geological Survey, Henry de la Beche, was one of them. Other processes might be involved in mountain-building than mere "up and down"—but what? He viewed with favour Elie de Beaumont's idea that the earth was contracting as a result of its slow cooling, mountains being thrown up as periodic "adjustments" of the crust to accommodate a diminishing circumference. "If we suppose with M. Elie de Beaumont," he wrote in 1834, "that the state of our globe is such that, in a given time, the temperature of the interior is lowered by much greater quantity than on its surface, the solid crust would break up to accommodate itself to the internal mass; almost imperceptibly, when time and the mass of the earth are taken into account, but by considerable dislocations." That "almost imperceptibly," combined with "considerable dislocations," is suggestive of revolution by stealth; of having his Lyellian cake and eating it, too.

Elie de Beaumont is commemorated by the mountain that carries his name in the Alps—the New Zealand Alps, not the Swiss. By geologists he is remembered—if at all—for the aberrations of his maturity rather than for the brilliance of his youth. Obsessed with finding a mathematical design in the arrangement of mountain belts, he developed what he termed the *réseau pentagonal.* He ascribed the directions of mountain belts to regular pentagons drawn on the surface of the globe—the pentagons themselves being derived from the intersection of fifteen great circles. His lines were "zones of weakness" along which mountain belts preferentially aligned. This, he believed, was where the subterranean controls on folding were expressed on the ground. It was a total global system, a theory of almost everything. It was also an idée fixe for Elie de Beaumont and he defended it to the last, and with his by then considerable authority. When anomalies in his system were discovered, he simply made it more complex to cope with them, until his intersecting lines drawn on the earth resembled a spider's web. To historians of science the situation is a classic one: it is reminiscent of the "epicycles" that had to be introduced to retain an earth-centred astronomy. To account for awkward facts, the system of orbiting planets was tinkered with until it collapsed under the weight of its own modifications. Similarly, the *réseau pentagonal* died with the great geologist. Of course, with the wisdom of hindsight, we might say that Elie de Beaumont was half right. As we shall see, there *are* zones of weakness where tectonics reawaken. He was more accurate in his reading of the Alps than was Lyell. And one day a great earth system would indeed be discovered. His ambitions had exceeded what was known of the earth at the time, and his grand mistake obscured his grander virtues. History sometimes cheats reputations.

Eduard Suess had more of the geology of the Alps at his command than any of his predecessors: after all, it was home territory for him. He was as averse as Elie de Beaumont to the notion of mountains as areas accountable to uplift alone. Everywhere he saw evidence of lateral movement, of rock formations piled on one another. He was familiar with much of the country we have explored, and he read it well. Suess never thought other than globally. Like Marcel Bertrand, he made connections between mountain chains, joining one to another like a cordon drawn across the continents, and he knew that such a global phenomenon required a general explanation. The Andes and the Rockies were all grist to his tectonic mill. By the end of the nineteenth century, it was accepted that there were many ancient mountain belts, and that the history of the planet had been punctuated by mountain-building events, or orogenies. We shall be visiting some of these old scars on the face of the earth later in this book. Whatever the cause of mountain ranges, then, it must be one that could stretch back through geological time. Exactly how much time that involved was far from clear when Suess wrote *Das Antlitz der Erde,* but it was certain that many millions of years were necessary to lay down the rocks of seas long vanished, elevate them into mountain chains, by whatever means, only to have them worn down to contribute to the next cycle of sediments. When Suess compiled his synthesis of the world, the geological scale of time periods based upon the succession of stratigraphy and fossils—Cambrian, Cretaceous, and the like—was essentially in its modern form. But much finer subdivisions of geological time had also been introduced. As an example, Suess went to great lengths to show that one small part of the Cretaceous, termed the Cenomanian stage, was a time when the seas advanced widely over the surface of the globe. What was not known was how many millions of years ago this event happened—the radiometric scale that could measure it was still to be developed. Still, the scientific basis for correlation of rocks was sufficient to make generalizations about when and how the past upheavals of the earth had happened. In Europe, three main mountain-building events—Caledonian, Variscan, and Alpine, respectively—served to divide the history of the continent. They were times when the earth convulsed, when Lyell's slow ticking of the geological clock was speeded up; between times, the earth was calmer. For Suess, summing up folding and thrusting in a dozen mountain chains, orogenic events happened when the earth shrugged violently and pulled itself together. After all, he maintained, the earth was contracting in volume: when the strain became too much, mountain ranges were the planet's way of coping.

Eduard Suess was a thoroughly political animal. He was a member of the Austrian parliament for twenty years. He knew how to make a case. *Das Antlitz der Erde* is more than a brilliant summary of how the world was made; it is also a disguised polemic for Suess' view of it. He first set out to establish geology's omniscience. An early chapter deals with the biblical Flood. The discovery in Mesopotamia of the clay tablets known today as the Epic of Gilgamesh, and to Suess as the Izdubar Epic, allowed for a new interpretation of the biblical event. The tablets in the Akkadian language described the flood as history, and in some detail. The decipherment of the tablets was a "state-of-the-art" discovery. Suess took the account on, point by point, to show that the Flood was a geological event: a tidal wave or tsunami approaching from the sea rather than the result of exceptional rainstorms coming from the mountains. It was the first attempt to demystify the Flood.* Geology was a great rationalizing force, and throughout his huge compilation, Suess could not resist providing a rational explanation for legends. Here he is entertainingly slaying a dragon: "*Trionyx aegypticus* lives in the neighbourhood of Beyrut, and it is a remarkable fact that the crocodile of the Nile still exists in the nearby estuary of the Nahr e' Zerka, or crocodile river, three kilometres to the north of Caesarea. Pliny knew of a town—Crocodilion—in this neighbourhood . . . These facts also throw unexpected light on the numerous and circumstantial accounts of the slaughter of a scaly monster by the knight Deodat von Gozon on the island of Rhodes in the first half of the fourteenth century." Curiously, Schiller, whose Wilhelm Tell was linked above with one of the classical geological sites of Switzerland, also wrote a poem about this "dragon," based on an account by one J. Bosio, published in 1594.

Suess was equally concerned about promoting his view of tectonics. "A complete revolution of opinion has taken place as regards the formation of mountain chains," he wrote, and it was his own ideas to which he was referring. When he wished to underscore a point he resorted to italics, so the reader is infallibly guided to the nub of the matter. "*The dislocations visible in the rocky crust of the earth are the result of movements which are produced by a decrease in the volume of our planet.*" Nothing ambiguous about that. The shortening of the earth's crust observed in mountain ranges was inescapable. Since it was also clear that such linear chains had a long history, perhaps extending back in time for hundreds of millions of years, there had to be a way to explain the inevitable telescoping of all those strata. A decrease in the earth's volume would provide it. But Suess regarded ocean basins as areas of subsidence, so there was no opportunity to reduce the area of the earth's surface there. It would take ocean-floor spreading to balance an apparently impossible equation, as we shall see. The ancient cores of continents seemed to be inviolable forelands onto which the mountain structures spilled, as would a wad of clay squeezed between two bricks. It had to be the mountain chains that took the strain, and initiated the next geological cycle of erosion and sedimentation in the long, slow dance of the earth. As for the Alps, the details of a single outcrop confirmed the larger scheme. Here is Suess' description of the kind of country we have explored in this chapter:

* A recent attempt, by the distinguished oceanographers Drs. Ryan and Pitman, makes a good case for the Flood legend originating in a huge marine inundation of the Black Sea. Clearly, geology can yield more than one answer.

> The Flysch has been driven like a plastic mass, and the great advancing sheet of Cretaceous limestone rears itself up, creeps upwards over the Flysch, assumes a vertical position and finally turns over backwards. Heim has termed this a turnover klippe . . . it is precisely these ends of the sheets which have directed attention anew to the diminution of the Earth's circumference.

Home territory is always the most persuasive. For "it is the number of sheets lying upon one another as seen in the Swiss Alps, which most impresses us with the magnitude of the phenomenon [i.e., shortening]." From Glarus to the world, indeed! The image that was often used to illustrate Suess' idea was a drying, and therefore shrinking, apple, whose wizened countenance and ridges and furrows seemed appropriate for the visage of the earth crossed by mountain ranges. The metaphor was taken so seriously that pictures of old apples actually appeared in textbooks. Perhaps it was secretly hoped that this apple would be to tectonics what Newton's was to physics. As recently as 1952, the Presidential Address to the Geological Society of London was on the topic of a contracting earth.

I like to think of a succession of observers contemplating a mountainside, perhaps the one across the lake from the hotel balcony by the Urnersee where Wilhelm Tell jumped to freedom. Old Kirchner would have seen a kind of girder reaching from the innards of the earth. Charles Lyell might have seen the consequences of ten thousand tremors. Elie de Beaumont might well have seen the same strata as a mere grace note on the *réseau pentagonal.* He might have smiled briefly to himself, recalling the beauty and completeness of his system, then frowned momentarily while recalling some critic or other. And Eduard Suess would have seen folding as yet more incontrovertible evidence of the incessant, slow shrinking of the earth. The image hitting the retina would be the same for all of them, but what is actually *seen* depends on the secret workings of the brain. Seeing is not believing; rather, it is belief that governs seeing. The modern observer? He might see little squirming strata on the back of the collision of Adria and Europe. No doubt the present-day viewer is closer to the truth than any of his predecessors, and the advances in understanding in the last half century have been enormous and permanent; but it would be foolish to suppose that an observer a century hence will see nothing differently. *Autres temps, autres yeux.*

By the time Arthur Holmes wrote his textbook, the structure of nappes was well understood. Heim and his successors' magisterial maps had been published and digested. In the fifty years after Suess, physicists had been to work on the idea of the secular contraction of the earth, which was originally based on the concept that the earth was a cooling body. The discovery of radioactive heating upset all the calculations: it just wouldn't work. We shall soon meet those who thought, on the contrary, that the earth was expanding. The poor old earth could apparently be pumped up and down like a balloon! The global pattern of mountain belts was well known, and Eduard Suess deserved much credit for outlining them; but, as Holmes remarked, the "assemblage of mountain systems combines into a world pattern of apparent simplicity which nevertheless masks a variety of details that remain perplexing and unexplained." There had to be a mechanism for the emplacement of nappes that did not depend on the contraction hypothesis. It must be a force capable of moving large chunks of strata laterally at low angles. Newton's pristine and unshrivelled apple might have its place in the argument after all, for gravity is such a force.

The higher nappes in the Alpine pile might, in this theory, have slid into place, on a lubricated sole of salt, or some other tectonic "oil." As they moved, they also folded progressively. You can get some impression of how it works if you lay well-puddled clay in a thin sheet upon a board and tip it upwards: if the consistency of the clay is correct it will eventually slither and fold under the influence of gravity. In the 1950s several serious academics, especially Hans Ramberg of Uppsala University in Sweden, spent a lot of time on clay models, and sometimes produced good simulacra of nature. Analogue models have developed considerably since those days, and are now used extensively in the oil industry. Curiously, the gravity sliding theory was in essence a revival in sophisticated dress of Charles Lyell's notions of uplift as the essential control on mountain-building. For you also need to elevate some region along the growing mountain belt to produce a "high" from which the nappes could slide. Such a line of elevation became known as a geanticline—an upward swell running along the axis of the mountain chain. As the geanticline rose upwards, nappes glided off like slicks from the back of an emerging whale. This might satisfactorily explain the higher nappes, but then you also need an explanation for the kind of nappes which we saw at the San Bernardino Pass—those that involve deeper (and older) rocks, metamorphosed under high temperature and pressure. How do you get them to rise up from the depths? It is a slightly difficult concept to grasp. Arthur Holmes explained it like this. The "deeper and hotter parts of the infrastructure . . . [were] where these [burial] effects are most intense, as in the core of an orogenic belt, [and hence] the zone rises to higher levels and may pass high into the superstructure." Deep, hot rocks "flow" upwards, deforming and creeping like a contrary glacier of the underworld. There was a model for this process in the salt domes of the Middle East. Deeply buried deposits of sea-salt dome upwards and pass through overlying strata, as a kind of intrusive lobe, eventually emerging at the surface—the rising tongue is called a diapir. I examined one of these salt domes in the Oman desert. It makes an improbable hump hundreds of metres across in the middle of a stony, flat wasteland: you can see it from miles away. A partly flowing, partly crusty hilly mass of salt is a curious thing to see in the fiery heat. Even odder is to find all sorts of blocks of rocks lying around that have been carried up with the salt as it cut through strata overlying it, including limestone blocks with fossils. Some of these salt domes have acquired commercial importance as oil traps.

The heyday of gravity tectonics came and went, as the contracting earth had done before it. Once more, detailed fieldwork provided critical evidence. In the first place, when nappes were very carefully reconstructed back to their original disposition (imagine refolding those sheets back on the bed) the evidence for any kind of downhill slope was very tenuous. It began to seem as if a "push" were needed after all: gravity alone would not do the work. Secondly, as the history of the various parts of the Alpine chain was unscrambled, it became clear that the emplacement of nappes coincided with the closure of ocean basins—precisely the circumstances when such a "push" might result from natural causes. The timing of the emplacement of the "packages" of nappes could be related to the history of the massifs and the sediments that accumulated upon and between them. Further research discovered microcontinents like the Briançonnaise—parts of "Europe" that, like Adria in relation to Africa, had a semi-detached history. It seemed a logical inference that the Alpine mountains were the consequence of the jugglings of all these major pieces of lithosphere, and not just a question of "up and down." That uplift was going on was undeniable—recall the Via Mala, that slit carved steeply into the earth—but this was as much a feature of the present history of the chain as an element in its early evolution. The time was ripe for the Alps to be integrated into the story of the plates, and that of the confrontation of old Africa and old Europe. When that happened, the green serpentinites in the Pennine nappes would suddenly make sense. But this is to anticipate.

. . .

We leave Switzerland by way of the Jura Mountains. The name of the mountains is famous in geology; it is the type area for the Jurassic system, 205–137 million years ago, which, as every child now knows, was when Dinosaurs Ruled the Earth. Rocks of this age are beautifully developed in the area, as you would expect of a type locality, although the ones I saw were all sedimentary rocks that had been laid down under the sea, and would therefore yield to the hammer the fossils of humble invertebrates like ammonites, brachiopods, snails, and clams rather than the bones of huge terrestrial monsters. The mountains run in a great swathe north-west of Neuchâtel—where Marcel Bertrand was a professor—following the direction of the Lac de Neuchâtel. All along the northern edge of the lake, the southerly facing hillsides rise in vineyard after vineyard. Beyond the vineyards a wooded ridge reaches to the skyline. The Jura Mountains comprise a series of great ridges like this one, running in parallel east-south-east–west-north-west. They are like the wrinkles on the forehead of a very old man. The high points of the ridges often exceed 1300 metres, so they are not negligible, even if they are dwarfed by the Alpine peaks south of the Lac de Neuchâtel. Between the ridges there are valleys with the same trend, so the whole countryside is one of densely wooded, rising ridges separated by intensively farmed depressions—corn, beet, and wheat—and this pattern extends well beyond the border into France. Cows have changed from the cute grey Swiss ones to a more everyday red-and-white breed. Most of the rivers faithfully follow the valley floors, but a few of them turn to cut directly through the ridges in steep-sided gorges. We follow one of these north of Neuchâtel, which traverses the Chaumont. Rocks are well exposed in the sides of the gorge. One difference from the Alps is immediately obvious. No longer are the strata thrown into elaborate contortions. They are tipped perhaps 30 degrees from the horizontal, but one stratum succeeds its predecessor in orderly fashion, younger upon older, as well-behaved stratigraphic sections should. An occasional small twitch in the bedding reminds us that these rocks, too, have not escaped tectonic influence, but, clearly, we are beyond the reach of the most violent of the Alpine convulsions.

The little town of Valangin guards the entrance to the gorge. It, too, is a famous name in geology, for the Valanginian is the label attached to the second earliest time division of the Cretaceous, following on from the Jurassic. The strata in the area record the passage of time from one period to another without a break. Old Valangin is built from the local limestones, which yield deliciously creamy yellow blocks. That the château was defensive in origin is obvious, for its massive walls boast round towers on the corners, and its entrance is heavily fortified. The block-like big house tucked away safely inside is functional first, luxurious second. This place was evidently built to protect the route running through the gorge. In Valangin you are reminded how appropriately villages that have been built from the local geology seem to lie in the landscape. They appear to grow from the hillsides. France, with its abundance of limestones, has hundreds of such towns and villages, from Provence in the south to Caen in the north. No matter if the buildings lack distinction—though they rarely lack individuality—for the harmonious whole is more than the sum of its parts. England, too, has dozens of villages in the Cotswolds with the same charm. Elsewhere, where natural building stone is unavailable, local bricks and tiles still grow out of the land, and add their character to the regional architecture. In Provence, I visited the source of ochre, which is a sandstone formation cutting across the vineyards near Apt. How suitable this colour wash now seems rendering walls in the Midi; like sunflowers—and much the same shade—ochre revels in the rich heat.

The structure of the Jura is comparatively simple. The great waves of hill and vale passing over the landscape reflect similar waves in the strata. The ground undulates in harmony with the folding of the Jurassic and Cretaceous limestones. The ridges are anticlines, meaning that the rocks are bowed upwards. The valleys between them are synclines, where the strata are bent downwards in the fashion of a cupped hand. The topography follows the bidding of the underlying structure almost slavishly. Everything else is a consequence of the geological underlay: the cultivated valleys; the wild, wooded hillsides; the courses of the streams; where vines will ripen grapes. Look carefully when crossing over a ridge and you will see the switch in the general dip of the limestone beds as you traverse the axis of the anticline. On the southern side of the ridge most of the limestone beds tilt to the south, northwards they are inclined to the north; a simple scheme only complicated by the inevitable minor undulations. The Jura is a geological switchback, a sine curve graven into the earth.

The form of the ridges was generated as part of the Alpine movements. Looking at the map of Europe it is easy to see the Jura as an outrider to the great weal of the Alps, flanking all that confusing disturbance, that most unruly and stirred-up pile. Imagine scrabbling up a rug into a mass of folds at one end, while at the other the material is thrown into a series of diminishing undulations. The connection between the Alps and the Jura was not always recognized. It was only with the growth of geology as a globally interconnected science that the marriage of the two regions was established. Here is Eduard Suess writing in the second volume of *Das Antlitz der Erde:*

> There was a time when every single anticline of the Jura was regarded as an independent axis of elevation; then it became clear that such a collection of parallel anticlines must have a common origin; next it was seen that there is a certain dependence between the Alps and the Jura; finally, the influence of the obstacle presented by the Black Forest was recognized . . .

Suess saw the Black Forest as part of the "foreland" against which the Alps were constrained. The Jura Mountains were the relatively minor folds produced as the Alpine movements lost their momentum against the immovable block to the north, a series of shrugs of exhaustion. The modern view is perhaps not so different in outcome, except that the cause is—at the deep root of things—the northward movement of Africa. The Jura is part of the marginal effects of the convergence between the European Plate and its great neighbour to the south. The folding is now considered to be the reaction of a thin "skin" of sedimentary rocks to the kick of continued Alpine movements, lubricated by a detachment in the thick Triassic salt deposits at depth. It is as if the folds in the rug were generated by changing the relief on the especially slippery floor on which it lay rather than by shaking the rug from one side. Seismic profiles made through this part of Europe suggest that the deep detachment extends southwards all the way to the Aar Massif. The vines and forest trees on the flanks of the Jura owe their particular habitat to events that happened several kilometres beneath their roots. Thus it is that the intimate details of natural history link downwards into unseen worlds. Everything connects.

At the top of one of the ridges there is a *Vue des Alpes.* Lying beyond the far shore of the Lac de Neuchâtel, the array of white peaks takes up the horizon, like ranks of shark's teeth. Distance scrambles them together, so it is hard for a foreigner to place them in order. They seem to erect an impenetrable barrier, yet we know that even in the late Stone Age there were skin-clad men who would brave the remote passes; one of them fell into a crevasse and was yielded up from his natural deep-freeze only a few years ago. Forensic palaeontologists have determined that he had ibex for lunch and that he may have been shot in the shoulder by a rival. Can it really be true that the Carthaginian general Hannibal Barca drove elephants over the passes 2200 years ago? We have the word of the Greek historian Polybius that it was indeed so. But before the era of the written word, whether the words we read are true or not, everything has to be established by inference. History envelops the past in uncertainty, like the mist obscuring the beech trees in the valley below me. The deeper the history, the more the outlines blur, the more inferences about the past are subject to change. In this chapter we have seen how different eyes contemplating the Alps each saw their own version of history. We have a vision for our time, but we can be certain that it will not be the last. Modern heroes have followed in Hannibal's tracks, but this time with map and pencil in hand, unscrambling the secrets of the Alps written in a code of writhing rock. We know that it is not faith that moves mountains; it is tectonics. Geophysicists today poke beneath the great pile with their seismic probes. We now know a great deal about the Alps: their structure, the role of plates in their genesis. There has been a revolution in our understanding over the last forty years, and the gains in knowledge are permanent. But we will never know everything, and that is as it should be. From the obscuring mist of the past, science has ensured that some of the mountains have emerged into clear view, but as soon as that happens the misty shadows of further peaks are glimpsed in the distance, rank upon rank: so many other heights to climb, so many mysteries to investigate.

5

Plates

There are lands of the imagination that cannot exist, but seem real; and there are lands that once existed that somehow seem remote and hard to credit. Perhaps their comparative solidity depends on the hand of a skilled writer. Who can doubt the reality of the countries beyond the sea that Jonathan Swift peopled so skilfully for his hero Lemuel Gulliver to visit, not merely to stimulate the imagination, but as a ruse to illustrate human frailties: puffed up and monstrous in Brobdingnag, or shrunk in Lilliput to petty proportions to match the triviality of their concerns? Yet to travel back in time to the land of the Gonds—Gondwana—or to try to grasp the reality of Pangaea 250 million years ago seems to require a greater leap of imagination. But these places existed, as solid as Africa is today.

Somewhere in a middling category is Atlantis, a familiar enough name, but one which is either myth, or myth ultimately rooted in reality. Determined researchers strive to place it on the map with the certainty of New York or Vladivostok. Plato described the kingdom in his *Timaeus,* locating it in the Atlantic Ocean—or at least "beyond the Pillars of Hercules"; but theories abound that prefer to place it within the Mediterranean Sea, near the cradle of the classical world. Whichever the preferred location, its catastrophic elimination as a punishment for the dubious ways of its inhabitants was a tectonic one, for Plato describes earthquakes and floods. A cursory search on the Internet will show the curious observer how many outré theories exist about the drowned kingdom. There are claims made for it being in the South China Sea, and, for all I know, on the moon. Probably the most plausible scenario relates the destruction of Atlantis to the eruption of Thera, now known as Santorini, seventy kilometres north of Crete. This was the biggest eruption that *Homo sapiens* has ever witnessed, at least in Europe. It has been estimated that thirty cubic kilometres of material was erupted—that is fifteen times more than in the famous eruption of Vesuvius in A.D. 79 that eliminated Pompeii. The eruption of 1700 B.C. also coincided approximately with the decline of Minoan civilization, and is therefore one of the crucial phases of western history. The Bronze Age could have had no better source for its subsequent legends. It must have been a most catastrophic blast, for it left a caldera six kilometres in diameter. It surely generated a vast tsunami, which would have drowned the vibrant coastal towns of Crete. On Thera itself, direct evidence of burial by pumice is preserved in the town of Akrotiri, where colourful frescos and wine jars testify to the good life before the Fall. The sky would have darkened with sheer volume of ash, the grapes would have withered upon the vine; misfortune would have fallen with the inexorability of volcanic tephra upon rich and poor alike. The plain-dwellers would have been punished with the mountain folk. That Thera *was* Atlantis is given some credence by Plato's description of the place (pre-eruption) as comprising several concentric belts of land and lagoons—a rather characteristic volcanic configuration. According to Robert Scandone, a vulcanologist at Rome University, the eruption of Thera may have inspired at least one passage in the story of Jason and the Argonauts. When they returned to Greece with the golden fleece, they passed by Rhodes to the eastern part of Crete; here they experienced a pall of darkness—as indeed they would have done if they were under a cloud of volcanic ash. When they fled northwards, a bronze giant called Talos pelted them with fragments of rock (this hardly requires further explanation). The eastern Mediterranean is replete with flood legends, which are not unreasonably linked with tsunamis in the aftermath of major eruptions. Today, what remains of Thera is the precipitous and sun-bleached port of Santorini and its flooded caldera, which shelters a small fishing fleet engaged in satisfying the demand for fresh-grilled sardines in *tavernas* catering for tourist cruise liners. It is a prosaic way for the Argonauts to be shadowed. An observant passenger on one of these liners might spot Nea Kameni, a small island gradually growing from lava extruded out of sight. Who knows if the volcano might once again wreak destruction? For now, we know that all that destruction was caused by Africa moving northwards into and onto Europe; for Santorini lies at the edge of the Aegean Plate. What we see at the surface, of course, is all down to the influence of the geological foundations. But how did this understanding arise?

For Eduard Suess the idea of foundered lands was meat and drink. Some Atlantis of the mind enchanted him. Before the fundamental difference between ocean crust and continental crust was appreciated, it was possible—preferable, even—to think of the oceans as places where continents had foundered. It would be wrong to think of that generation of geologists a century ago as reluctant to admit all change. Rather, they would admit the kind of change they wanted and deny other processes. They appreciated some of the evidence for what we would now call plate tectonics. Gondwana* should be taken as the prime example, because it was the acceptance of the existence of this vanished land that set the seal on modern geology. Suess coined the word—the "Gonds" were an ancient Indian tribe. He saw very well connections between the geology of Africa, South America, and peninsular India, and appreciated that they were bound together by more than coincidence. Like finding fragments of an old treasure map, he saw that broken lines must connect to make a picture, and that the picture was a key to fundamental truths about the earth.

When I set out to read through Suess' dense prose, I had not realized how intelligently he appreciated the evidence of the rocks. Rocks do not lie. They do, however, dissemble as to their true meaning.

The most straightforward example is probably provided by tillites. In the nineteenth century, most geology in remote regions was of no more than reconnaissance standard. Exploring geologists were probing the innards of Africa for the first time, and the dark continent yielded its secrets reluctantly. Reports were sent home to whatever European country had the appropriate colonial interests, there to be published in the journal of the national geological society. Suess hoarded the insights dispatched by these trail-blazers. A rock formation from South Africa was christened the Dwyka Conglomerate, and reported by one pioneer, J. Sutherland, in the *Quarterly Journal of the Geological Society of London* for 1870, under the title "Notes of an Ancient Boulder Clay of Natal." This insouciant title embodies an extraordinary idea. A boulder clay is the deposit left behind as a glacier retreats. You can see deposits broadly of this kind dumped at the foot of Alpine glaciers shuffling backwards in their valleys away from the warm breath of global climate change. The boulder clay is a mucky kind of thing: dumped boulders of various sizes all mixed together and sealed by a sticky brown clayey gum. This is debris dumped unceremoniously by the

* I was once ticked off for using the term "Gondwanaland" rather than "Gondwana": my critic pointed out that "wana" means "land," making the inclusion of the English equivalent tautological. Both forms can in fact be found in published accounts of the evolution of the continents, and they appear to be interchangeable.

A block of tillite from the Permian strata of South Africa, showing included pebbles of glacial origin. Natural History Museum.

melting glacier, its dirt alongside its cargo of rocks, unsorted, chaotic. It is like the contents of a badly made, old-fashioned bag pudding. Look closely and you will see characteristic scratches on some of the bigger pebbles where they were scraped along at the base of the glacier. When such a clay becomes hardened and annealed by time it is called a tillite ("glacial till" is the proper term for the deposits of ice, and the suffix "-ite," as usual, designates a rock). But the scratches on the boulders are preserved, and so is the potpourri of different kinds of rock types all jumbled up, just as they were picked up and carried promiscuously on the back of the original glacier. So, here in South Africa, in the Great Karroo, was a rock identified as being of glacial origin, in a countryside where succulent plants now abound. It was an original idea, to say the least. Nonetheless, the signature of the fossil action of ice is easily read: there had evidently been an ancient ice age in South Africa. These boulder rocks were followed by what Suess called the "upper Karroo Sandstone." When some of these rocks were split they were found to contain fossil leaves having something of the shape of bay leaves

Leaves of the distinctive Permian Gondwana tree Glossopteris, *whose remains serve to unite the ancient supercontinent.*

but more tongue-like, which is why they were called *Glossopteris* (from the Greek for "tongue").

Then, in peninsular India, very similar conglomerate rocks were discovered in what Suess called the "Talchir Stage." Of these beds of boulders he remarks: "the resemblance to the Dwyka . . . is very striking." Evidently, the ancient glaciation extended beyond South Africa. As if to confirm this judgement, leaves of the fossil tree *Glossopteris,* and its associate *Gangamopteris,* were found in overlying strata in India, just as they had been in Africa. Suess goes on to describe many Indian basins bounded by fractures in which the "lower Gondwana coal" derived from these fossil trees can be exploited. *Glossopteris* is unknown in Europe or North America. It was a peculiarly Gondwanan tree. Nowadays, we could enumerate dozens of plant and animal fossils with a similar Gondwana signature. We can put a Permian (about 275 million years) age on many of the *Glossopteris* fossils, and we know them well from Australia. Evidence for glaciation is everywhere on Gondwana. I walked along a wadi in Oman in 1995, in the midst of the Arabian Peninsula and in a bone-dry desert, where the sides of the valley were lined with glacial boulder beds as blatant as any I have seen in the Arctic. The floor of the wadi looked as if some gigantic dragon had been sharpening its claws there; the grooves were gouges left in the rock floor by an ice sheet. Thus evidence is piled on evidence until the signature of Gondwana—its Permian life, its ice age—seems as graphic as the artefacts that define ancient Egypt. But the outline of it all, the facts that lay at the *fons et origo,* were available to Eduard Suess. As he said: "An undeniable resemblance exists between the structure of South Africa and that of the Indian Peninsula." He surmised that the former continuity of Gondwana was the appropriate explanation,* the Indian Ocean only subsequently separating the "fossil tongues" that spoke to him so eloquently: "Then came collapse. A new ocean was created and the continents assumed other forms . . . Out of the abyss of the ocean arises the great island of Madagascar presenting all the characters of a horst." As to adducing evidence from other sources, Suess remarks: "Eminent zoologists have been led to imagine the existence of an ancient continent on the site of the western half of the Indian Ocean to which the name 'Lemuria' has been given." There are indeed similarities between the animals of Africa and India that were accounted for by the existence of a bridge of land joining the two areas, or by a foundered continent. In either case, Africa and India as we know them were, in the past, exactly where we see them today. So the world apparently all makes sense, with the facts tucked away into a neat bag of theory.

But of course the description was all wrong.

It is worth reflecting upon what we would now see as an obvious mistake. Most theories are of their time, and Suess' was no exception. More facts are acquired, and, in the end, a mistaken theory is jettisoned, and then forgotten. The discovery of what made the ocean floors—basalt, and more basalt—is one obvious fact that falsified the subsidence theory: no foundered tillites there, nor *Glossopteris* fossils. However, that the foundering idea was *capable* of such falsification still qualifies it as a scientific theory, at least, if one accepts the criterion of science that Sir Karl Popper has explained in *Conjectures and Refutations:* the scientist has a licence to be wrong. The theories that cause much more trouble are those that can twist and turn in a breeze of new facts without ever fracturing completely. In fact, fixity of continents was already being challenged not long after Suess'

* The theory of the original continuity of South Africa and India can be found in earlier sources, mostly little acknowledged. See, for example, an article by Stow in the *Quarterly Journal of the Geological Society of London* (1871).

magnum opus was translated into English. It is one of the famous stories in science how a German meteorologist, Alfred Wegener,* proposed "continental drift" many years before it was widely accepted that the continents were neither fixed nor the residues of a swathe of Atlantises. Wegener's first account was published in German in 1915, and was largely dismissed. It was translated into English in 1922, and probably only came to be on the "serious" scientific agenda by the time of the French edition of 1936, *La genèse des Continents et des Oceans,* although even then it was regarded by most earth scientists as distinctly wacky. Wegener pointed out—and he was not the first—the "fit" of the continental profiles: the western coast of Africa and South America; Madagascar, far from showing "all the characters of a horst," seemed instead to wish to snuggle back against the East African coast, into an indentation that closely matched its western outline; and India and Africa were once *joined.* Gondwana was a united entity over which an ice sheet might move, or a forest might spread. The continents must have moved to their present positions. You could shuffle them back into their original contiguity by solving the jigsaw puzzle of their coastlines. Drift South America back towards Africa and—see!—they lock together in a happy marriage. The ancient glacial deposits made sense if there had been an ice sheet centred upon this Gondwana; but it must have occupied a different latitude at that time, nearer the pole, to support such frigidity. *Glossopteris* and its allies spread over the cooler parts of this united world, and there were early reptiles that could stroll unencumbered throughout the same ancient glades.

The idea of continent fusion was taken still further by Wegener. Perhaps *all* the continents were once fused together into one "supercontinent." At this distant time, all dry land was annealed: terra firma and sea were most perfectly parted one from another, for a single vast ocean counterbalanced one huge continent. North America and Europe/Asia were conjoined in the same way as South America and Africa; the North Atlantic Ocean had yet to appear. According to Wegener, in late Carboniferous times this huge northerly continent was bound to Gondwana as well. The current geography of the earth was claimed to be a legacy of a marriage of all continents, consummated more than 250 million years ago. The current continents resulted from the break-up of this continental behemoth. They

* Tony Hallam has ably summarized Wegener's intellectual prehistory (in his *Great Geological Controversies,* 1989). There were several earlier writers who suggested parts of the continental drift hypothesis. "New" ideas usually prove to have had progenitors.

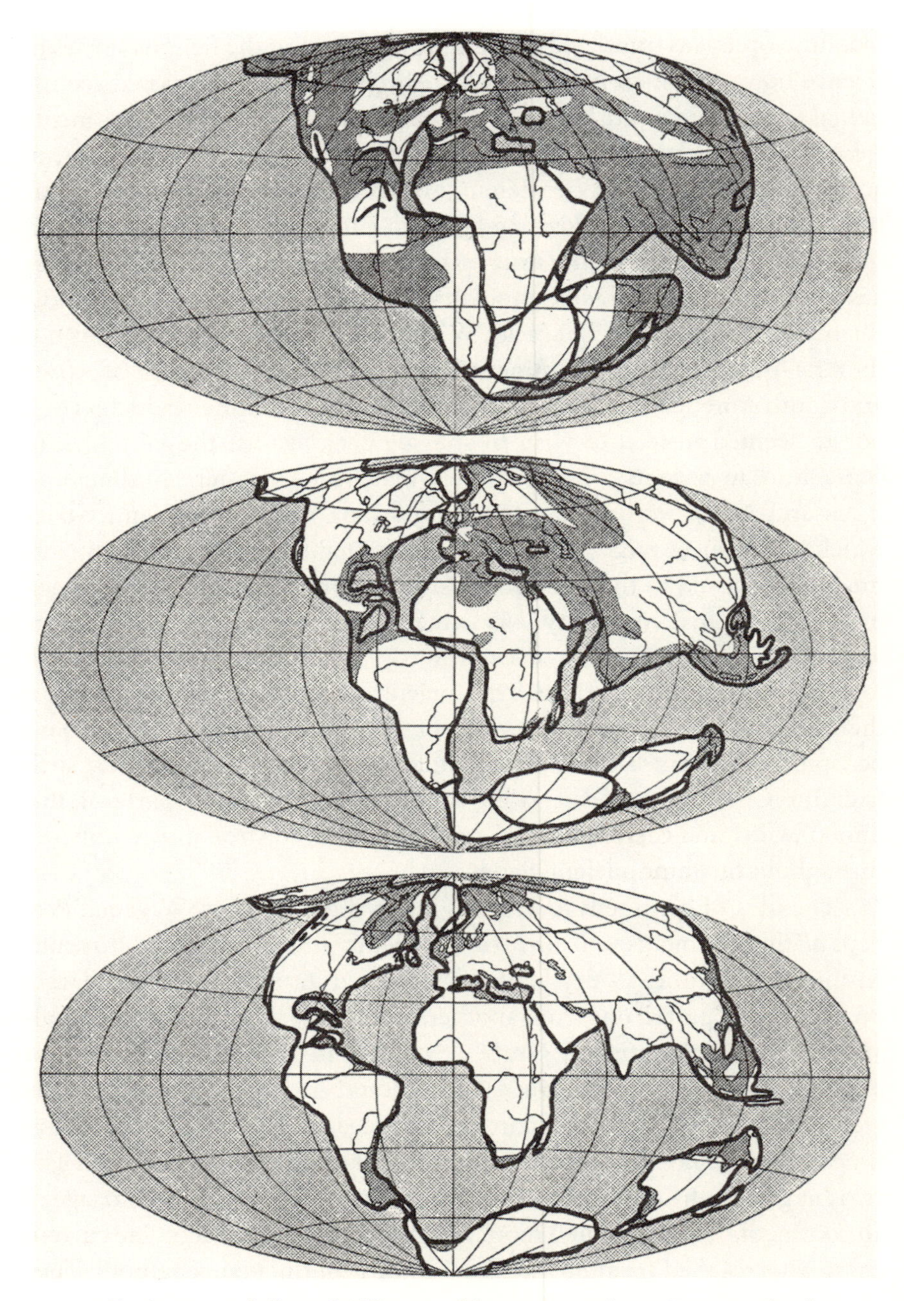

Alfred Wegener's "pre-drift" map of the supercontinent Pangaea (top), and its subsequent fragmentation.

had moved apart—drifted—to their present positions over many millions of years. This was, by any standards, a profound reorganization of knowledge. Gondwana was the name applied to the united India–South America–Australia–Antarctic portion of the supercontinent; Laurasia usually refers to the northerly part, the marriage across the North Atlantic. The whole is known as Pangaea, from the Greek for "universal world." I shall concentrate on the Gondwana portion, if only because the evidence for or against it figured so prominently in the history of the debate about the origin of the shape of the world.

Originality may be the capacity to look at the same facts and see new explanations. Much of the same evidence that Suess had adduced was paraded again by Wegener to draw an utterly different map, and to make predictions as different from drowned continents as could be imagined. And just as Suess' theory died by its predictions, so the notion of Gondwana, and of Pangaea, finally lived by virtue of facts that were still to be uncovered.

The drift hypothesis explained to Wegener the crumpling up of mountain chains. The movement apart of the continents entailed the elevation and folding of ranges at their further edges: the rucked western edge of South America; the huge arc of the Himalayas, pushed upwards as India collided with Asia. To see the world as it should be seen, he implied, the world must be looked at as a totality. The fate of mountains was bound up with the fate of oceans.

Perhaps if Wegener had had the eloquence of Dean Swift, or John Bunyan, then Pangaea, or even "Gondwanaland," might have rapidly achieved the cultural currency of fictional Lilliput or the Slough of Despond. Instead, the concept lurked on the fringes of respectable science, like some eccentric location for Atlantis. More than forty years elapsed before a version of Wegener's surmise hardened into something like a fact. But there were champions who kept the idea alive. Through the 1920s and 1930s these were primarily field geologists who worked particularly in the southern hemisphere, where the evidence was most striking. The South African Alex du Toit summarized a mass of information in *Our Wandering Continents* (1937). He went far further than Wegener in identifying strong connections between the Gondwana fragments. There were South American rock formations that trended towards the eastern coast of that continent, where they were more or less abruptly truncated at the Atlantic Ocean. He claimed that their natural continuation could be found in southern Africa, and beyond into Australia. Close the Atlantic and Indian oceans, place the continents back together, and the match was perfect. No shattered Minoan urn from Santorini could be repaired with greater confidence. Wegener himself had described the logic behind such observations: "It is just as if we were to refit the torn pieces of a newspaper by matching their edges and then check whether the lines of print run smoothly across. If they do, there is nothing left but to conclude that the pieces were in fact joined this way."

hampions
gain but
nly if . . .
Council t ax rise is
'highway r bbery'
abour
ury as
county
rove
ap
13.4%
incre
Crossing appro ved
to a

hampions
again but
only if . . .
Council tax rise is
'highway robbery'
Labour
fury as
county
approve
13.4%
increase
Crossing approved
to alter
HENLEY
TOYOTA

The principle of reconstructing continents is not so different from rejoining the headlines from this torn newspaper.

Most geologists today would accept such evidence without demur, but it was still "fringe" science when du Toit was publishing. Geologists didn't travel as easily then as they do today, so sceptics would not go to the trouble of taking a steamer halfway around the world just to disprove some crackpot idea that was bound to crumble under scrutiny. Geologists met at the International Geological Congress in some capital city every four years or so, as they still do; but not everyone could afford to attend, and the best funded, and therefore the most numerous, attendees were mostly Americans, who included the most convinced anti-drifters. "What's all this nonsense?" you can almost hear them ask. "How can a continent motor across an ocean? We all know that the resistant forces of rocks could not accommodate any such motion."

The opponents included some of the intellectual heavyweights of the day. Harold Jeffreys was a brilliant geophysicist who wrote the standard textbook, published in 1926. He summarized Wegener's theory as an "impossible hypothesis" and concluded that "the assumption that the earth can be deformed indefinitely by small forces, provided only that they act long enough, is therefore a very dangerous one, and liable to lead to serious error." In a word, there was no mechanism for continental movement. That same year, the American Association of Petroleum Geologists' symposium in New York was the first convened to discuss the subject of "drift": virtually everyone was against it. I wish I could report that those of my own persuasion—palaeontologists—were more positive, but it was not generally so. Charles Schuchert of the Smithsonian Institution in Washington favoured land bridges across the Indian Ocean and elsewhere as the conduits of biological similarity. Where had they gone? Why, they had foundered in Suessian fashion. It may be significant that Schuchert had made his name studying fossil brachiopods (an important group of "sea-shells" unrelated to molluscs), of an era much older than Pangaea. The world was differently arranged again 500 million years ago, and Schuchert's prejudices may have been forged by an even earlier world in which, indeed, continents were once more separated. The credit of the palaeontologist is somewhat restored by one of my own former colleagues in the Natural History Museum in London, the palaeobotanist A. C. Seward. After he had studied the distinctive flora associated with *Glossopteris,* he found the assumption of the existence of Gondwana more probable than any of the alternatives. He said as much in several papers published in the 1920s and 1930s. Those who studied the fossils of the Indian subcontinent in the field followed suit, notably the remarkable Birbal Sahni, who founded a botanical institute in Lucknow that still carries his name. Other palaeontologists on the "anti" side invoked drifting logs and seeds—the distinguished mammal expert George Gaylor Simpson called them "sweepstake routes"—to explain far-flung similarities between species. Reports of lizards clinging to floating stumps in the mid-Pacific were eagerly garnered in support. What a lizard could do, *Glossopteris* could do better.

Many tectonicists were on the opposition side, too, and those with thunderous reputations carried their less extrovert colleagues and students with them. Hans Stille was a formidable advocate of the importance of vertical movements in tectonics. Not for him the slithering, sideways movements of drifting continents, where all was push and thrust; no, he was a stabilist. German professors of that time were like God, only more frightening. There were, besides, other things pressing on the German national psyche at the time; perhaps the fate of mountains was low on the agenda. In the United States, there were other stabilists almost as formidable. But they did not have it wholly their own way. Emile Argand's nappe theory of the Alps saw mobility of the crust everywhere, and evidence that rocks could indeed flow was accumulating. But in spite of the efforts of the few, the consensus carried the day. For more than thirty years the idea of mobile continents lurked out of sight of a scientific establishment predominantly certain it was untrue. Continental drift was an appealing distraction, a siren song that needed to be resisted by those grounded in the certainties of physics: if it can't move then it didn't move. To acknowledge the reality of Gondwana would have had the effect of changing just too many things in the geological status quo. One is reminded of the lines of Hilaire Belloc on the fate of Jim:

. . . always keep a-hold of Nurse
For fear of finding something worse.

But what of Arthur Holmes?

Holmes was, in fact, one of the early pro-drifters. It is clear from his correspondence that he appreciated the explanatory power of the new theory. He was an "Africa hand" as well: his first geological employment had been on a mineralogical survey there in 1911. This was pioneering exploration in Mozambique, in the most remote country imaginable. His diaries reveal a buoyant enthusiast whose powers of expression did not yet quite match his powers of observation. He was a young geologist in love with his trade, and any scientist will recognize his intemperate zeal to be up and doing, out and discovering. Academic instruction is like fitness training: its point is only revealed when racing over the ground. What Holmes saw in Mozambique was not immediately concerned with the problems of Gondwana, but exposure to field conditions could help but prime his mind to the scale of things. Some years later, Holmes had a position in the University of Durham. He had tenured employment, and his work on the age of the earth had earned him respect in the geological community, even if he had made a few enemies as well as many friends. In December 1927 he read a paper to the Edinburgh Geological Society where he revealed his pro-drift colours. It was an extraordinary presentation and anticipated so much of what was to come. Not only did he accept that "drift" had occurred, but he also suggested its elusive mechanism. Holmes' work with radioactive elements had made him aware of the possibility that they could be a source of heat. He proposed that the differential effects of such heating created convection currents in a deep "substratum": solid, yes, but sluggishly flowing over millions of years, in an analogous way to the proverbially slow creep of glaciers. The upward limbs of the convection cells reached the lithosphere and then parted in opposite directions, much as the surface of simmering pea soup develops a pleat between boiling vortices. The drag of the convection cell provided the motor that moved continents. Where the cell arose beneath a continent it could split it and create a new ocean; where a cell turned down again it might pull ocean crust with it to produce an "ocean deep." The account is almost clairvoyant in its anticipation of future discoveries. His heat-based scheme was summarized by Holmes as "a purely hypothetical mechanism for 'engineering' continental drift." This is almost like referring to Darwin's *Origin of Species* as a "modest proposal possibly germane to the appearance of new forms in nature." However tentatively couched, it did not go down well with the leading anti-drifters, including Harold Jeffreys, who was obdurate about this force being as inadequate as any other for transporting continental slabs (recall that his own textbook was but a year old). The gap between reading a paper and its publication is one of the running complaints that scientists have, and Holmes suffered very badly with this seminal work—the relevant volume of the *Transactions of the Royal Society of Edinburgh* did not appear until 1931. It is worth remembering that arguments about who made the critical discoveries of the AIDS virus have hinged on matters of days (even hours) in recent years. Priority of discovery has become a crucial passport to scientific glory. This has occasionally led to jiggery pokery, as one research team has attempted to nudge a nose ahead of another, and publication priority is crucial. The leading journals, *Nature* and *Science,* keep discoveries under strict embargo until the very day of publication. Less scrupulous scientists sometimes speak to the press before they speak to their colleagues. Perhaps it was as well that continental drift was so profoundly unfashionable in the 1920s and 1930s. Holmes' paper was seen by his critics as being merely the latest folly in a minority delusion. As one of them wrote: "I believe that we need to apply elementary physics and mechanics to the continental drift problem to show how impossible drifting would be." Arthur Holmes would have to wait more than thirty years to establish his priority.

But Holmes did not lose faith in mobile continents. Rather, he was sympathetic to the evidence subsequently accumulated by Alex du Toit and the palaeobotanists—perhaps he appreciated better the hardships under which such evidence had been acquired than did his chair-bound critics. He believed that the rocks, ultimately, would speak the truth, and that appropriate theory would follow. When the first edition of his *Principles of Physical Geology* was published in 1944, continental drift was there in the last chapter. It was a bold move for the times. For the price of thirty shillings, or seven dollars, the reader could find a summary of the evidence for "Gondwanaland." Problems in the imperfect "fit" of the coastlines were aired, too. Critics had been quick to point out that Gondwana was a jigsaw puzzle that had to be fudged to fit. Du Toit's reconstructions had improved on those of Wegener, but neither was perfect. It was realized later that the true edges of the continents lay not where the shorelines happened to be, but at the edges of the continental slabs themselves, below sea level—the rims of the building blocks of the earth. If the 100-fathom contour was taken to represent the outline of the continent, a much better fit of the pieces of the puzzle was obtained. This was another example of the interplay between sea level and terrestrial geology with which this book started in the "Temple of Serapis." The sea is fickle as to the height it laps on shore: geological reality often operates at a different scale.

Holmes' book sold very well despite its stolid title. The first edition eventually reprinted eighteen times. The reviews were good, but some of the reviewers would have much preferred the last section to have been left out. For them, the mixture of fact beautifully explained with speculation almost smuggled in at the end was disquieting. Some followed the geophysicist's view that what is obviously impossible can never be the case, no matter how the evidence appears to favour it. (I am reminded of those irritating stickers that get posted on office walls: "Don't confuse me with facts, my mind's made up.") But, as any newspaper proprieter knows, the bottom line is sales. A new generation of students brought up on "Holmes" (the book became synonymous with the man) would encounter the idea of Gondwana in a less fettered way than their teachers had. The young mind is often a more open one. The post-war years were optimistic and innovative times, a time to jettison received ideas. Technology developed during the Second World War was beginning to make an impact on pure research: ocean depths were there to be read for other reasons than the pursuit of enemy submarines; decoders for intercepting enemy signals were ready to become computers. Thanks to jet engines, air travel to remote lands became commonplace, and so it became easier to organize field trips to establish, test, or clarify geological facts. In the mid twentieth century, mobility of continents began to seem like a theory with a chance of being properly tested. Physicists, who had been with the sceptics in that war of ideas about continent mobility, were about to become crucial to the campaign on the other side.

It is interesting to follow the change in the respectability of continental movement theory in the later edition of "Holmes." Like all such compilations of second thoughts, the revised book got longer (and some would say it lost a little of its punch). The bulky 1965 edition is the one I encountered as a student: twenty years of second thoughts. It is still nothing in bulk compared with the five volumes of Suess, but it is a dozen times more read-

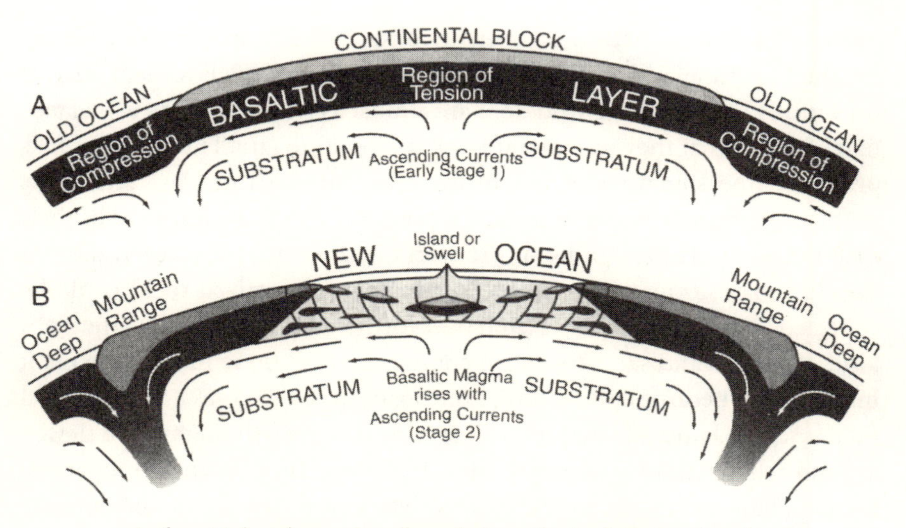

Arthur Holmes' prescient "purely hypothetical mechanism for 'engineering' continental drift" published in the Journal of the Geological Society of Glasgow, *1931, redrawn by Cherry Lewis in her biography of Holmes (2000).*

able. The final chapter on continental mobility is still there, as is the section on "Gondwanaland." Of course, there was no mention of plates—that was for the future. But there *was* a section on polar wandering.

The earth behaves like a vast magnet. The magnetic field streams from (magnetic) pole to pole. From time to time the field reverses—north becomes south, and south becomes north. There is a small number of minerals that readily become magnetized. Appropriately enough, the commonest of these, a simple iron oxide, is called magnetite. This is the lodestone, whose interesting properties have been known since antiquity. It is present in several common rocks, of which basalt is the most ubiquitous. There are also plenty of sandstones that contain tiny grains of magnetic minerals. When a basalt cools from molten lava it passes through a threshold known as the Curie Point, at which the magnetite (or a related mineral) takes up magnetization from the earth's field as if it had been freshly minted from a magnet factory. These natural magnets point at the poles like a compass (you have to record the orientation of the rock in the field very carefully, for reference). Because the quantity of the iron mineral is generally rather small, any magnetometer recording this "fossil" magnetism has to be very sensitive. Technical improvements in these instruments made in the 1950s meant that magnetization of ancient rocks could be examined critically for the first time. This allowed for determination of "palaeomag," as it is universally known; those who do it are almost as widely known as "palaeomagicians." What they did was, after all, a kind of magic. The ancient rocks preserved the signature of where the poles of the time lay. The rocks pointed their time-frozen finger at the pole as it was at the moment of their magnetization. If everything in the world had been fixed forever, then so should the positions of the poles. Fortunately, magnetized rocks could be collected from many sites scattered over former Gondwana where the right kind of basalts and sandstones are common. From these measurements it immediately became apparent that the position of the poles had moved. At first, it seemed possible that the whole crust of the earth might have slewed around relative to the poles, like the "skin" on a thick broth when you tip the pan. However, soon enough it was clear that the tracks of ancient pole positions from different continents took quite different paths: this would not be expected if the whole crust had behaved coherently; instead, each continent appeared to have a will of its own. And they were indeed paths. The ancient poles did not jump around here and there, hither and yon; instead, progressively older poles in Africa, for example, shifted away from pointing at the present pole along a distinct track, a journey through time towards a different place. There was sense in the signal, and it pointed to Gondwana.

By putting all this evidence together at last, there was a sea change in the respectability of continental movement theory. Precision of physics and delicacy of instrumentation had finally joined cause with the geological hammer. When the apparent polar-wandering curves for the different components of Gondwana were compared, they only made sense if there had once, more than 200 million years ago, been a united supercontinent. Suess' ancient land had finally been explained, and the explanation was accepted by a majority of younger scientists. As Gondwana split up, its component continents embarked on their own tracks, and each "saw" the poles in a signature fashion as one geological period succeeded another. Lavas erupting on the travelling continents became magnetized, freezing the details of each journey as a diary written in aligned atoms. But all the journeys started in the same place; that was now evident. Then, too, the fossil and ice-age details that had been known for more than half a century fitted logically and convincingly into the same scheme: for "palaeomag" also indicated that the South Pole in the late Palaeozoic ought to be in the midst of Gondwana—just where an ice sheet might be expected to grow, and where tillites might be left behind to record the legacy of frigid times.

Here was something resembling a vanished Antarctica. And after the ice age had waned, cool forests prospered and spread, with their tongue-shaped leaves sheltering special animals that liked it there. It all added up, and the sum of the facts was more than the parts, as is often the case with momentous scientific shifts.

It is difficult to cite a date, a particular moment in this intellectual drama, when the consensus in the geological community changed. Such transformations are somewhat mysterious; the *Zeitgeist* switches in ways which are not necessarily logical, although the principal players in the drama often like to present it that way. Nor does everyone shift together: the later "Holmes" pointed out that a Russian textbook published in translation as late as 1962 still berated "the total vacuousness and sterility of the hypothesis" of moving continents. The crucial work on palaeomagnetism was already under way less than ten years after Holmes' first edition. Keith Runcorn, one of the pioneers of these methods in Britain, was certainly a geophysical convert before 1956. The proceedings of a U.S. symposium published in 1962 in the *International Geophysics Series,* under the explicit title "Continental Drift," show how far views of the more free-thinking scientists had shifted by then, and there were already inklings of the next phase in the great global tectonic synthesis. On the other side of the newly liberated Atlantic Ocean, a symposium at the Royal Society early in 1964 was also overwhelmingly pro-drift. By this time, Americans were among the most radical of the theorists, a position they maintain today. As St. Luke's gospel puts it: "the truth will set you free," and once the continents themselves had been untethered so, too, scientific imaginations were liberated to pursue their dreams. Mountain chains could now be seen as the consequence of continent shifts and collisions; mineral plays might prove to be related to continental movement, too. The truth of the world would be rewritten, and who would be first with the new version? Some scientists were laggards, while others surged forwards. Some geology departments took several years to adapt to the new world. But there was little doubt where the "cutting edge" of geological discovery now lay: it was in the study of the consequences of continental mobility. What Robert Herrick applied to poetry might apply equally to geology:

To print our poems, the propulsive cause
Is fame, the breath of popular applause.

The northern hemisphere equivalent of Gondwana—the marriage of Eurasia and Laurentia—was already familiar in Holmes' second edition. These continents, too, had split apart at the same time as their antipodean counterparts. Evidence for that greater fusion—of Pangaea, the unification of all the continents—followed naturally enough. Both traditional geological and new geophysical facts were employed: it would take a longer book than this to list the detailed evidence, and it would perhaps be nugatory to do so since the arguments are in many ways similar to those rehearsed for Gondwana. There are still differences in detail about the constitution of Pangaea, and for how long it was united at the end of the Palaeozoic. Early computers were used to produce "best fits" of the continental shelf margins, but the labour required to produce a pre-drift map was still prodigious. In the mid 1960s, the latest equipment was in the Geophysical Laboratory on Madingley Rise, just outside Cambridge, where Sir Edward Bullard presided over the attempt to use the latest technology to solve the fit of the continents. In 1965, the computer fits of the continents around the Atlantic Ocean were published to "the breath of popular applause." The originality of Wegener's vision had been successively endorsed, and everyone in "earth sciences" felt that they were living in exciting times.

But what of the oceans? To concentrate attention on the mobility, or otherwise, of the continents leaves out two-thirds of the world. At the same time as evidence for continental movement was firming up, the distinctiveness of the oceanic crust was being demonstrated by improved sampling of the ocean floor. We have seen how new technology after the Second World War also allowed a clearer picture of the topography of the deep-sea floor. It was possible to look at the whole earth for the first time. We followed the fate of a tiny piece of it—the Hawaiian chain—in some detail to show how movement of an oceanic plate accounted for so much of the local features. How was the link made between what went on in oceans and what was now accepted as happening to continents? The crucial step was the appreciation of the structure of the mid-ocean ridges as the place where new crust was manufactured. The relevant observations were made by sea floor sampling and mapping in the 1950s. One of the sea's principal explorers, H. W. Menard, who worked at the Scripps Institute of Oceanography, has described in *The Ocean of Truth* (1986) the kind of discoveries that were made. These were not just details: they were major features, such as the Pacific Fracture Zone. The clearly linear course of the ocean ridges emerged from semi-obscurity; the deepest parts of the ocean, the trenches, were mapped as great lineaments, too. Off South-east Asia, fringing the Philip-

ABOVE: *Alpine magnificence: Weissbad, Canton of Appenzell, eastern Switzerland.*

BELOW: *Geologists inspect the base of the Glarus Nappe at the Lochseiten section in Switzerland. The author is in the middle, and Geoff Milnes is to his left.*

Dark grey Flysch produced roofing slates at Engi, near Elms, in the Sernft Valley, Switzerland.

Spanning two continents: the bridge that crosses the Insubric Line, a major fault marking the junction of "Africa" and European plates near Pianazzo.

ABOVE: *View of the Glarus "Thrust" from Flimserstein, with St. Martin's Hole (Martinsloch) on the skyline.* BELOW: *From the depths of the earth to the heights: highly metamorphosed gneissose rocks on the summit of the old San Bernardino Pass, between Italian and German Switzerland.*

OVERLEAF: *A mountain chain dies away: Lake Geneva, with the highly deformed Bernese Alps to the south and the wrinkled chain of the Jura Mountains to the north.*

ABOVE: *Fragments of ocean floor transported to the continents: the Oman ophiolite makes the dark hills on which little or nothing will grow.*

BELOW: *Where plates are subducted, volcanoes arise, fed from deep sources.* South Wind, Clear Sky, Red Fuji *(c. 1830), one of Hokusai's* 36 Views of Mount Fuji: *a cone which is almost the definitive volcano, and a profound cultural influence.*

ABOVE: *St. John's, Newfoundland: a geological fragment of Europe. A boat passes through The Narrows at the oceanic edge of St. John's, backed by grim late Proterozoic hills.*

BELOW RIGHT: *Almandine garnets in a metamorphic rock, schist. This striking mineral only begins to form under conditions of high temperature and pressure.*

BELOW: *Bell Island near St. John's—a piece of Ordovician Avalonia. The sandstones and shales forming the cliffs yield trilobites and other fossils closely similar to those from the other side of the Atlantic Ocean.*

ABOVE: *A piece of Ordovician Laurentia: limestones that formed in shallow, tropical waters which lapped over the western coast of Newfoundland.*

BELOW: *Thick beds of limestones which tumbled off the edge of Laurentia alternate with shales laid down in deep water, in the Cow Head rocks of western Newfoundland. All these rocks have been tipped vertically by earth movements.*

ABOVE: *An ancient mountain belt continued into Appalachia: the Blue Ridge Mountains.*

BELOW: *An Ordovician trilobite fossil,* Pricyclopyge, *a species which typically lived off the edge of the Gondwana continent of the time.*

pine Sea, the "spot depths" already observed by *Challenger* were linked together as the huge Japan-Bonin-Nero Trench system—in their own way these trenches are more dramatic than the mid-ocean ridges, somewhat like the Himalayas in reverse. The exploratory process at sea was similar to a version of the child's game of "join the dots," whereby a picture emerges as more and more numbered dots are joined. You have to be certain that there is a picture to discover in the first place, and, for a while, prejudice about what the finished picture might be can mislead; but eventually a likeness is the outcome, albeit a jerky and disjointed one; and the more dots you join, the better the picture. The image is still being filled in today. Modern methods of surveying, such as multi-beam sonar, simply provide a much more accurate and immediate view of the whole scene.

The breakthrough in understanding can be summarized in one short phrase: "sea floor spreading." The concept grew out of research by two American geologists: the expression was coined by Professor R. S. Dietz, but the pioneer publication was written by Harry H. Hess. This research paper had, like Holmes' thirty-five years before, a publication lag of two years. It had been available in unpublished manuscript in 1960, but did not appear in print until 1962. The earlier typescript is what scientists like to refer to as a pre-print, a kind of moral claim to an idea. If the idea turns out to be a seminal one, a pre-print can be dangerous for the author, to the extent that it tempts others to try to jump the gun. In fact, Hess' ideas spread indecently fast. The 1960s were a time of extraordinarily rapid progress in the understanding of continental mobility. Hess' paper explicitly stated why the most obvious thing about the Mid-Atlantic Ridge was, indeed, so: it was median "because the continental areas on each side of it have moved away from it at the same rate . . ." This is not exactly the same as continental drift. The continents do not plough through the oceanic crust propelled by unknown forces; rather, they ride passively on mantle material as it comes to the surface at the crest of the ridge and then moves laterally away from it. Continents are part of processes that entail the oceans, too. Thus, at last, the criticism of the physicist Harold Jeffreys was answered about the impossibility of the continents sailing across the basalt oceans like huge, granitic barques. They didn't. They were at one with the oceans. New oceanic crust was added at the ocean ridges—where it was known that heat flow was high and that there were submarine volcanics. A moving continent and its contiguous ocean crust moved in harmony. Thus South America moved westwards with its associated slab of ocean crust extending as far as the Mid-Atlantic Ridge, while Africa moved eastwards with *its* slab of oceanic crust. The increasing width of the Atlantic Ocean was accommodated by new crust erupted—ultimately derived from melted mantle—and inserted at the ridge, like caulk in a leaking ship. It was a wonderfully simple way of putting so many disparate observations together: volcanic ridges, which were buoyed up by hot magma, even the rifts at their peaks, were an expression of the urge for separation of plates. Earthquakes below ridges were discovered to be only 10 to 15 kilometres deep; they become deeper outwards—up to 100 kilometres—just as would be expected. Geologists could go to Iceland and stand astride the Mid-Atlantic Ridge itself to see evidence of the processes on the ground. The irony was that at virtually the same time as the pro-drifters finally triumphed over the stabilists, demonstration of the mechanism of ocean widening rendered the very term "drift" obsolete. "Continental drift" had to be expunged from our vocabulary, to be substituted by "sea floor spreading."

Knowledge and theory move forward together in a kind of uneasy, shuffling collaboration. In an account of the kind I have given above, it is easy to forget those who were part of the process but who lurk on the margins of history, because they do not fit into a narrative of brilliant theory anticipation (Wegener, Holmes) or experimental confirmation (Bullard, Hess). They get forgotten; perhaps they had espoused an idea that never had its day. Had you been at a conference in the 1960s, though, these figures might have glittered as brightly as any of the stars that history remembers. In particular, I think of Professor S. Warren Carey, of the University of Tasmania. Carey was one of those pro-drifters who kept the faith through the 1940s and 1950s, when to do so required courage and determination. His ideas were taken seriously, for example, by Arthur Holmes. Carey, though, believed in an expanding earth. His notion was that the increase in diameter of the earth had caused the split-up of Pangaea and was the deep cause behind most of the geological phenomena with which this chapter has been concerned. Continental crust had once almost enwrapped the earth, it was claimed, but with the earth's expansion it split and buckled and parted company, removing the continents to where they are found today, with the ocean basins "filling" between. No wonder the continents fitted together, just as the pieces of a peeled orange skin fit together! The idea is not so strange. As we have seen, in the nineteenth century, and in the early part of the twentieth, there had been a strong school of geologists who attributed tectonics to a *shrinking* earth. After all (so they supposed), the earth had cooled from its hot "primeval condition"; thus it would shrink, and, as it did so, mountain ranges would result. Suess evidently believed in such a mechanism when he contemplated the intensely folded rocks of the Swiss Alps. He realized that the recumbent folds there indicated that the surface must have contracted, or shortened: such folds "have directed attention anew to the diminution of the earth's circumference," as he wrote. Recall, too, that early estimates of the age of the earth were based on the idea that it was cooling progressively. Discovery of radioactive "heating" turned the tables. There was now a possibility that the world was actually heating up and expanding, fuelled by radioactive elements in its interior. Carey later turned his attention to changes in the gravitational constant as the cause. For a while, the expanding earth was in with a chance. Serious scientists, like the oceanographer Bruce Heezen, believed it. However, the recognition of sea floor spreading made sense of continental splitting without having to invoke earth expansion. It seemed to be a more economical explanation. More lethal was the recognition that oceans had existed *before* the assembly of Pangaea; it was only a phase in the history of the planet, after all, and not necessarily its primitive state. If the earth had been smaller still then, how could there be room for such oceans? The expanding earth theory had some trouble accounting for those mountain ranges, like the Andes, that abutted oceans, though it could cope well enough with those marking where continents had collided, as in the Himalayas. Still, Carey gamely defended the theory. I went to a lecture he gave in the early 1970s, and a fine, flamboyant affair it was, a charismatic display of bravado. You could not help admiring his chutzpah, as you might that of a pianist who carries on playing while the ship is sinking. However, he had made a convert in the Natural History Museum in London—my colleague Hugh Owen. Hugh continued to pick holes in the continental reconstructions favoured by computers for the next fifteen years or so, and published an alternative set of maps in which some earth expansion was incorporated. Both Carey and Owen were absolutely convinced of their arguments, and if some twist of knowledge were to prove them right in a hundred years or so, they, too, might gain iconic status. For the moment, this seems improbable.

The proof of sea floor spreading was in the heating. If the theory were correct, hot magma welled up at mid-ocean ridges. When the lavas cooled they would pass through the Curie Point, and their magnetic minerals would become magnetized. If sea floor spreading theory were correct, such cooled lavas would have the same age symmetrically on either side of the mid-ocean ridge, as their respective pieces of crust migrated away from the line of genesis. So far, so logical. This is where the other property of the earth's magnetic field comes in: its periodic propensity for reversal, for south and north poles to switch. If such a reversal occurred, its record, too, should be symmetrically disposed in the basalts to either side of the centre-line of the mid-ocean ridge. You might expect a strip of oceanic crust on either flank of the ridge that gave a reversed signal, and then, beyond it again, a strip with normal magnetization, and so on. Colour normal white and reversed black and the sea floor should be painted with symmetrical stripes of black and white running parallel to the ridge. The ocean crust was like a tape recorder spooling away on either side from the mid-ocean ridges, with the narrative recorded in magnetism. A sensitive magnetometer was needed to make these measurements, and the right kind of mid-ocean ridge where a good density of samples might be obtained. Two Cambridge University geophysicists, Drummond Matthews and Fred Vine, got the evidence they needed on the Carlsberg Ridge in the north-western Indian Ocean. The proof was published in the journal *Nature* in 1963. This paper is always referred to in the trade as "Vine and Matthews," and is one of those rare classics that provides a benchmark in the progress of science, like the determination of the speed of light or Planck's Constant.

My description might give the impression that this was one of those "Eureka!" discoveries, where prediction met result in a happy consummation. But that is not an accurate portrayal. Fred Vine was far more diffident than one might imagine about the likely acceptance of his paper. He had doubts. He has even recalled saying to his colleagues, "If they publish that, they'll publish anything." Nor was the discovery as clear cut as the publication date indicated. As so often in science, when the time is ripe for an idea, more than one person has it. Lawrence Morley worked as a palaeomagnetist for the Canadian Geological Survey and had reached similar conclusions to Vine and Matthews, but his letter to *Nature* that same year was rejected. He lowered his sights a little and submitted the paper to the *Journal of Geophysical Research.* In a notorious riposte, it was rejected *again* with the judgement: "this is the sort of thing you would talk about at a cocktail party." In other words, it was not to be taken seriously by serious scientists. Morley was thus to be denied "the breath of popular applause." The paper was published in modified version a year later by the Royal Society of Canada, but not one in a hundred who read *Nature* would have seen this more obscurely published version, and a year is a long time in science when a subject is "hot." Science, like life, is often unfair. Then, too, it has to be admitted that "Vine and Matthews" was incomplete; they had not really demonstrated the required symmetry of the "stripes," and there was, as yet, a very imperfect chronology for the magnetic reversals. You need to know just how old a reversal is to approach an estimate of the *rate* of sea floor spreading, for you require an independent timepiece to calibrate that slow, basaltic conveyor belt. Fred Vine himself was admirably free of hubris, and remained to be convinced by better evidence from better ridges. But, by now, this was "big science," which meant that the well-funded oceanographic laboratories from the United States provided the next big push. These included the Lamont Geological Observatory, and the U.S. Geological Survey at Menlo Park, the big boys with big bucks—ocean coring does not come cheap. Within three years of those first, prescient words written in Cambridge, new and more detailed descriptions had been released to the world of the patterns at other ridges that proved to have clearer "signals," including the Reykjanes Ridge, south of Iceland, and the Juan de Fuca Ridge, off Vancouver Island. When the reversal stripe pattern was proved again on the Pacific Antarctic Ridge, it would have taken a sceptic of great pusillanimity to deny that a truly global phenomenon was being recorded. By 1967, it had become material for future textbooks.

The detailed study of ridges revealed one other fact that proved essential to what would soon be known as plate tectonics. As the topography of the ridges was revealed, so it became clear that they were not as straight as was originally believed. In places, the ridges were strikingly displaced, as if they had taken a "hop" sideways. Something else was going on here that could not be attributed to sea floor spreading. The geologist Tuzo Wilson realized that the ridges were being moved along faults, but these were faults of a special kind that did not require one or another flank to move up or down. Instead, chunks of ocean floor were slipping past one another. Put two planks side by side on a table and draw a line across them: slide one of the planks, and you will get the idea, as one half of the line is displaced. Tuzo Wilson called them "transform" faults.

The dating of the magnetic reversals required experts in radiometric dating of rocks. As so often in geology, once the need was perceived, the end was achieved by a refinement of technology. The goal was accurately to date the lavas in the reverse magnetic "stripes." Did the reversals in different parts of the world coincide, as they should? The short answer is: they did. The longer answer is that a new mass spectrometer was constructed at the University of California at Berkeley that could measure smaller amounts of radiogenic argon than previous instruments, and hence could date the rocks of a given "stripe" more accurately. Coupled with the new cores derived from recently explored spreading ridges, the way was open to test the question of the age of magnetic reversals. A prominent reversal interval was discovered in several of the profiles and in 1966 was christened the Jaramillo Event. It has proved to be as identifiable as a fingerprint throughout the world. The rates at which new ocean floor is grown can now be calculated. Even this longer answer is a travesty of a detailed story of discovery that has been related by W. Glen in *The Road to Jaramillo* (1982). In this fashion, knowledge begets questions which beget new technology which provides answers—which in turn beget questions. This is the implacable carousel of research.

Fossils also came to play a part in confirming spreading rates and dating ocean floor. Tiny shells of planktonic animals rain down from the surface waters of the ocean on to the deep-sea floor. Among these are the delicate remains of single-celled organisms, particularly foraminiferans and radiolarians. The former are often tiny spirals, or collections of chambers looking rather like minuscule popcorn, while the latter are glassy silica nets or spheres, delicate as lace. They evolved fast, changing shape in subtle ways; hence their various species can be used as miniature chronometers to measure the passage of geological time. Eocene foraminiferans are very different from Pliocene ones, and even *within* the Eocene there are many fine time divisions that can be recognized using foraminiferans. It is actually much quicker to examine a slide of fossils to get a geological age than to go through the technically complex business of extracting isotopes, so the fossils have an advantage. Logically, you would expect the very first shells to fall on newly minted ocean crust to give a good indication of its age, almost as if it had been date-stamped. By the same reasoning, you would expect the oldest ocean crust to be furthest away from the mid-ocean ridges, the newest, closest. So it proved to be. Of course, old oceanic crust was often covered by progressively younger fossil deposits—it had to await the development of coring on the deep sea floor to produce good evidence. The results were exhilaratingly consistent with sea floor spreading theory, and provided concrete evidence that although the ocean floors were of many different ages, none was older than the Jurassic, since when it had spread into existence. The sea floor was far, far younger than the earth itself.

Some people feel a certain nervousness about mathematics. The world is so wondrously complex that to reduce it to a collection of formulae excites suspicion in many individuals. If the formulae are arcane and impressive it seems to stimulate further anxiety, of the kind that people feel when they are at a party where they know nobody, or where everyone else is talking a foreign language. We all like to feel that we know what is going on, and few are polyglot. At the time Eduard Suess was writing it seemed impossible that the world might be boiled down to numbers. As he wrote: "It is with extreme distrust that the geologist regards all attempts to apply the exact methods of mathematics to the subject of his studies." The implication was that the world was too messy to knuckle under to figures. But knuckle under it has. Arthur Holmes was less nervous about quantification, especially since his work on the age of the earth demanded the use of grindingly primitive calculators, but at heart he was still a visionary. In the story I have related it does seem that vision has always preceded calculation, with inspiration preceding measurement. The lesson of Sir Harold Jeffreys and his unyielding (but honest) and careful (but bogus) mathematics impeding the drift hypothesis is one that every gung-ho theoretician has taken to heart. However, there has been a complete change since Suess. Today, the way much geological science is done entails proposing a mathematical model, and then going to the field to see whether it is supported in the rocks, or perhaps conducting an experiment to suggest what we should be looking for in the rocks in the first place. This preamble is necessary to explain the final transformation in the story of moving continents: plate tectonics.

Plate tectonics spliced together all the different aspects of the world we have met within this book into one grand design. What went on in the oceans was linked to what went on in the mountains. The dipping and rising in the Bay of Naples was part and parcel of the fiery trail that led to Hawai'i, and that too could not be disentangled from the heave of the Alpine nappes. Even the genesis of a single crystal in a particular rock might be a response to the bidding of greater forces in the earth's fundamental engine. It was nothing less than a unifying theory of all the world.

And it was once again a visionary who fathered the new ideas. Tuzo Wilson was an unusually imaginative man. We have already met him once, and will do so again. He always seemed to leap ahead to something new, largely because, like Holmes and Suess, he saw the world as a whole. I met him in London when he was a Grand Old Man a few years ago at a lecture at the Royal Geographical Society. There he reviewed his intellectual progress, in a matter-of-fact way. It seemed a little flat, because what he said was now part of the way we all thought. I was reminded of a time when the secret of a magic trick was explained to me: to recapture the sense of wonder you have to stuff the rabbit back into the hat and forget all you know. Tuzo Wilson had invented the trick. He had minted the idea of plates in a paper published in *Nature* in 1965. Within a few years, plates were everywhere.

But first I must add ocean trenches to the plot of the story. As previously noted, these had been recognized since the days of the *Challenger* from spot depths, but their profile had only become known in detail following the surveys of the sea floor undertaken after the Second World War. The trenches were great gouges in the sea floor—but not holes; rather, arc-shaped cuts. The greatest series looped off Japan and the Philippines. It is as well to get their profile correct, as trenches are usually portrayed with a greatly exaggerated vertical scale on classroom maps: this may add to their drama, but it detracts from the truth. But why were they there at all? The explanation proved to be linked to a series of observations made on earthquakes that happened on the landward side of the trenches. Any seismic map of the world shows that a concentration of earthquakes lies along the length of Japan and southwards through the islands of South-east Asia. This part of the world has the serious shakes. From seismometer readings it is a comparatively simple matter to calculate the epicentre and the depth of origin of any quake, and in Japan there is no shortage of data. What was revealed got into the guts of the earth. There was very little activity in the trench and seawards of it; but towards the islands the quakes increased. As more data was gathered, it was clear that there was a pattern to the distribution of the quakes. Close to the trench they were shallow, while they increased in depth towards the island. In detail, they lay along an almost straight line, which dipped at an angle away from the ocean and underneath the Japan Sea towards the land. This must, surely, be a major dislocation in the crust of the earth, a line of failure.

Here lay the answer to the obvious question: if new crust was being added at the mid-ocean ridges, where was it being destroyed? The downward-dipping earthquake zones were the burial traces of plunging ocean crust. The trenches were dragged downwards: warped tracts of the ocean floor where the crust dived towards the mantle from which it had once emerged. The "scrape" of the moving crust against the adjacent buttressed margin engendered earthquakes. Crust was born at ridges and died in deep graves at the edges of continents. Where it plunged downwards it would tend to melt, admixing with partly melted rocks of continental type, perhaps, to feed subterranean chambers of liquid rock. Magma would find its way to the surface in explosive volcanoes: hence the "ring of fire" that marks the Asiatic margin of the Pacific Ocean. Some of the greatest eruptions of recent years have occurred there: Mount Pinatubo in the Philippines awoke in 1991 after sleeping for 400 years, sending eruption plumes twenty-five kilometres into the atmosphere.

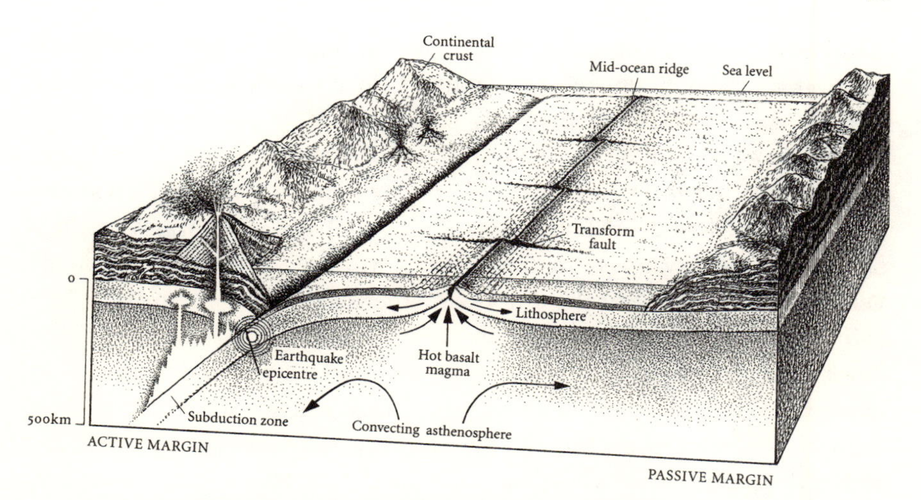

The modern view: a simplified cross-section through the oceans and continents, showing upward limbs of convection and creation of new crust at spreading ridges, consumption of oceanic crust at subduction zones, island arcs, and the generation of magma.

So the edge of the Asian continent coincided with the line where the rim of the adjacent oceanic plate plunged downwards. The ocean crust is said to be *subducted,* literally "led under" to oblivion. Even before the topography of the ocean floor was fully known, Eduard Suess had realized that a crucial boundary lay in the deep seas off Japan. Once more, he regarded subsidence as the key to understanding. He remarked that "all marine abysses which sink below a depth of 7000 meters are foredeeps in a tectonic sense, and indicate the subsidence of the foreland beneath the folded mountains" and also that they "mark the eastern boundary of the Asiatic system." He was, as so often, half right. Where Suess saw deep subsidence of continents, a modern observer sees the plunge of basaltic plates. What we see is partly a function of what we believe we see: our truth is constrained by the times in which we live.

The mathematical description of plates required these few assumptions: creation of crust at ridges; its subduction at destructive margins; and transform zones where plates could slip past one another. All were recognizable in the real world. And because that world is not expanding, the overall creation of new crust should be balanced by its destruction elsewhere. It is a matter of reconciling the equation. As we have seen, all the ocean floor today has been manufactured since the Mesozoic began: subduction has removed all that was older. Hence, it became possible to describe the world in terms of a few rigid plates that moved relative to one another over the surface of a sphere. The mathematics involved is a spherical geometry based on Euler's theorem. Leonhard Euler was a brilliant eighteenth-century Swiss mathematician, who became professor in St. Petersburg at the time of the Empress Catherine. His ideas came into their own nearly 200 years after his death.

Imagine! To describe the geometry and structure of the world by the movement of a dozen or so plates: Suess' ineluctable complexity reduced to motion, direction, and spreading rate. How to picture it? I like to think of the different plates as distinctly coloured swathes dividing the earth's surface, growing apart at ocean ridges. Imagine that the oceanic parts of the plates are darker, the continental parts paler. You can visualize the plates jostling one another, like crowded ice floes. In places, two differently coloured "floes" slide past one another with minimal interaction. But where spreading oceanic crust meets the rim of a stable continent at the edge of an adjacent plate, a subduction zone appears along the join. The darker-coloured oceanic crust will then disappear into the zone of destruction, plunging down to obscurity. Volcanic islands—a Hawaiian chain perhaps—may be plastered on to the neighbouring continent, drawn in willy-nilly by the subduction process. Inexorably, the two continents themselves will begin to approach one another as the oceanic part of the moving plate vanishes into the depths. Thick continents cannot dive under one another; instead they collide, and where this happens the earth suffers, buckles, thickens. A mountain belt arises over millions of years: a Himalayan or an Alpine system. The ocean that once ran before the Indian subcontinent has vanished beneath the Himalayas; now, it is the turn of the continents to meet face-to-face. The mountain belt grows on a line that marks the demise of the ocean that preceded it. Its linear shape betrays its origin. Forced together, the two continents mark their conjunction by squeezing lathes of land in mutual accommodation; great slabs of crust are folded and ejected, turned over, sliding, slithering, twisting. The light continental crust thickens on collision, then bounces back to rise rapidly. Snows and icefields gather on the jagged peaks. Rivers gouge downwards to reduce the impertinence of altitude. Deep beneath the mountains the isobars, depressed by the squeezed pile, return to heat the roots of the ranges. Rocks melt, but selectively. There are granites, there are gneisses.

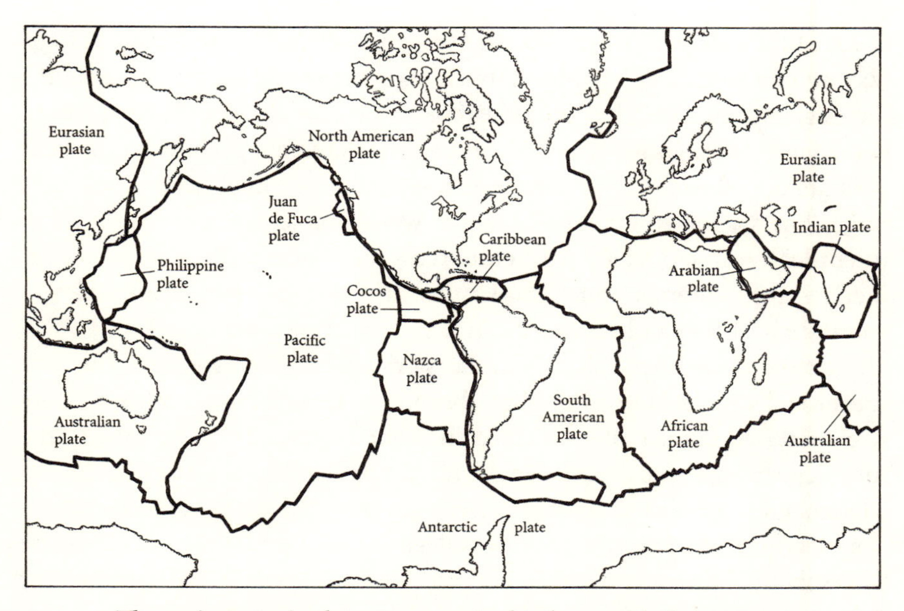

The main tectonic plates as recognized today—with their names.

Deep in the earth there is a cooking extravaganza, a hundred rock types born of collision. Minerals and metals are squeezed into the interstices of modified country rocks. This marriage of two continents is also a refinement, a distillation of rare elements into ichor that can be infused into secret places. The modern mineralogist sniffs out these places with his instruments; he now knows the knitted continents for what they are. They are the true repositories of the earth's alchemy.

In the original formulation of the theory, a handful of plates encapsulated the rocks of the world, permitting enough permutations to accommodate our apparently inexhaustibly varied planet. This may seem almost an impertinence, in view of Suess' caveat about numbering the world. But this is how the plate tectonic revolution wrought a transformation. At the present time, there are many more plates recognized, but the principles remain valid. "We shall all be changed, in the twinkling of an eye"—and so we were, at least relative to the fifty years it had taken for the idea of mobile continents to become respectable.

I had the curious privilege of being a bystander during the period when formal plate tectonic models were being developed in Cambridge in the late 1960s. There were two inspired scientists who worked on the mathematical treatment of plates at more or less the same time but on opposite sides of the Atlantic Ocean—Jason Morgan at Princeton and Dan McKenzie at Cambridge. Cambridge University is organised into colleges, and I belonged to King's College, named after its founder, the devout King Henry VI. King's is famous for its choir and its chapel. The chapel lies on one flank of a quad, and two of the other three sides are walled in by academic buildings. The visitor will notice little stairwells like secret tunnels. Some of them lead to comfortable and cluttered rooms occupied by famous academics and writers. I met E. M. Forster for tea in one such room: he was very old and very polite, and did not say anything memorable, but I was awestruck nonetheless. "Mr. E. M. Forster" a painted label said at the base of the staircase, as if he might have been in charge of the plumbing. There were fellows of the college located in other rooms who were solving the problems of the global economy, or the fundamentals of matter, or how to understand the difference between "ego" and "I," or things even more important. I was preceded in my college life by Dan McKenzie, while he was solving the problems of the construction of the world. He was a year or two ahead of me, and I was always invisible in the glare generated by his brilliance. After the first degree came the Ph.D. thesis. Dan McKenzie's on applying Euler to plate motions was instantly famous. At a college function he explained that he had printed *extra* copies to sell to the leading laboratories in the United States. The usual form in those days was for the student to take a bank loan to pay for the typing of the great work, which would then sit, magnificent but untouched, on a shelf in the University Library. The notion of people queuing up to read your thesis was bizarre. Within a year or two Dan was leading a double life, being flown out to California from Cambridge on a regular basis, like a film star. To the rank and file, all this was impossibly glamorous.

But it was all deserved. There are those who believe that theories have their time, that they arise with a kind of inevitability because of what has gone before, and as a result of new facts added to the store of knowledge. Notwithstanding, it remains true that there are few people who can grasp the historic moment. It always requires boldness, sometimes a certain unscrupulousness, and it is risky. Everything could be wrong, or half right in the manner of Warren Carey. So credit justly belongs to those who take the first bold leaps. That said, there are usually a few intellectual sprinters who take to the field at the same time, like Morgan and McKenzie. Ideas lie half concealed, awaiting their historic moment for discovery, at which point there could be several gifted people who independently realize what, in retrospect, seems inevitable. By 1970 plate tectonics was not, yet, a "theory of everything," but there was scarcely any aspect of earth history that plate theory would leave untouched. Plates provided a new way of looking at the world. Most immediately, there was a great boost for the Lyellian method. Geologists could go now into the field and look for analogies with the present-day world in ancient rocks, their search focused through the understanding that plate tectonics brought. In the Himalayas, could this, perhaps, be the remains of an island arc like Japan enmeshed in the mountain-building as two continents collided? In the Alps, might these be the deposits laid down in deep water on a sea mount? Can we find fossil equivalents of the down-warping crust in front of the Himalayas? Can this particular mineral only grow where the conditions were exactly right during the closure of an ocean? The questions were endless, and answers were often forthcoming.

Take ophiolites first. Eduard Suess had known that there was a characteristic association of rocks that could be found in many mountain belts. These rocks had been studied by an Alpine geologist, G. Steinmann, in the early years of the twentieth century. He had noticed a consistent association of dark, basaltic volcanic rocks, often forming "pillows," as around Hawai'i, with richly green and speckled serpentinite, topped by beds of the dark, fine-grained siliceous sedimentary rock known as chert. Sometimes the latter included fossil remains of tiny radiolarians, those single-celled oceanic plankton mentioned previously. So common was the association that it came to be known as "Steinmann's trinity." For instance, in the eastern Alps, Suess recorded that "the green rocks (gabbro, serpentine, diabase) are frequently associated with deep sea radiolarian chert of middle to upper Jurassic age." They were called "ophiolites" from *ophis,* the Greek for "serpent," because of their serpentinite component. Fifty years later, Arthur Holmes was to note that ophiolites indicated eruption in deep water in "geosynclines." Plate tectonics would show that they were much more significant.

In 1994, I visited the Sultanate of Oman in search of trilobites. To reach my destination in the southerly desert I was obliged to cross the Oman Mountains from Muscat, the capital city. Like many countries in the Middle East, Oman has adopted western styles but has attempted to make them compatible with the region's cultural traditions. This has resulted in some curiosities. The new outskirts of Muscat are cheerily like western suburbs, although the white-walled villas are tiled for coolness rather than lined for warmth. But in the middle of roundabouts, or in public spaces, there are new monuments. I particularly liked an enormous painted coffee pot, complete with matching cups. In another site, a vast dagger in its ornate sheath commanded attention. My assumption was that these were precious items in the tents and caravans that were formerly the common habitat.

You are out of new Muscat soon enough and pass into the foothills of the mountains. There are steep-sided valleys—wadis—with fig trees and almonds in irregular groves wherever there is a little silt on which they can grow. Little brown goats fastidiously pick at spiky bushes. The wadis are dry now, but the large cobbles lying on the dry stream floor tell you that they must be able to turn into torrents when the rains come: how else could such boulders be transported there? Soon, you pass higher into the mountains. The few villages have all grown up around springs and seem to be ancient: uniformly buff or pallid brown, each with a fortress that is all steep walls, punctured with tiny square windows in ranks. There are groves of date palms, making rows of neat canopies. Near the water source are irrigated squares of vegetables looking improbably green against the overwhelming brown. Higher still, and you suddenly pass on to the moon, or so it seems. Here there is nothing and nobody. Even the scrawny acacia bushes seem to have given up the struggle for life. The hillsides, completely barren, are composed of beds of rock, and have an extraordinary heaviness about them, an oppressive blackness crusted with umber brown. It is not just that the rocks are dark: they seem to suck out the very energy from the sunlight, which is brilliant at this height. They absorb everything and give nothing back. They might indeed be from another world. Then you notice that some of the rocks show pillow structures, as if some underworld giant had laid out a whole field of cushions for a hellish party. They are identical to lavas erupted on the ocean floor. Perhaps you might see a few patches of chert, lying in the joints on top of the pillows. Elsewhere, there are black dykes running like ribs over the hills. A coarser crystalline rock glints darkly in the sunshine, and proves to be almost entirely composed of dark green olivine, reminiscent of the green-sand beach in Hawai'i. This rock is called dunite; nearby, there are lumps of dark green and reddish mottled serpentinite scattered about on the ground. No doubt about it, Herr Dr. Steinmann would have recognized this part of Oman as an enormous ophiolite complex.

In a sense it is quite appropriate to think of these mountains as being derived from another world. It is the world of the deep sea. As more was learned about the structure of the oceanic crust, it became clear that ophiolites actually *were* slices of the deep sea floor wrested from the abyss.

In the eastern Mediterranean the island of Cyprus is centred on the

Troodos Mountains. The Troodos are another slab of ocean crust that has moved into the sunlight, and the area has become the classic of its kind. Professor Ian Gass and his colleagues showed that rocks recording the whole profile of the ocean crust are preserved in these mountains. Careful mapping revealed a "fossilized" anatomy of the lithosphere of the sea floor. Pillow lavas that had been erupted under the ocean depths from liquid magma were merely the top of the pile. Beneath the pillows there was a layer of dykes arranged in steep sheets. These were the "feeders" that supplied magma to the ocean ridge where sea floor spreading occurred. These dykes were the very struts of creation. Beneath this layer again there was dark, crystalline, igneous rock—gabbro—which underlies the whole of the ocean. In places, there were black seams of the mineral chromite (source of the element chromium) looking like tarry stripes. No wonder plants eschewed these places, for chromium is poisonous to many of them. Crystals of this heavy mineral must have settled out in the magma chamber, like beans falling to the bottom of a soup tureen. And in places, beneath the gabbro, there were patches of the rocks that lay at the base of the lithosphere, like dunite and lherzolite: we met pebbles of these rocks, as messengers from the deeps, where they had been brought to the surface in some of the lava flows on Hawai'i—dragged up from their natural home at the margin of the mantle. In these ophiolites the same rocks were hijacked wholesale from the depths of Neptune's (or should it be Pluto's?) domain. The whole profile could be more than ten kilometres thick. So this huge pile was out of place, lifted bodily from the ocean to form the backdrop to the traditional home of Aphrodite, goddess of love, whose temples are so numerous in Cyprus. Coins from Paphos show Aphrodite's emblem, a sort of cone, though it would be fanciful to imagine it represented any kind of volcano. But it is not fancy to suppose that Pluto's kingdom had somehow been levitated to the realm of love and sunshine. No wonder those similar mountains in Oman looked so unearthly, darkly glinting in the brilliant light.

Surely, though, this should not be? Ocean crust is destined to be consumed at subduction zones, not displayed layer by layer as in Oman and Cyprus. There was only one possible explanation. In places, slivers of oceanic crust were pushed *upwards,* rather than downwards. They were squeezed out from the tectonic vice, like pips from a crushed orange. In some places, a piece of ocean evidently came to rest on continental crust—as a displaced chunk, a tectonic refugee. This process is called "obduction," a kind of anti-subduction. Plate tectonics completely altered the perception of these puzzling morsels of the earth, and Holmes' observations of their deep-sea origin were explicable in a different way. Even the little radiolarians made sense, because such plankton drifted on the surface of the seas overlying the depths, and were preserved upon their death in the sediments above the pillow lavas. Later, these little fossils became invaluable in dating the ages of ophiolites and cherts, because they changed with the passage of geological time: a fossil chronometer.

In a nutshell, ophiolites were the pieces of oceanic crust that got away. Steinmann's old "trinity" proved to be three times as interesting. Within the Alps, too, there must have been other "pips" squeezed out as ancient oceans closed, when Africa and Europe met. The modern view is that ophiolites are commonly associated with the closure of smaller ocean basins—between an island arc and an adjacent continent, for example. It has become clear that subduction is not entirely destructive. Wedges of sediments may become plastered onto the edge of the continent as subduction proceeds. These form prisms of sedimentary rocks, gummed on, or as we should say, "accreted," to the continental margin. Once more geologists went into the field to decipher plates in action and found evidence of this process on the ground in the Makran Desert of Pakistan. But nobody would have heard the message of these rocks without the plate tectonic stimulus.

It could be said that entire cultures are under the influence of the geological underlay. As in Jung's original concept of the Collective Unconscious, we, the surface dwellers, are moved and bound by deeper things. Japan has now, at least superficially, adopted the western capitalist mode. Earthquake-proof technology has allowed for the construction of buildings in Tokyo that are as routinely modern as those in other capitals around the world. However, the classical era of paper-and-wood constructions was entirely appropriate to living on the quaking edge of a great subducting region. Maybe the Shinto religion with its welter of gods provided a way of propitiating the uncertain earth. Iwo-Jima is a small volcanic island 1150 kilometres (715 miles) south of Tokyo. The shore that gave brief haven to the survivors of Captain Cook's fatal voyage of 1779 is now 40 metres above sea level: the height of a church steeple. The island lies in the midst of a caldera hidden beneath the sea that is 10 kilometres across and that was formed explosively some 2600 years ago. Magma is evidently rising again and being injected into the crust. There may yet be another cataclysm. Mount Fuji has been stable for hundreds of years, but it is only sleeping. The wonderfully symmetrical cone is a typical stratified volcano, built from successions of pyroclastic flows that would have destroyed everything in their path. The explosive lava type is a particular product of its position where plates meet, so this symbol of everything Japanese is in fact a product of the underworld. Even in these secular times the mountain is dotted with shrines, some of them dedicated to Sengen-Sama, goddess of the mountain. There is no question that its presence in Hokusai's inexhaustibly inventive set of portrayals, *36 Views of Mount Fuji,* is more than merely decorative. The mountain centres the world of his art. In Shinto days everybody made a pilgrimage to the top of the mountain, which at 3776 metres (12,388 feet) is the highest in Japan. The sun goddess Amaterasu required it. Today people still feel drawn to watch the sun rise, looking towards Tokyo 100 kilometres away, or perhaps towards the Pacific Ocean where the unseen plates move. The American journalist and author Lafcadio Hearn, who married a Japanese woman and took her nationality, was one of the first westerners to try to explain the virtues of Japanese culture. A hundred years ago he described his emotions on seeing the sun rise from the summit of Fuji in these terms:

> But the view—the view for 100 leagues and the light of the far faint dreamy world and the fairy vapours of morning—and the marvelous wreathings of cloud: all this, and only this, consoles me for the labour and the pain. Other pilgrims, earlier climbers poised upon the highest crag with faces turned to the tremendous East, are clapping their hands in Shinto prayer saluting the mighty day . . . I knew that the colossal vision before me has already become a memory ineffaceable—a memory of which no luminous detail can fade till the hour when thought itself must fade.

Lush prose, perhaps, but you are left in no doubt as to the importance of the occasion, and the importance of Fuji in Shinto cosmology. The Shinto religion may not have the grip it once had, but there are still shy little offerings placed in wayside shrines on the way to the summit. Just in case.

Geology dictates the lie of the land, and climate controls how the design of the world accommodates to life. But climate itself is in thrall to geology. A landmass over the poles permits ice sheets to grow, and this mediates the sea levels of the world. There have been warm times when much of the land surface has drowned, and such times will come again. Mountain ranges modify weather systems, specify where there shall be deserts, and steal rain. Then, too, oceans are great climatic modifiers. Think how the coast of Europe is so ice free while the icebergs drift off frozen Labrador, at the same latitude. The North Atlantic Drift (Gulf Stream) moves warmth northwards—but only for some. There are climatologists who believe that the warm pump might be turned off in the twinkling of an eye, geologically speaking. Then long, frozen days might turn England's hedge-fringed green fields into birch and conifer scrub. The shape of the ocean basins is the stuff of climate: deep gyres transfer cool water and nutrients about the world. Yet ocean and mountain are no more than a consequence of the geological foundation: the arrangement of plates in this mosaic of our earth. Change the plates, and you will rearrange everything else. Mankind is no more than a parasitic tick gorging himself on temporary plenty while the seas are low and the climate comparatively clement. But the present arrangement of land and sea will change, and with it our brief supremacy.

6

Ancient Ranges

Newfoundland is a curious island. It hangs off the eastern coast of the North American continent as if it doesn't really belong there. The natives call it "the Rock," and it is an apt name. There is certainly no shortage of geological exposure on the island, although most of it crops out along its rugged and indented shores. The coastline is 10,000 kilometres long if you add in all the numberless bays, where fishermen have established their "outports." Inland, Newfoundland is a mass of rambling rivers, and small lakes, called "ponds." The cover is a scrubby forest, mostly under-sized conifers, but relieved by shaking aspens, small alders, and birch trees. Many of the conifers have a sickly look, and are crowded together, a working example of the struggle for existence. It is an endless kind of landscape, mile after mile of the same, for hundreds of miles. There are few roads away from St. John's, and if you are foolish enough to leave the beaten track and wander into the bush, you can become disorientated within seconds.

In spite of the uncompromising terrain, from an early period the island was colonized by Europeans, who displaced the native Micmac Indians. Even the Vikings found their way to L'Anse sur Meadows in the Northern Peninsula, at a time when the climate was more equable than it is now. Hundreds of years later, fishermen from the English West Country and from Ireland followed the first wave of immigration into North America. The coastal villages they settled were isolated from the world, and from each other. Some of the place names—Trepassey, Tor's Cove—would not

Newfoundland, a bleak land. At Bauline, with ice in the bay and a typical outport wooden construction.

be out of place in Cornwall. Others—Blow-me-down Brook, Fogo, Goose Tickle, Heart's Desire—have a unique charm all their own. The extraordinary Newfoundland accent may have preserved something of its original Tudor twang. Certainly, it is most unlike that of the island's Canadian neighbours. To the "incomer" it sounds Irish at first, but then other harmonics emerge—a hint of Cornish, perhaps. Aitches are dropped and added; and words that are obsolete elsewhere are preserved here in common usage. The island only formally joined Canada in 1949. Before that, it was the oldest British colony, founded in 1583. It remains fiercely proud of its own identity.

The once thriving cod industry has collapsed, ruined by non-indigenous factory ships over-fishing the Grand Banks. Nowadays, the distinctive two-storey wooden houses that line the outports are painted brighter than they have ever been: there is little for a beached fisherman to do other than redecorate the house. The cod "flakes" along the shoreline on which the fish were once split and dried have all but disappeared. There are still lobsters to be had, and squid some years, but the great days are over. Newfoundlanders annually receive the vilification of animal rights activists because of their participation in seal culling. But to these men it is a tradition, and clubbing is the most humane way of dispatching a proportion of baby harp seals for their skins. The year I was living there I saw the Bishop of St. John's bless the fleet before it sailed. This was my first lesson in moral relativism.

Newfoundland is a geological textbook laid open to the skies. The traverse east to west across the island is a pilgrimage that many famous geologists have made. There is something there for everyone: for the palaeontologist, for the tectonician, for the just plain curious. The rocks embrace an appreciable chunk of geological time—several hundred million years—especially spanning the later Precambrian, Cambrian, Ordovician, and Silurian periods, 900–419 million years ago. It is a big slice of earth history spread across a big stretch of land, making it difficult to select just a few places that reveal the secrets of the Rock—many students have spent their whole lives untangling the complex geology of small parts of the island. It becomes necessary, perhaps, to imitate one of the huge seagulls that haunt the Newfoundland shore, alighting briefly on one rocky outcrop before flapping off to another distant cliff.

St. John's lies on the eastern edge of the island. It is built on folded, resistant, Precambrian sedimentary and volcanic rocks that form spectacular, sheer cliffs, ribbed and pleated, from the top of which you can watch pilot whales sporting offshore. The old town around the splendid natural harbour is picturesque enough in a scruffy way. Steep terraces of painted wooden houses run down to Water Street, which was once a bustling centre for shopping and shipping before the edge-of-town malls and supertankers took their toll. Grand Victorian houses in their own grounds that belonged to the merchants lie uphill. Those whom Newfoundlanders still call "marchants" were the beneficiaries of the fishing wealth of the island: it is said that there were more millionaires a century ago in St. John's than anywhere else in North America. They kept the fishermen in a state of perpetual debt by an iniquitous system of "loans" advanced on the year's catch. Older inhabitants still remember running around barefoot through the snow. This is a tribute to their sheer toughness. Because the Gulf Stream passes Newfoundland by, the climate is, to be tactful, frequently inclement. You would never guess it was on the same latitude as Paris. Spring seems to be delayed forever, and a dump of slushy snow down from the north is not unknown even in June. Fog obscures the delightful sea views for what seems half the year; and the mournful lowing of foghorns through the Narrows provides an eerie accompaniment to winter walks. Ice floes brought down on the Labrador Current accompany the brief early days of summer. For geologists, it means that the field season is decidedly limited, and waterproofs are indispensable at all times. When the sun comes out, it is totally delightful, except for the mosquitoes, and the blackflies, and the deerflies.

Bell Island is a small island in Conception Bay, a few miles west of St. John's. It is more or less encircled by cliffs comprising dull yellow or brownish sandstones, much less folded than the rocks around St. John's. The cliffs are quite accessible in a number of places, and a brisk breeze usually keeps the biting and bloodsucking flies at bay. These sandstones are the kind of rocks that were originally laid down as sediments in quite shallow seas. A very lucky hammer blow might reveal fossils that look like small, black hacksaw blades, laid out on "leaves" of intercalated shale. These are the remains of extinct, colonial, planktonic animals called graptolites. Every geologist likes to find graptolites because they can be used to date rocks quite precisely. They evolved fast, and spread widely around the ancient oceans in which they floated—making them ideal biological chronometers. The ones on Bell Island are no exception: the local expert, Henry Williams, identified them precisely. They indicated an Ordovician age; more exactly, a part of the early Ordovician called the Arenigian. There were other fossils there as well, including some of my own favourite animals, trilobites. Superficially more complex than the graptolites, trilobites were "bugs" with jointed legs—arthropods—that swarmed in the seas of the Palaeozoic era. Sadly for all biologists, they are now extinct, but there were once thousands of kinds of trilobites, individual species of which were confined to the seas surrounding different continents of the ancient world. These ones on Bell Island were old friends: I knew their names as well as I know my children's birthdays: *Neseuretus* and *Ogyginus.* Both of them had been yielded up to my questing hammer in strata cropping out along streams and roadsides in Wales. I knew them also from France and Spain; I had even been sent a sample of one of these beasts from the remote deserts of Saudi Arabia. Much more common on Bell Island were scratches and burrows left in the sandy sediments by animals on the Ordovician sea floor. Some, called *Cruziana,* were probably made by the same trilobites that you could hold in your hand. They look like braided plaits laid on the rocky slab, as distinctive in their way as many more substantial fossils. Exactly the same kinds of *Cruziana* could also be found in France and Spain and North Africa. It is beginning to look as if the sandstones on Bell Island are trying to tell us something.

Move further westwards to Random Island and it is a similar story. More fossils, a little older, just like those from Europe. On the way you pass over a cheerless, bleak, and empty region, where even the hardy conifers have given up trying to take root. These are the "barrens." The underlying rock is unforgiving granite, and the soil above it is so poor that little seems to flourish. Little, that is, except a fruit called the partridge-berry that clambers over the ground on inconspicuous vines. It is said to be a good source of vitamin C and is resistant to frost. Newfoundland families used to keep a stash of these purple-red berries the size of rabbit droppings under water in a jar by the door, and a few would be taken every day over the interminable winter to keep scurvy at bay. In tough places everything seems to have a use. The eastern part of the Rock comprises two broad lobes of land trending north-east–south-west connected to the rest of the island by a narrow link of land that is almost an isthmus. This eastern part of the island

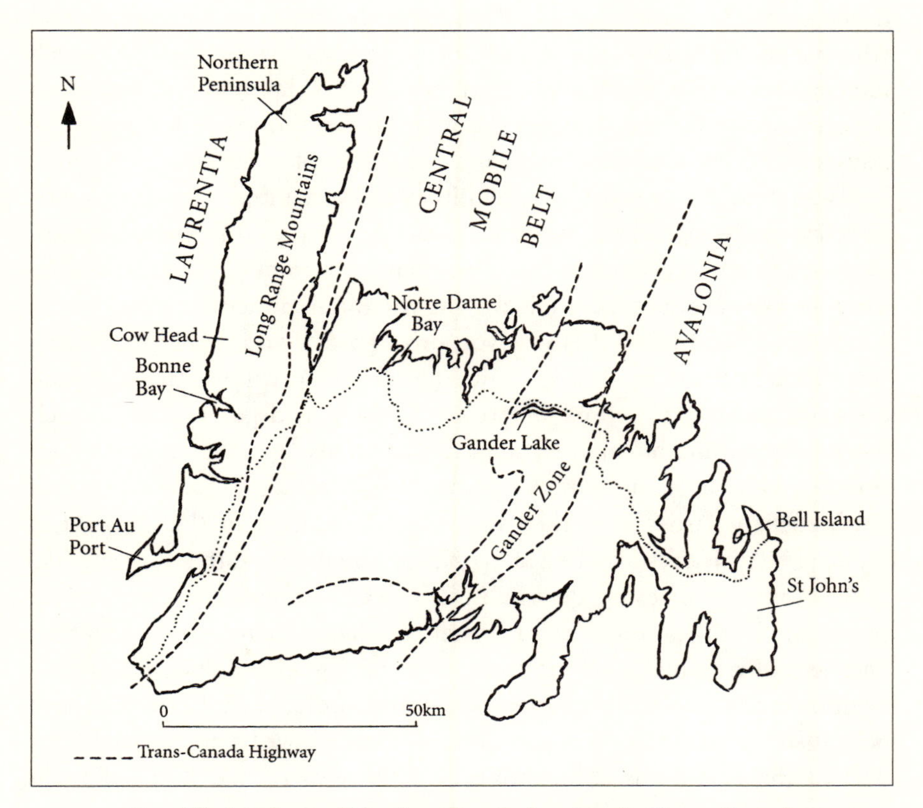

Simplified geology of Newfoundland, showing the Trans-Canada Highway running west from St. John's, and locating some of the places visited.

is known as Avalon. As so often in geology, understanding of the area was dependent on making maps of the rocks, and the unscrambling of Newfoundland began in Avalon, with the pioneering geologists Alexander Murray and James Howley publishing the first map of the Avalon Peninsula in 1881. The accounts for 1871 fieldwork submitted to the Geological Survey Office note that 100 pounds of bacon cost £5, as did a new tent. Murray was able to hire four men for four months for £120, but his own salary was £490. Those geological maps have been under a continuous process of refinement ever since: no map is ever truly finished, simply because you cannot write down all the truth of the world on a sheet of paper.

Move westwards again. You will find yourself driving along the Trans-Canada Highway—there is no other way to go. Several hours pass before you get to Gander, a remote settlement with nothing to it except a hotel and an airport. Then, when you reach Gambo and look at the rocks cropping out in cuts along the roadside near the village, you notice that something seems to have changed. Gone are the simple sedimentary and volcanic rocks that we saw around St. John's and on Bell Island, which there are laid out in clear layers, even though those layers might have been elevated to the vertical. Instead, the rocks at Gambo present a bewilderingly various appearance: striped and twisted, in places they show a crude layering, though it seems to wander hither and yon. Pinkish layers a foot or so thick can twist and pinch out on a rock face a dozen metres in length. Quite often you see smaller, pale veins that seem to be scribbled onto the rockface like graffiti. If the sun is shining you might notice that many of the rocks glisten and flash briefly as you drive past. If you manage to find a place to get off the road to examine the rocks you will see that the pink ones are mostly composed of creamy-looking feldspar crystals, which impart the dominant colour. Then there are veins of quartz, which sometimes wriggle through the rock like creamy worms, or cut across its grain like petrified flashes of forked lightning. The shine in the sunlight is induced by little crystals of mica, with perfectly flat faces that respond like diminutive mirrors to incident light. Coarse and irregular layering is often picked out by drifts of dark mica crystals. In places, the rocks look as if they had been stirred by a gigantic spoon. They are gneisses. Such rocks have been changed by baking in a terrestrial pressure-cooker. Then they have been disinterred: kilometres of overburden must have been removed. This part of the country is comparable with the pass of San Bernardino in Switzerland, despite its different geological age. Nowhere near Gander are there marbles, such as those that enclosed S. Gennaro's relics in Naples in such splendour. This is gneiss country, boggy and impenetrable. Just as marbles started life as limestones, many gneisses started out as shales, before they were "cooked." More specifically, many of these rocks are described as migmatites, in which the pink veins seem to blend insensibly into a kind of swirling granite. Sometimes it is difficult to be quite sure whether you are looking at gneiss or a banded granite. If you take a diversion off the Trans-Canada Highway, along Bonavista Bay as far north as Deadman's Bay, there are huge masses of wholly unambiguous pinkish granite plunging direct to the sea. Some of the coarse gneisses contain garnets. Jewel-quality garnets are pretty, plum-coloured stones, much beloved by the Victorians for silver settings. Most of the ones you find in metamorphic rocks are a disappointment; the best you can hope for is to find one about the colour, size, and shape of a partridge-berry. They are interesting to find, though, because they only form under particular temperatures and pressures. Almandine, the variety found here, is stable at temperatures of 540° to 900° C at a pressure of 200 GPa. This has been proved by experiment under controlled conditions in the laboratory. So it is possible both to visualize and quantify something of the suffering the rocks in this area endured in the great vice of the earth. No fossils could be expected to survive such treatment.

Between Avalon and Gander, therefore, it is evident that a major geological threshold must have been crossed. The threshold is marked by a great fault cutting through the geology, the Dover Fault, which runs north-east–south-west near the eastern side of Bonavista Bay. The Avalon Platform lies to the east, back to Bell Island and St. John's, while to the west lies the Mobile Belt, forming the central part of Newfoundland, a land of complexity and problems. This is where the striped and redoubled gneisses, and many other rocks besides, all run together and are twisted and faulted up. It is mobile, not just in the sense that there must have been movements of strata, faulting, and geological mayhem, but also that whole chunks of the world may have reached where they now lie from some other place.

Past the village of Gambo the road runs along a pond that is big enough to be called a lake—Gander Lake, where there are folded, metamorphosed sandstones (quartzites). The rocks of the apparently impenetrable area north of the lake were mapped in the 1960s by a celebrated Newfoundland geologist, Harold Williams, universally known as "Hank," a native of the Rock and wise in its ways, with an accent that puzzles those not used to its idiosyncratic inflexions. Hank Williams worked out much of the geology of the Mobile Belt the hard way, by foot and Indian canoe; latterly, helicopters made life easier, if more dangerous. Like Murray and Howley before him, he produced geological maps—but what maps! Unscrambling the Mobile Belt was like solving a cryptic crossword, one in which even the clues were anagrams. The resulting maps are speckled and daubed like a painting by an abstract expressionist, but even through the extravagance of colours you can clearly see how many features have a north-west–south-east trend governing the shape of the bays and headlands. The design of the country is clearly written in the geology. Hank employed many students, too, and one of his criteria for their selection was the ability to play a musical instrument. There was always a band in the Geology Department at Memorial University in St. John's, with Hank on banjo or fiddle, and with assorted bassists and dulcimer players, all making a fine racket with their jigs and reels, of which there seemed to be an endless selection. When the fog was thick on the north shore, I am told that Hank used to sit in the bow of the leading canoe furiously strumming the banjo, to guide the party onwards to the next offshore island. He thought nothing of dumping a student in some remote area until the field season was over, with only a violin and a field assistant for company. But Hank was much more than a decoder of complex geology; he built up a picture of the whole of the Newfoundland central fold belt and extended it far, far beyond the wild shores of the island—finally, it provided an interpretation of the whole eastern side of North America from the Adirondack Mountains to the Appalachians. But here I am getting ahead of my story.

It is clear that there are "fossil" volcanic islands within the Mobile Belt—around Notre Dame Bay, on the northern coast, for example. Volcanic rocks are easy enough to recognize, even when they have been baked or folded; a number are like those I have already mentioned from Hawai'i, including pillow lavas. A few ancient islands still stick up out of the sea as rock islets today; disinterred, they reproduce their anatomy of 460 million years ago. The detailed chemistry of the lavas can be studied in modern laboratories, and some show the kind of "trace" elements that are characteristic of oceanic islands at the present day; others are like those from island arcs close to subduction zones. These elemental ratios are like fingerprints that identify the magma source of the ancient lavas. This geochemistry is a wonderful application of Lyellian principles, and a boon to the geological interpretation. Not all the rocks have been as heavily metamorphosed as those near Gander. In places there are beds of rock composed of cobbles (conglomerates) that you can just imagine tumbling off the side of the unstable volcano. Some of the islands had shallow seas around them, and, just once in a while, fossils of some of the shelly animals that lived in those waters have survived all the vicissitudes that nature has thrown at them. These speak of Ordovician dates. It is curious to think of seals now basking where trilobites once crawled. Discovering these critical creatures in these volcanic-derived rocks was an heroic accomplishment; for once the needle and haystack analogy can be used without blushing. So this part of the Mobile Belt is a place where all the branches of geological science meet and converse.

There are ores and minerals in the Mobile Belt, too, and many of them are associated with the volcanic rocks. Every few years there are rumours of gold finds, and indeed in a few localities there are significant amounts of the precious metal—and rather more silver. Sulphide ores have been important enough to give rise to working mines for copper, manganese, and lead-zinc in other parts of the island. For wherever the earth moves, metals are concentrated.

A friend, John Bursnall, was assigned an area around Notre Dame Bay as the ground on which to work up his doctoral thesis. He returned from fieldwork each year with his brow progressively creased, and with thick notebooks full of sketches. The geology was, he remarked from time to time, a real mess. Everything on the ground seemed to be in a state of flux. The rocks he had been assigned are known as the Dunnage Mélange. Poor John was obliged to share an office with me, where I noisily dug out trilobites from their limestone matrix whilst he attempted to put some shape into a hideously complicated piece of Newfoundland. He sometimes had to retire with a headache. A mélange is, I suppose, a polite word for a mixed-up mess. The principal feature of the Dunnage Mélange consists of huge blocks of rock. These now form islands and shoals in the grandly named Bay of Exploits. If you are accustomed to think of rock successions as deposited in a logical order—oldest first, then progressively younger—it is not too difficult to envisage how such successions can get folded, or even, as in the Alps, turned completely upside down. It is not so impossibly hard to conceive of these rocks subsequently becoming metamorphosed by heat and by pressure, and changing their character accordingly, in the grip of the earth's inexorable movements. But in the Dunnage there are places where it is almost impossible to make sense of the relationships between rock units in the field. These enormous blocks—some more than a kilometre in diameter—of oceanic-style volcanic rocks rest on a kind of groundsheet of shales, which must have originally accumulated in the deep sea. Some of the blocks have evidently tumbled or slid downslope until they came to rest in a confusing array. Fortunately, those useful fossils, graptolites, have been found in the shales to tell us that these events happened in the Ordovician. But then the confused mass is itself folded, even overturned, thus superimposing complexity on perplexity. To make matters worse, there are places where igneous rocks, cooled from magma, have been injected into the shales while they were still soft muds, so there are other kinds of rocks swirling around in the mixture. And then the whole area has been cut with steep faults, and pushed bodily north-westwards along thrusts.

Hank Williams and his associates divided Newfoundland into a number of zones, each one with a distinctive geological signature. By now we have moved progressively westwards from the Avalon Zone, through the Gander Zone into the Dunnage Zone. As geology always turns out to be more complicated than was thought at first, the Dunnage is now subdivided in various ways, the details of which would further tax the imagination. Each of the major zones is separated from its neighbour by a major structural line, often a series of faults that cut deep into the crust. The Grub Line separates the Gander Zone from the Dunnage, while the Baie Verte Line separates the Dunnage complexity from what happens further west. Although they wander somewhat, each major fault follows the grain of the land from north-east to south-west. Everything on the surface is in thrall to the deeper order. It is almost a relief that by the time we arrive at the west coast, some of that bewildering complexity is lost.

One important road crosses the Trans-Canada Highway in the west; it runs northwards up the Great Northern Peninsula, a great protuberance that sticks up on the west side of the island like an optimistic thumb. This highway runs to the sea at the Gros Morne National Park. The western part of the Rock retains vestiges of French influence in place names like Port au Port and Port aux Basques—after all, it does face Quebec. But it is still defiantly Newfoundland in spirit, even though the "mainland" is only a ferry ride away. The Gros Morne Park is spectacularly disposed about Bonne Bay, a "drowned" valley flanked by relatively lush forests with cliffs comprising mostly Cambrian-age sedimentary rocks plunging steeply to the sea. Down at Norris Point there is a little ferry that will take you across Bonne Bay to Woody Point: on a clear day the crossing is a delight, with the sun flashing the tops of the waves and picking out tree-lined coves up the Bay. Woody Point is protected from the winter weather, so it has mature trees, as well as immaculate clapboard houses flanking the hillsides to which some of the more affluent islanders choose to retire.

Following the road upwards from Woody Point towards the outport of Trout River, you suddenly encounter a different world. All the trees have disappeared, as if removed by blight. There is little up here but sphagnum moss and a few hardy plants that seem able to survive on nothing at all on the open, rounded hillsides. The whole landscape has been transformed into a dull brown wilderness. The umber rocks suck out any vigour from the sunshine, while on an overcast and misty day you might easily imagine that you were entering Mordor, J. R. R. Tolkien's blasted and evil kingdom, for so little thrives here. The rocks are clearly of a kind on which terrestrial life is ill at ease. Looking closely, it is clear that there are pillow lavas among them: we have seen those bulbous forms before. Then there are some stratified rocks: their dark and crystalline character shows them to be igneous, and some of them are of the kind associated with mantle material, whilst a few of the dark grains might be the mineral chromite. Elsewhere, there are densely packed dykes—sheeted dykes—that closely resemble the stacked legacy of ocean-floor spreading. Suddenly, it is obvious that the whole desolate area is nothing less than a piece of ancient ocean floor: an ophiolite suite. Its alien bareness is akin to the landscape I encountered under cloudless skies in the Sultanate of Oman. Small wonder that trees cannot flourish on the malnourished hillsides. Rather than becoming obliterated in the plunging mills of subduction, this naked knob of land on the western coast of Newfoundland has been obducted—squeezed upwards rather than downwards, and thereby saved from destruction. Since further westwards lay only the great continent of Laurentia (North America + Greenland), there must have been an oceanic basin to the *east* from which this slice of crustal history was derived. How could it be otherwise? To this we shall return.

The story continues further northwards up the Northern Peninsula. From here you can see the inaccessible Long Range Mountains that form the "spine" of the peninsula. These are composed of metamorphic rocks of Precambrian, more particularly Grenville, age, far older than anything in the Mobile Belt: these speak of earlier ages, other dramas. They were annealed before the first rocks around Bonne Bay were laid down under the Cambrian seas. The village of Cow Head is excluded from the park, so it is an agreeably ramshackle kind of place. Youths hang about on the main street, and even the arrival of a geologist is worth a glance. You used to be able to buy a plate of cod's tongues in the Shallow Bay Motel—and surprisingly good they were, too—but the fishing ban probably put paid to that delicacy. I got to stay in a mobile home, and in one of Payne's Cabins. There only seem to be two family names on the coast hereabouts—the Paynes and the Crockers—so I imagine marriages do not cause much taxonomic surprise. Cow Head itself is a prominent headland connected to the settlement by a natural causeway, or "tickle" as the Newfoundlanders prefer it. The rocks running along the shore are extraordinary. From a distance it looks as if some Titan had emptied out sacks of massive boulders along the cliffs. Closer to, it appears that the rounded boulders are allied to dipping rock-beds, cemented together by smaller pebbles and fine limestone: in fact, everything—boulders, pebbles, and all—is made of limestone. These are conglomerates of a very distinctive kind. Then you notice that individual conglomerate beds are separated by something altogether softer that weathers back. This rock comprises thin beds of limestone and shale; you may have to grub with your geological hammer to claw out bits big enough to break. But—wonder of wonders!—the shales are full of useful graptolites. Following one bed of shale to the next above, these fossils show a succession of typical species that prove that the earlier half of the Ordovician period is represented by strata of conglomerates and shales (in fact, similar rocks carry on down into the Cambrian). The shaly rocks were originally laid down under deep water. Then, when you bring a bigger and stronger hammer to bear on the limestone boulders themselves, you discover that they, too, yield fragments of fossils—trilobites. I spent more hours than I care to remember persuading tough limestones to give up their shelly treasure trove. On clear days, such work was nothing but a pleasure, with gentle breezes bearing the aromatic smell of pinewoods. On wet days it was necessary to keep reminding myself of the nostrum that genius is 99 percent hard work; this optimistically based on the classical deductive mantra: "all genius is 99 percent hard work; this project is 99 percent hard work: therefore, I am a genius." However, the labour was worth it, since the kinds of fossil animals recovered showed two things very clearly: they were species that lived in shallow water; and they were utterly different from their near contemporaries on Bell Island, where this journey began. The conclusion is inescapable that the limestone blocks originated in shallow seas, and then slumped or tumbled down to a shale sea floor in massive flows, carrying with them their cargo of trilobites into the domain of the graptolites. At Lower Head, nearby, some of the blocks are the size of houses.

Journey's end is a little to the south and west: the Port au Port Peninsula. This is a comparatively open part of the island, with many grassy fields, in which a cow or two watches the world go by. It was once very poor, and is still very Catholic. Small square wooden houses sport the longest washing-lines in the world. They are hung with trousers and shirts of various sizes, graded according to age. When there is a wind blowing they look like bunting. As usual, the rocks are exposed along the shore: limestones again, but ones that appear different from the spectacular conglomerates on Cow Head. These are regular limestones, in regular beds of rock. The strata are gently tilted, so you can climb up or down the rock succession much as you might go up and down a staircase, respectively ascending or descending through geological time. This is about as simple as geology gets; it is known as "layer cake stratigraphy." Compared with the Mobile Belt, only the gentlest of earth movements has disturbed these rocks. In some places, on the surface of limestones it is easy to see coiled shells of fossil snails, which tell the palaeontologist that—again—the rocks are of early Ordovician age, 470 million years. But on Port au Port there are no deep-water graptolite shales. All the evidence from the limestones points to their accumulation in shallow water environments during the Ordovician. Here and there are beds of oolites, for example, made of little rounded grains—like millet seed—that form only in agitated warm waters, such as you might find off the Bahamas today. Elsewhere, there are fossil algal "mats" that grew close to the inter-tidal zone: they look like rucked paper tissues in cross-section where the sea has polished them through. Or the limestones have been replaced by dolomite, which is something that happens in hot lagoons. There is no question about it: not only were these limestones laid down in shallow seas, but those seas lay under a tropical sun. There were calcareous lagoons and tidal flats and oolite shoals where now the gulls keen over a choppy Atlantic sea. Similar tropical limestones crop out further north in western Newfoundland, right up to St. Antony at the tip of the Northern Peninsula. Further afield, they are widespread over much of North America and Greenland. This was a time when sea levels were high world-wide, and shallow seas penetrated into the interior of continents. The whole of Laurentia basked under a tropical sun in the Ordovician—a vast area of shallow seas known as a carbonate platform. However we fashion the Ordovician world it is evident that we are obliged to place Laurentia close to the equator of the time. And the trilobites? Some of the species found in the tumbled blocks at Cow Head are found also in situ in the contemporary platform limestones. So this is where the blocks must have come from: they slumped and tumbled off the edge of the carbonate platform at the rim of the Laurentian continent as it was in the Ordovician. Into the deep sea, and into history.

One final fact: careful geological mapping showed that both the Cow Head conglomerates and the Bay of Islands ophiolites were piled on *top* of the carbonate platform limestones. This happened along a series of thrust faults, a shove of the earth that propelled huge tracts of land towards Laurentia. Unlike the platform rocks on which they lay, the Cow Head rocks and ophiolites were tipped steeply, folded, and faulted; they had suffered as they moved.

How are we to interpret the geology we have seen as we bobbed like a restless seagull from east to west over the Rock? Plate tectonics finally supplied a satisfactory answer, but understanding came piecemeal. It is necessary to avoid thinking of the progress towards what we now know as a kind of moral improvement. It must *always* have felt as if the tectonic truth were just around the corner; but the confident assertion of today is all too often the retraction of tomorrow. To the pioneering geologists making their first maps it must have seemed enough simply to lay out the ground into its constituent rock formations. They would have been delighted to find sufficient fossils to put some dates on the rocks of the Avalon Peninsula, just as on the other side of the island as early as 1865 the Canadian palaeontologist Elkanah Billings was uncovering trilobites from the same limestones I had chopped with my hammer more than a century later. Eduard Suess was certainly aware of the integrity of the ancient continent of which those limestones formed a part: "That vast region of North America which is formed on ancient rocks overlain by horizontal Cambrian strata has received the name of Laurentia." This expresses the idea clearly enough. He noted that Cambrian trilobites overlay the older "schists" of the Northern Peninsula in Newfoundland, making that part of the island a sure component of Laurentia. And he was insistent that Greenland belonged to the same continent, too: "*Greenland is a part of Laurentia*" was an assertion important enough to be worth italicizing. It was also clear that the geology of Newfoundland connected south-eastwards with that of eastern Canada, and beyond into Maine, Vermont, and New York State, and thence to the southern states. The whole grain of the Appalachian mountains followed the same northwest to south-east trend, just like the faults separating the main geological divisions. The Blue Ridge Mountains traced the identical line, as did the

Alleghenys. Then there were the metamorphic rocks, those schists and gneisses that gave such clear evidence of having been caught deep in the vice of the earth, but which were now disinterred. They, too, could be followed from Newfoundland through the Appalachians close to the edge of the great Laurentian continent. Surely these rocks proved that a range of mountains more elevated and dramatic than today's scenic remnants had once towered above the Laurentian interior. Hundreds of millions of years of erosion had brought them down, exposing their innards to the snows and rains of the twenty-first century. To be sure, there are places where the mountains are still splendid and wild, especially where granites have bolstered their durability; but there are others where silvery schists crop out along urban roadsides under the sumach trees, or are dug out to make the foundations of superstores. Geological time is enough to wear down the mightiest peaks, and lay low former Everests.

It was appreciated early on that there was a contrast between the successive Cambrian, Ordovician, and Silurian sedimentary rocks that accumulated under the sea on the Laurentian platform and those that were laid down at the same time along the Appalachian line. James Hall was a great geologist who prepared the first official accounts of the fossils and strata of New York State. In 1859 he observed that the sedimentary rocks of the Laurentian platform were ten to twenty times thinner than their time equivalents in the folded mountains—where sedimentary piles could be several kilometres thick. It was evident that linear troughs followed the trace of future mountain chains, which were steadily subsiding over millions of years, simply to accommodate all that sediment. In 1873 James Dana, whom we met earlier, called them "geosynclinals"—literally, a sag in the earth—and Suess used the same term, though the alternative designation "geosyncline" is the one that later stuck for more than half a century. For a while, geosynclines were very popular. Forty years ago the crumpled and complex Mobile Belt of Newfoundland would have been considered a typical result of a deformed geosyncline. Many different kinds of geosynclines were recognized in the 1950s, most notably by the "prince of geosynclines," Marshall Kay of Harvard. He tacked all kinds of prefixes ("taphro-," "poly-," etc.) on to geosynclines to designate what he considered their distinct geological circumstances. Marshall Kay is principally remembered—rather affectionately—by those who met him as the most garrulous man who ever lived. But he did point to Newfoundland as an ideal place to study the genesis of ancient mountain chains, in which regard he was absolutely correct.

The problem is that the mere fact of a thick pile of Appalachian sediments alone does not explain how they subsequently came to be folded and metamorphosed—in short, to have all the features of a mountain belt. A great thickness of comparatively light sedimentary rocks accumulating in a geosyncline would ultimately "rebound" because of isostasy—like a depressed rubber duck bouncing back upwards in a bath. The lower parts of the sedimentary pile might have been temporarily subjected to sufficient heat and pressure to become metamorphosed, but the effect would not have been adequate to account for the extreme conditions required to produce some of the minerals found in schists and gneisses—nor for the observable tectonic mayhem and thrusting such as forced the ophiolites in western Newfoundland over the "platform" limestones. Arthur Holmes struggled gamely with applying geosynclinal terminology, but found it inadequate. Clearly, as the detective in "locked-room" mysteries is wont to say, it was time to consider the facts in a new light.

The acceptance of mobile continents provided more than merely new illumination: it supplied a floodlight. The 1964 edition of "Holmes" includes a diagram showing what Europe + America looked like when you closed the North Atlantic Ocean, prior to the breakup of Pangaea. An extraordinary convergence happens. The whole Newfoundland–Appalachian mountain chain suddenly joins hands across the (non-) Atlantic with another ancient mountain chain—the Caledonides. Caledonia was, of course, the Roman name for Scotland:

O Caledonia! stern and wild,
Meet nurse for a poetic child!
Land of brown heath and shaggy wood,
Land of the mountain and the flood

declaimed Sir Walter Scott, who celebrated its romantic embodiment in *The Lay of the Last Minstrel.* The Caledonides, however, do not just include Scotland: the ancient mountains have a compass as generous as that of the Appalachians. Like that range, they comprised a complex of Lower Palaeozoic rocks variously folded, faulted, or metamorphosed: wild country, for the most part, mountains and brown heaths, indeed. The chain ran south-west to north-east through Ireland, embracing hilly Wales and the Lake District, much of rugged Scotland, and thence to include the whole of the Scandinavian coastal mountain chain running through Norway, from Stavanger in the south to Hammerfest in the north—this section alone spans more than 10 degrees of latitude. Close the Atlantic Ocean, and the Caledonides run on naturally north of Newfoundland. The whole chain when put together snaked its way through the midst of the ancient supercontinent of Pangaea. Had we been able to peer from a satellite in Permian times, the Appalachian–Caledonian line would have been one of the most easily recognizable signatures written on the face of the globe: a seam ancient even then. Clearly, the opening of the Atlantic Ocean had ripped apart two segments of a single geological monument—not neatly so, either, but raggedly, like an incriminating photograph roughly torn in two by a jealous lover.

Eduard Suess had not recognized the northward continuation of the Appalachians for what it was; for him, Newfoundland was the local end of the structure. If anything, he traced the continuation of the Appalachian chain into what he termed the "Altaides" of Europe and Asia. Marcel Bertrand and, later, Arthur Holmes, had spliced the Caledonides and Appalachians together. By 1964, when the second edition of "Holmes" was published, the "drifters" had finally achieved respectability. The recognition of such an extensive ancient range could be viewed as another consequence of the acceptance of Pangaea.

The Caledonides themselves had a tradition of study that was both more historic and more intense than was the case for Newfoundland's geology, largely because these ancient mountains crossed a major part of western Europe, where there were many geologists. After all, Britain was where the rocks of the Cambrian, Ordovician, Silurian, and Devonian periods had first been discriminated one from another during the nineteenth century. The very names still embody country (Cambria: Roman Wales), county (Devon), or ancient tribes (Ordovices, Silures) that were part of the human history of the British Isles. This is where the "type areas" for rocks of these ages still reside. Pioneer geologists did fieldwork in areas they could reach by train or on horseback, or by foot. Improvements in public transport, particularly expansion of the railway networks, made formerly inaccessible areas readily available for study. And how those pioneers could walk! When Ben Peach and John Horne exquisitely mapped vast areas of the Highlands of Scotland at the end of the nineteenth century, they thought nothing of covering thirty miles a day of "brown heath and shaggy wood" across some of the most strenuous territory in Britain. As so often, geological maps drawn by heroic foot-sloggers preceded tectonic enlightenment.

Hard on the heels of the first maps were furious geological rows about the interpretation of particular features. In the north-west Highlands there is a geological line running from Durness in the north to the Isle of Skye in the south that marks out the contact between metamorphic rocks—gneisses and schists—to the south-east and what I have called platform carbonate sedimentary rocks, to the north-west. The latter are of Cambrian to Ordovician age, and virtually identical to the rocks that underlay the Port au Port Peninsula in Newfoundland. They even have the same fossils. In a book that records the true density of historical detail accompanying discovery, David Oldroyd has written excellently about the controversy surrounding this single contact, the meaning of which occupied some of the most famous geologists of the nineteenth century. On modern maps the outcome of these intellectual battles is recorded by a mere two words: Moine Thrust. The metamorphic rocks have been bodily pushed north-westwards over the platform sediments in a thrust movement comparable with that around Bonne Bay in Newfoundland, at the western edge of the great mountain range. The rocks under the thrust were ground to paste under the weight of vanished mountains, a paroxysm of the earth now marked by no more than a bluff where a few sheep graze and the wind tickles the cotton-grass. Charles Lapworth, who was as instrumental in working out the truth of the Moine as any geologist, described it thus in 1882: "Conceive a vast rolling and crushing mill of irresistible power . . . shale, limestone, quartzite, granite, and the most intractable gneisses crumple up like putty in the terrible grip of this earth engine." It is hardly possible to describe such an historic drama in the dry language of science.

Southwards, the Highlands themselves are a metamorphic geologist's delight and a palaeontologist's despair: even if the rocks were once full of tiny "bugs," what could survive? Schists, gneisses, and quartzites rule, making for wonderful scenery and no fossils. In fact, most of the rocks were originally deposited in the Precambrian, so you would not find large fossils there anyway. You are obliged to remind yourself that today's steep hills are but a remnant of the grand alps that once towered here. It still seems like hard uphill walking, particularly on wet and windy days (which constitute most of the year). Patient mapping by officers of the Geological Survey of Great Britain has shown that the metamorphosed strata are twisted into great overfolds and nappes, comparable with those of the Alps, that are now concealed beneath bogs or partially exposed along the flanks of the wildest hills. Kilometre-long slabs of the earth have been overturned hereabouts. The crust must have been thickened by the "earth engine" as one fold piled on another as the mountain range rose up, and then the great pile was subjected to heat and pressure deep within the earth. As the rocks cooked, new minerals grew. Which particular mineral appeared depends on the conditions of temperature and pressure that the buried rocks experienced. There are minerals with simple chemical compositions, like aluminium silicate, which produce different minerals according to how hot and pressurized they became. One of the most famous diagrams in geology is a profoundly simple one that shows the different conditions required to grow one of three mineral varieties of aluminium silicate. Enough pressure and you find kyanite; enough heat and you get sillimanite; less of either and you get andalusite. It is almost like whether you get fudge or toffee when you boil sugar. This mineral cookery can stand in for dozens, if not hundreds, of experiments carried out in the laboratory that simulate natural conditions in "pressure-cookers" to find out which mineral is stable at a particular temperature and pressure. Natural circumstances are often more complicated—for example, the presence of even small quantities of water as an extra component can change the outcome—but the scientists who like these kinds of rocks can now "map out" the temperatures and pressures in these metamorphic belts much as field geologists map out rock boundaries. It is a way of peering into the intimate thermal history of mountain belts. The Highlands were one of the testing-grounds for this kind of science, where the idea of metamorphic "grade" was refined: the higher the grade, the more profound the tectonic suffering. It is even possible to plot out the subsequent cooling history of the rocks, as a high temperature mineral "reverts" to a lower temperature one as it cools, whilst leaving its original signature in the crystal form preserved in the rock. The

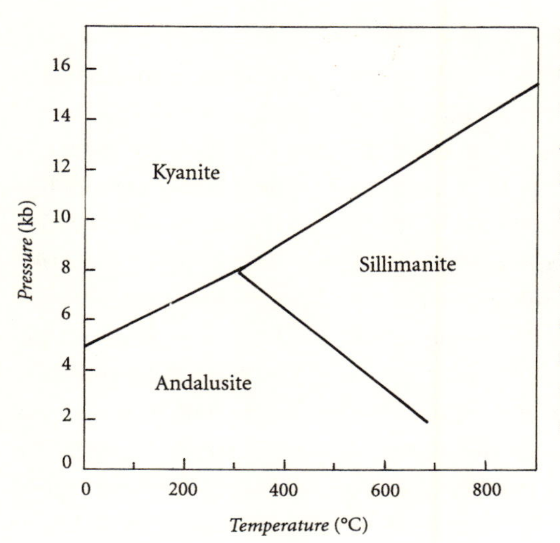

One of the most famous diagrams in determining conditions of metamorphism. Andalusite, kyanite, and sillimanite have the same aluminium silicate chemical composition, but different crystallography. Experiments determined the pressure (P) and temperature (T) under which each species could exist in nature.

Highlands have seen it all: heating under pressure, cooling, uplift, and deep erosion. These rocks make for thin soil and poor drainage. Then there are faults. The Great Glen follows one of those profound displacements with a north-west–south-east trend that we have seen in Newfoundland. Long, thin Loch Ness merely fills the gash.

The southern part of Scotland has not been heated so much. Endless bleak, rounded hills composed of shales and ribs of tough "greywackes" make up the Southern Uplands—all of such rocks were originally deposited under marine conditions. Up on the coast at Ballantrae there are volcanic rocks that compare with the ophiolites in Newfoundland. For the most part, these rocks are folded and thrust, in places furiously concertina'd and rucked. Were it not for the friendly graptolite fossils, this land would have been beyond decoding. As it is, careful mapping has disclosed its secrets. Following stream beds, delicately chipping at the dark shales, the little silvery films of graptolites betray the ages of this mass of similar-looking rocks. It is not merely that Silurian rocks are concentrated towards the south, and Ordovician to the north. The wild, open country is divided into north-east–south-west segments and bounded by thrusts, as if a gargantuan machete had butchered the whole of it into slices; and within each slice the strata rucked and repeated. It is obvious that if all the folds, and all the salami slicing of the crust, were straightened out and restored to their original disposition then this must be a region where everything has been compressed—to use the jargon, there has been "crustal shortening." What was once spread out on the ocean floor has been compressed, and stacked and folded in the process. It might seem impossibly complicated, but a plate tectonic explanation made it seem simple at a stroke.

Southwards again—across the Solway Firth into England, and the Lake District. Those who have queued to have a look around Dove Cottage at Grasmere may well have regretted that William Wordsworth and his friends made the "Lakes" so popular. But you have only to walk up a hill like Skiddaw to get back to wildness. There, the sight of other walkers a few hundred metres away from you serves only to put a scale upon the open landscape and the dark, bald hills in the distance. The rocks themselves frequently resemble the apparently endless marine shales and grits of the Southern Uplands of Scotland, although they are less crumpled. They have been heated enough to impart a delicate blue-green hue, which is because of the abundance of the "low-grade" metamorphic mineral, chlorite. The geology of the Lake District is still being worked out; only five years ago it was realized that huge Ordovician submarine slumps or slides, like those in

Newfoundland, had affected considerable areas. Fossils other than graptolites are very rare. It requires patience and luck to find trilobites in the early Ordovician slates, and, when you do, they are not beautiful—they, too, have suffered with the rocks. But enough have been found to tell us that they are similar to species found from Wales, France, and Morocco. Thus they link also to Bell Island, in eastern Newfoundland, but not to Laurentia.

It grieves me to have to pass over Wales in a few sentences, for I have spent years working there, and have grown fond of its open hills and fern-choked streams. Fossils have helped unscramble its geological complexities, and my work has helped fill in some of the details. In Wales, you always have the feeling of following in the footsteps of great geologists, sometimes revisiting the same quarry as they did, for the same reasons. The rocks are often folded, but there is little of the alteration that has afflicted the Highland strata. To be sure, some of the shales have been changed to hard slates by being squeezed in the inescapable vice of regional compression, particularly in North Wales; but enough has survived to permit the discrimination of strata. Cambrian, Ordovician, and Silurian fossil animals were disinterred in order, even though there were vigorous disagreements among the pioneer collectors in the nineteenth century about exactly how the geological time embracing this succession of "organic remains" should be divided. By the time of Eduard Suess' global compilation, however, such problems of definition were largely resolved: geological time had labels for the periods on which to pin our understanding; "Cambrian" described a time interval as well as a region on the map. The classical rock successions of the Principality are usually described as comprising the "Welsh Basin," which sounds a rather domestic term for a thickness of several kilometres of overwhelmingly shaly rocks. (Of course, this sag in the earth had also been described as a geosyncline in its time.) In the midst of the "basin" are the deposits left behind by Ordovician volcanoes, forming the untamed heart of the Cambrian Mountains, including Snowdon, where you feel as if the glaciers of the last ice age melted only yesterday. Small wonder that these wild fastnesses preserved the Welsh language from extinction while cultures came and went in the softer countryside of England. The fossils, too, have been sequestered safely. On the western tip of southern Wales there is a tiny blob of land, Ramsey Island, home to many sea-birds and a few sheep. On the top of the windy cliffs on the northern end of the island you can grub out pieces of a rather tough, sandy rock, while far below you the sea boils and seethes against the cliffs. It is the kind of place where you routinely lose your hat. Persist hatless, though, and you will recover specimens of the trilobite *Neseuretus,* the same beast that occurs in the early Ordovician rocks on the far side of the Atlantic Ocean, on Bell Island. Not far from Ramsey Island, on the mainland coast in the pretty little inlet of Abercastle, you can find its companion *Ogyginus,* and the *Cruziana* tracks that are so common in Newfoundland.

Twenty-five years ago I took the coastal steamer along the length of the Norwegian Caledonides. The cruise took more than a week. This is fjord coastline *forma typica.* During the Pleistocene—indeed, until a few thousand years ago—a huge ice cap on the mountains fed glaciers that nosed their way to the sea along the length of Norway and carved out their paths deep into the country rocks. When the ice age came to an end the glaciers retreated, but left their steep-sided gouges behind. As the polar ice caps melted, the sea level rose and flooded the valleys, so that the coastline is today deeply excavated by sinuous marine inlets. Even when the open ocean rages—as it often does at these latitudes—the fjords can be spookily calm. The mountain slopes rise steeply and implacably above the narrow shores. As in Newfoundland, the fishing communities led isolated lives in the coastal villages; different dialects recorded the separation of the populations. You cannot take issue with the tourist blurb: "spectacular fjord

Which way had the nappes of the Scandinavian Caledonides come from? Professors Törnebohm (left) and Svenonius (right) attempt to push their own points of view. Cartoon originally by E. Erdmann (1896).

scenery . . . picturesque fishing villages . . . unrivalled vistas . . ."—that kind of thing. I do, however, recall an excess of trolls. In every fjord it seemed that some troll or other had left his mark. "On your left," the guide would remark, "you will see the famous Troll Cave." In the next fjord a looming lump of rock with vague breasts would be described as the famous Troll Wife; a day later a pinnacle would be the illustrious Troll Castle; and so it went on. When our boat put briefly into harbour at the small towns dotted up the coast, it was something of a relief to escape the trolls and look at the rocks. The random samples presented to us were metamorphic, often high-grade gneisses like those in parts of the Highlands. There were also igneous rocks of several kinds. It was clear at a glance that this part of the Caledonides belonged to a most highly deformed part of the great range. It is enormously difficult country to understand geologically, even though the rocks are so well exposed along Trollfjord and on Troll Mountain. Swedish, Norwegian, and more than a handful of British geologists have spent their lives unravelling the structure of the ancient ranges; that they have got so far is a major achievement. It has been shown that the Norwegian Caledonides comprise a series of great nappes, somewhat on the Alpine model. A marvellous cartoon shows two Scandinavian professors of geology fighting over which direction the nappes came from—east or west? Professor Svenonius attempts to push them one way, while Professor Törnebohm tries the other way, both of them puffing like bad-tempered trolls. The answer is now clear: the root zone lay to the west. Eastwards lay folded rocks of the region around Oslo—an area famous for its Cambrian and Ordovician fossils; eastwards again lay the "platform" rocks of Sweden and the Baltic States. No, the huge overfolds had to come from the seam to the west that was revealed when the present Atlantic Ocean was "closed." Not only that, subsequent work has proved that there are four major nappes, piled one on another, like some tectonic pick-a-back on an enormous scale. Imagine! The whole country is like a club sandwich of tectonic slices. Furthermore, they are stacked in order: the highest nappe has travelled the furthest from the west; those lower in the pile have travelled progressively shorter distances. Fossils even helped with this discovery, since there was one corner of the highest nappe that escaped the regional metamorphism, and in one corner of that corner trilobites and brachiopods were discovered. Some of them were like species from the "platform" limestones of western Newfoundland.

. . .

Such is the briefest of sketches of the Appalachian–Caledonian mountain chain. When the Atlantic Ocean opened as Pangaea broke up into pieces, it evidently did so approximately following the line of this ancient range. But there is an enormous time gap. The opening of the "new" ocean happened well over 200 million years *after* those ancient mountains had been elevated to Himalayan proportions. The vast time interval in between had been occupied by the ineffably slow grinding down of the mountain range, for whatever presumes to height will eventually be laid low by erosion. Ice, wind, and rain are inevitably attracted to mountainous regions, and by the same token guarantee their eventual levelling. Many Devonian-age red rocks record the "waste" (molasse) from the Appalachian–Caledonian chain, and still survive along the flanks of the old mountain system: in Scandinavia, Greenland, Scotland, Wales, and New England. These freshwater sedimentary deposits are of more than passing importance in the history of our planet, because they contain graphic fossil evidence of the transition of life from water to land. This is where four-footed animals began—but that is not our story. From the geological point of view, the salient fact is that the widening Atlantic Ocean followed an ancient seam, an old memory revived, a weakness exploited. But the opening did not follow the old seam *exactly.* Consider that in Scotland, north-west of the Moine Thrust, the trilobites are like those of Laurentia. This is a fragment of earth's crust that really belongs on the other side of the ocean: when the Atlantic Ocean opened, this fragment was, as it were, left behind on the wrong side of the widening seas. Contrariwise, the Avalon Peninsula in Newfoundland truly belongs with Wales, Spain, and North Africa—for that is what those trilobites *Neseuretus* and *Ogyginus* told us—even though it now dangles off the eastern coast of the "Rock." You could say that this piece of Newfoundland has been marooned on the Laurentian side: small wonder, perhaps, that it feels so like Cornwall. You cannot trust the permanence of geography in a world on the move.

Now we come to a crucial question. As the Atlantic Ocean slowly widens at the present time—at the pace at which fingernails grow, according to a familiar analogy—no new mountain belt is being generated along the old Appalachian–Caledonian line. This is because both sides of the Atlantic Ocean are passive margins. No subduction happens, no volcanoes, no mountain-building; just slow drifting apart. This leads you to wonder: how did the Appalachian–Caledonian mountain chain get there in the first place? Mountain belts, like the Himalayas, are associated with *active* margins. Surely, then, it follows that at one time there were active margins with subduction zones running along the length of the Appalachian chain. This was the conceptual breakthough that changed our way of understanding how the world works.

Simple to state, perhaps, but hard to conceive of as an idea. The crucial scientific paper was published as a short note to the journal *Nature* in 1968, under the title "Did the Atlantic Close and Then Re-open?" The author was Tuzo Wilson, whom we have met previously as one of the crucial players in developing plate tectonic theory. The idea he had was straightforward. Five hundred million years ago there was a great ocean—call it the "proto-Atlantic"—separating Europe and Laurentia. That oceanic separation accounted for the very different kinds of Ordovician fossils found on either side of it—the differences between what you could collect on Bell Island and the Port au Port Peninsula in Newfoundland, for example. The ocean broadly followed the line of the present North Atlantic. That ocean then closed, probably by the late Silurian–Devonian period, involving the usual processes of subduction, and it was this closure that threw up the Appalachian–Caledonian mountains: they were the Alps or Himalayas of their time. This was when Charles Lapworth's great "earth engine" ground away at its work. By 400 million years ago the process was completed. Several hundred million years later, the present Atlantic opened, passively, along the weakened and eroded line of the ancient mountain chain. Simple as that. The fact that Tuzo Wilson had put a question-mark at the end of the title of his paper signalled that his conclusions were speculative. Top-notch scientific journals do not usually like speculations, but the note survived the editorial review process, no doubt because it came from such a well-known and innovative thinker, and made it into print: a few hundred words that changed the way we look at the world.

Tuzo Wilson's paper revealed a wholly novel way of applying Lyellian methods to very old mountain belts. Plate rules applied in the distant past. You could now look for evidence of the same kinds of tectonic and volcanic processes as happened today—around the Pacific Ocean, perhaps—in rocks that had been caught in earth movements hundreds of millions of years ago. A corollary to the methods that had begun in the "Temple of Serapis" could now be played out on the bleak shores of Newfoundland, or in the Arctic Caledonides. Of course, Wilson's model was far too simple, although right in essence; but it was the changes it wrought in the way geologists *looked* at familiar rocks that was important. The world may have stayed the same, but the perceptual glasses through which it was observed had changed. At such crucial moments, we have already seen that there are those who "run with the ball" while more cautious souls wait to see who the crowd will follow. So it was on this occasion.

John Dewey was a young lecturer at Cambridge in the 1960s. He was quick to appreciate how the whole Appalachian chain—and particularly the magnificent cross-section seen in Newfoundland—could be viewed in the light of the vanished ocean. He worked at a furious pace to be ahead of the game. He occupied a room adjacent to mine, and seldom left it during the crucial period, except to go into the field or to conferences. A kind of metabolic speediness makes John Dewey speak at twice the pace of anyone else I know, as if his tongue were struggling to catch up with the pace of his thoughts. He can give the impression that he is not listening to what you have to say, but I have concluded that he actually absorbs it, processes it, and assesses it as uninteresting, all within a couple of seconds. The only truly disconcerting thing about him is, or at least was, a propensity to do handsprings impromptu, and other such athletic feats. As a field geologist he is superb. When his whole concentration is focused on the rocks before him, he has the knack of apparently instantly fitting the detail he has seen not merely into a fold, but into a whole mountain range. To hold the totality of the Appalachians and half the Caledonides in his brain came naturally to him. Dewey had worked in Ireland, and had been to Newfoundland with Marshall Kay; he was primed to make the plate tectonic reinterpreta-

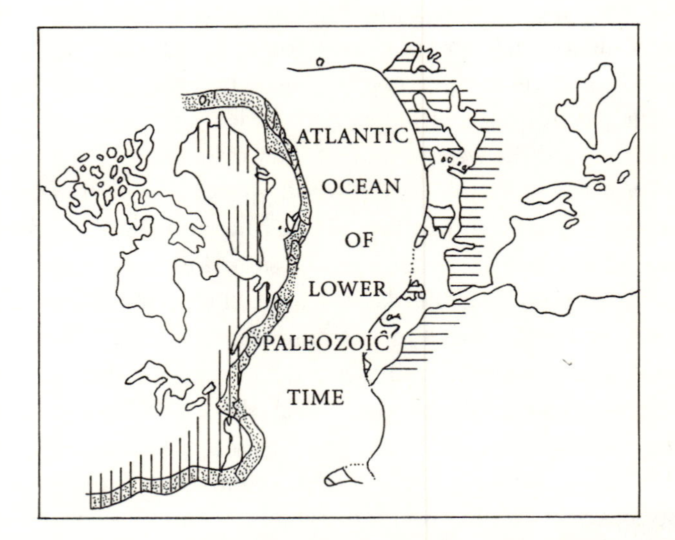

The simple diagram, redrawn from J. T. Wilson's original paper, posing the question: Did the Atlantic close and then re-open?

tion. Hank Williams was open to the new paradigm, too, and in his case the emphasis was on what he was best at—making maps. From maps of small and difficult areas around Gander, he graduated to maps of the whole Appalachian chain interpreted in the light of plate tectonics, but based, ultimately, upon what could be seen in Newfoundland.

As to complications, Wilson's simple sketch showing the two continents could have implied that their closure was almost like whacking together two blocks of wood. The real world is much more various and complex. In particular, there are volcanic island arcs—perhaps several—that will collide well before the "main event," which happens when continent–continent collision eventually occurs. They are outriders, if you like. Basins before and behind such arcs collect sediments that become entrained in the whole tectonic process. When one continent *does* eventually engage with its opposite number, the relentless pressure and crustal thickening where two thick crustal blocks meet induces the highest grade of metamorphism, making it likely that nappes are then "squeezed out." This is the furious centre of orogenesis. But it is also possible to have a much "softer" approach of continents—for example, if the opposing sides are concave. You can try an experiment to illustrate this that is valid in spirit, if not to the letter. Make a small, book-sized block of modelling clay. If you press it hard between the flats of your hands it will buckle intensely—in some tests, part of it will "mushroom" upwards. But if you use cupped hands instead you will get much more modest crumpling—and more of the original width will be preserved. The latter case is more like Newfoundland, which is why geologists love it. Less has been obliterated: you get a real chance to find out what happened within the vanished ocean. Then there are major faults. It is rather tempting to assume that closure of an ocean was like automatic doors slamming shut. It does not have to have been like this. Some of those faults may have been the site at which huge blocks slid past one another—transcurrent movement—rather than moving "up and down." This opens up the possibility that rocks that are neighbours now across faults may have had a very different relationship in the past. Everything, it seems, is capable of being somewhere else.

"Proto-Atlantic" is not a very appealing label—too indefinite—and the vanished ocean soon acquired a better one: Iapetus. It is surprising how the provision of a proper name confers respectability on a concept. In a curious way, naming it makes it *real.* Paradoxically, but appropriately, Iapetus was an entirely mythical name: a Titan, the son of Uranus by Gaea, the Greek earth goddess. Since he also fathered Atlas, which identifies a range of mountains, and Prometheus, provider of fire to mankind, you might say the geological credentials of the name are impeccable, and geologists were happy to accept it.

So we can now examine our trip across Newfoundland in the light of Dewey and Williams' reworking of the island's history. It became the story of the life and death of the Iapetus Ocean. Bell Island, and the whole Avalon Peninsula, lay on the eastern* side of a wide Iapetus in the early part of the Ordovician period. The trilobites (like *Neseuretus*) that scuttled over the sandy sea floors were happy scratching out a living in cold water, for at the time Avalonia was at high, even frigid latitudes. This can be confirmed by "palaeomag." We have seen already that these particular animals lived over an area that embraced much of Europe and northern Africa, which is where the South Pole lay at the time, and helped map out the former Gondwana continent. If we now jump to contemporary rocks on the other side of the former ocean, we arrive at the limestones of the Port au Port Peninsula. These rocks were laid down in shallow seas close to what was then the equator. The snails and sponges and trilobites that are found encased within them are all completely different from those on the eastern side of Newfoundland—not just different species, but even different families. Well, of course they are different: they were living in the tropics on a limy sea floor. We would not expect to find the same seashells living today in the cool waters off the "Rock" as in the blue lagoons of the Caribbean. In their own way, these tropical fossils were just as distinctive as the cool-water ones on the east of the island. Many of the species are found widely distributed over the Laurentian continent—that much had been recognized even before Eduard Suess. So, in the early Ordovician, Iapetus was wide enough to have one side in high latitudes and the other in the tropics: a massive ocean, indeed.

Remaining for a while on the western side of the island, the spectacular boulders and their intervening shales around Cow Head can now be seen for what they are: debris that collapsed off the margin of the Laurentian continent, where there was enough of a slope to prompt instability. Once an earthquake set a few large, tumbling blocks in motion, they would gather force, picking up other fragments on their dramatic slide as the whole slurry rushed down the slope into the deep water beyond. Mud may

* In all accounts of changing global geography "east" and "west," orientations will also inevitably change. To simplify matters, I refer to these directions as they are on current maps, whatever orientation they may have had in the past.

have lubricated the process. It is no wonder that the boulders look as if they had just been dumped out of a gigantic sack. Then quiet conditions returned for a while, and shales were slowly deposited on top of the boulder beds, and graptolites gently came to rest there to provide us with our yardstick for the passage of time. It must have been a good environment for them. I collected dozens of specimens from thin shales around St. Paul's Inlet, laying them out in the sun something like 470 million years after they had first come to rest on the sea floor.

As for the dark and treeless country above Woody Point, this was a slice from the ancient sea floor itself. We have seen the evidence that would be needed to prove the existence of an ocean to the east—and these brown hills comprised a piece of it miraculously preserved. Pillow lavas, dykes in sheets, even the deeper layers of mantle-type rocks, like lherzolite—the whole cross-section of oceanic crust was there. Out of the depths of time and the depths of the ocean, history was preserved in this ophiolite by the bleak road to Trout River. Mapping showed that both the Cow Head rocks and the ophiolite were stacked on top of the limestone platform—thrust into place. Furthermore, the ophiolite lay on top of the Cow Head rocks. As in the Norwegian nappes, the highest slice had travelled the furthest distance. We can imagine a massive shove from the east, first slicing up, and then pushing over the Cow Head rocks on to the platform, and then—heave upon heave—a sliver of ocean crust from far away being piled on top again. This small part of marginal Iapetus was pushed upwards and outwards rather than being doomed to destruction, for which redemption all geologists are grateful. It all proves that there was plunging down—subduction—of ocean crust in a westerly direction in the Ordovician. Some part of the floor of Iapetus was diving beneath the rigid mass of Laurentia. It was being consumed, just as the Pacific floor is destined for destruction along the edge of Asia. The ophiolite escaped the general destruction but was propelled in the same direction as the subduction zone. Now, at last, the contrasting rocks that lined the shores of Bonne Bay and the Northern Peninsula made sense. What might seem like baffling complexity when you are standing on a single outcrop of rock, scratching your head at a particular crumple, can be simply explained in a paragraph once you set the ancient plates in motion.

The age of the thrusting event was not, perhaps, what you might expect. It stands to reason that the date of the thrust has to be *after* the youngest platform limestones over which they are driven. Since the latter are of mid-Ordovician age, that sets a lower limit. Nothing in the Cow Head rocks goes any younger than that age either. At the very top of that set of rocks there is a peculiar green sandstone full of the kinds of minerals you would expect to be derived from a volcanic island; but there are no such islands to the west, so these must have been derived from the east, towards the open ocean. These features can be explained if a volcanic island arc was approaching from that direction. The major obduction of the ophiolite occurred in the middle part of the Ordovician when that island arc encountered Laurentia, even as it was being carried relentlessly westwards by the motor of subduction. This event was long recognized southwards in the Appalachians, where the folding associated with it is known as the Taconic orogeny. The implication is that the closure of Iapetus was much more interesting than just the collision of two continents. This crunch was heralded by several other tectonic events as island arcs, and possibly isolated oceanic islands, were all carried alike on the conveyor belt of oceanic destruction. In the Ordovician, Iapetus was still wide, but closing. Folding and destruction at the active margin happened over a long time period before continent–continent collision during the later Silurian. Some of the major faults separating Hank Williams' "zones" might well mark boundaries between these different events in Iapetus' history of closure.

Now it is even possible to make something of the Mobile Belt. Moving eastwards over the Long Range, it would not be too gross a simplification to describe the Dunnage Zone as the "squidge" region of Iapetus. This is where residues of the great ocean are crushed together in perplexing ways. In other parts of the Appalachian–Caledonian chain equivalents of this region have been squeezed or subducted out of existence. It is good fortune to have it preserved in Newfoundland because of its "soft" closure. In places, the chaos seen on the ground is exactly the turmoil you might expect, where pieces of ocean floor get mixed up with pieces of volcanic islands, and all blended with sediments that accumulated off their flanks. The latter tend to "give" more easily, and can be twisted around like so much black lasagne. It is small wonder that some parts of the Dunnage are so difficult to interpret. The zone includes volcanic islands that have survived to the present, and as many places in which blocks of volcanic rocks have slid down into the surrounding sedimentary shale basins to make that perplexing mélange. The detailed geological map of the area shows just about every rock type you can imagine. The fine-grained silica rock known as chert is worth mentioning because it is often associated with volcanic rocks and with what Marshall Kay would have called a "eugeosyncline." But there are sandstones, too, some of them the kind that originally rushed downslope as sand in suspension in the kind of slurries that have been termed turbidity currents. There are even blocks of limestone containing fossils: these were probably laid down on the margins of the islands before tumbling into the depths, where they were preserved. A great variety of igneous rocks add to the richness of the geological potpourri. Once upon a time, geological observers who pottered in their dories around the northern bays of Newfoundland were as confused as the rocks, but now we can see that the confusion itself is the outcome of explicable processes.

As for the Gander Zone, the heart of mountain-building lay there. It was the site of a thickened pile of sediments and volcanic rocks that experienced the full effects of Gondwana colliding with Laurentia: the crunch zone, if you like. Initially, the rocks between the two great masses accommodated to the pressure by crumpling, and then by sliding and pushing over one another, shortening the crust but also thickening it. Initially, perhaps, the thick pile might have depressed the geothermal gradient deep in the crust. But when the heat gradient recovered, the vast folded pile then heated up at depth, and the minerals composing the rocks became progressively uncomfortable. They needed to change in harmony with the new ambient conditions of heat and pressure surrounding them: metamorphism took over. The mineral legacy of this time is "frozen" in the rocks today, after tens of millions of years of exhumation from the deep cauldron of the earth. The garnets and the micas speak eloquently of the strata's ordeal. They can be read like fossil thermometers and pressure-gauges. In some places the original strata and folding of the rocks can still be discerned through the transforming prism of metamorphism, like faded snapshots in which a vanished scene can still just be made out. In other places, the rocks have been so transformed that any banding they show is only a reflection of the way minerals grew as they adapted to their clenching by the earth's interior. Those pink rocks I saw not far from the Trans-Canada Highway—migmatites—were so transformed that they began to melt, to sweat out a magmatic juice that was not far from granite in its constitution. Quartz in solution was squeezed out, subsequently to infiltrate surrounding rocks, filling fissures and pods. Finally, granite magma nudged into the centre of everything. And all the while the mountain range grew. For this is how the world bends, with neither a bang nor a whimper. The crust at first winces and buckles, then gives before inexorable tectonic forces; then piles up and dislocates; as mountain-building proceeds it heats up and changes its composition in an attempt to adapt to its profound new circumstances; finally, it may melt. It is hard to avoid dramatic imagery, even in the most sober account of such continental concatenation. There is a shortage of words to embrace all that buckling and thrusting and roasting.

So a journey across Newfoundland is a journey back to the time of Iapetus. Stand on the clifftops at St. John's and the early stages of this ocean can be imagined, a time when it was wider than the Atlantic. Pause under the fragrant conifers on Cow Head and you can visualize its subsequent destruction. All the evidence is laid out upon the ground: hardly a pebble on the beach that could not somehow be related to an ocean that vanished hundreds of millions of years in the past. Now another sea washes upon the old ocean floor, or polishes rocks that once were forged anew in the bowels of the earth. You can pick your way over a shoreline that was once the site of massive slides. Barnacles now encrust pieces of foundered volcanic islands where trilobites once hung on to a precarious existence.

Newfoundland provides but one profile of many across the former ocean, and to explore them all would take a book much longer than this. A brief review is all that is possible here. Hank Williams was able to map out his tectonic zones southwards all the way down the Appalachian chain. Avalon continued into eastern Canada, a connection that had been noted as early as 1887 by Marcel Bertrand, "who, with bold hand, traced the connecting trend lines," as Suess remarked. The other zones expanded or contracted all the way to the Carolinas and beyond. In some parts of New England the metamorphic belt expanded, and fresh mapping revealed huge nappes. Old observations were dusted down and examined in the new light of ancient plates. Eduard Suess himself had recognized lateral movements in the Appalachians and had surmised that they were propelled from the west. He, too, had wrestled with the problem of how to account for such long, linear features as the Blue Mountains. He had to have a global mechanism and opted for a contraction of the earth's interior, which caused the crust above to slide and buckle along zones of weakness. It was ingenious, but ingenuous, and there has never been subsequent evidence for a contraction within. But all those geologists who looked again at the Appalachians found the plate tectonic interpretation to be nothing less than a revelation. Disparate observations all seemed to fall into place: this bed of volcanic rock was arc-related; that piece of chemistry was oceanic. Science proceeds by a thousand small advances when a basis for understanding is in place. Once the big picture is drawn, the colouring-in can begin. And the geosynclines? They simply became expressions of different sites along continental margins. The thickness of strata accumulating off the edge of the Laurentian carbonate platform was not so remarkable on a continent edge. It was to be expected that the more oceanic sites included thick layers of volcanics and shales and cherts, especially in basins related to island arcs. Geosynclines became an "epiphenomenon" of plate tectonics.

We can move northwards into the Caledonides, where much the same enlightenment took place. There was no shortage of old field data that could be recruited into the new way of seeing: no less than 150 years of the tramp of boots and the tapping of hammers. England and Wales were part of Avalonia: the fossils that I had collected in stream bed and quarry told no lies. The limestones of north-west Scotland were a part of the opposing side of the present Atlantic, a stranded fragment of Laurentia. They were just the same as the rocks on Port au Port, down to the identical little snails curled up on the bedding planes. Between Wales and Durness, therefore, lay the wreck of Iapetus. There was an ophiolite in Scotland, too, albeit a much more modest affair than its Newfoundland counterpart. This comprised the pillow lavas and serpentinites and cherts—Steinmann's famous "trinity"—on the coast at Ballantrae. Delete "Ballantrae volcanics" from the geological map; substitute "Ballantrae ophiolite"—such is the mundane end of a great revolution in thought. Then there were the folded and faulted rocks of the Southern Uplands that form such uncompromisingly bare hills, beloved only of sheep and masochistic walkers. They, too, could now be looked at in a different way. That the country was sliced up into north-east–south-west trending "packages" was already recognized: but could they represent an accretionary prism? When a subduction zone plunges downwards, the sediments lying on top of the oceanic slab may be "scraped up" against the continent in front rather than destroyed. Jeremy Leggett from Imperial College, London, showed that it was possible to "read" the Southern Uplands in this fashion. He compared the Scottish rocks with those of the Makran in Pakistan, where more recent oceanic events had accomplished just the same kind of accretion as part of the construction of the Himalayan chain. In Scotland, the slices of graptolite-bearing shales and grits were piled up in slivers in exactly the way that might be expected. This meant, of course, that a subduction zone would have to lie to the *south* of the Southern Uplands, along the present line of the Solway Firth. When deep geophysics investigated this area, the "shadow" of a northward-dipping line at depth provided just the evidence that was required: a "fossil" subduction zone. Southwards again, the Lake District had an Avalonian stamp from its Ordovician fossils: the main part of Iapetus thus must have disappeared along the Solway Line that separated the Southern Uplands and the Lakes and followed the familiar Caledonian north-west–south-east trend: a short distance now, but the requiem for an ocean. The sliding and slipping of the Lake District rocks, which have recently been recognized by the officers of the Geological Survey of Great Britain, have echoes of the tumbling blocks on the other side of Iapetus at Cow Head.

The Highlands include the highly metamorphosed belt. Toughened up by gneisses and granites, the geology still allows for respectable mountains, even though they are but shadows of their former greatness. It is clear that the original sediments that were variously baked and squeezed in the Highlands originally lay off Laurentia, since unmistakeable remnants of that continent survive at the edge of the Midland Valley (where younger rocks predominate) and in the Southern Uplands. The whole area divides into two: highly metamorphosed Moine rocks to the north and less maltreated Dalradian rocks to the south and east. Some of the Highland rocks were folded into nappes and have been mapped out by a generation of devoted field geologists. The Tay Nappe and its ilk will be their permanent memorial. The general "push" of the structures was towards the north-west, just like the Moine Thrust itself. The history of the Highlands is much more complex, and extends over a longer period of time, than that sketched out for the rest of the ancient mountain system. Still earlier phases of history conflate with the Caledonian continental crunch, and whilst radiometric dating has helped unscramble some of this complexity, disentanglement is still in progress.

The Norwegian Caledonides are also generally metamorphosed, as I learned many years ago on that coastal cruise. In the context of plate tectonic history we can now begin to understand its great pile of nappes. They had originated from the west—where Iapetus lay. Squeezed out during orogeny to accommodate closure between opposing continents (rather than "soft" closure as in Newfoundland), their direction of movement contrasts with the heave towards Laurentia that we have met to the south: instead, the Norwegian nappes were directed eastwards. I have mentioned that there is good evidence from fossils that the furthest travelled—and highest—nappes in the pile had originated from the Laurentian side of

Iapetus. This is explicable now that we know the closure story: squeeze upon squeeze propelled masses of rock from the centre to the periphery, one after another, the most distant last. All that remains of the ocean itself are little pods of ophiolites entrained in some of the nappes; they must have been like apple pits squeezed out from a cider mill. This part of the great chain still has an alpine feel to it today, which is appropriate to the geology.

This short résumé of the history of a whole mountain chain appears to have the neatness of a "whodunnit," at the conclusion of which all the loose ends are tied up and the perpetrator is unmasked. It seems almost a pity to introduce further complexity. But nature has a habit of never being quite as simple as you first thought. On reflection, it might seem rather surprising that any mountain system that spans a hemisphere could be simply described as the meeting of two massive slabs. It did, indeed, prove more complicated, though the core of Tuzo Wilson's intuition remained intact.

The first complication concerned the ancient continents themselves. Further work served only to confirm the integrity of Laurentia. But the eastern side of Iapetus was another matter. An area centred on the Baltic Sea comprises a number of countries on the present political map: Norway, Sweden, Finland, the eastern part of Russia, and the old states of Estonia, Latvia, and Lithuania. The ancient geology of this whole area ignored our man-made borders. The Ordovician rock succession is very uniform in character, so much so that it has been claimed that you can trace a single Ordovician stratum throughout much of the region. The whole of it was underlain by very similar and very ancient "basement" rocks—the Baltic Shield. Not surprisingly, perhaps, the name Baltica has been applied to this area as a whole. In the earlier part of the Ordovician—the same time as the strata on Bell Island were being laid down—much of Baltica was covered by a shallow sea. The legacy of this time is thin limestone beds full of fossils; my own favourites, trilobites, are common, but there are other shells such as brachiopods in equal abundance. The remarkable thing about them is that they resemble neither their contemporaries in Avalonia, nor those in North America. They are their own thing, entirely. The limestones, too, are peculiar, being of a kind typically laid down today in temperate, rather than tropical, latitudes. None of the trilobites (remember *Neseuretus*?) that were found in Avalonia were found in Baltica, yet we have seen that they were common all over central and southern Europe and northern

Africa. What can be going on? After all, with Baltica continuing southwards into southern Europe, there seems no reason why the same animals should not have thrived over the entire eastern seaboard of Iapetus.

The answer was that there was a *second* ocean separating Avalonia from Baltica during the early Ordovician (p. 201). Robin Cocks and I christened it Tornquist's Sea in 1982, after a geologist who had recognized a tectonic line running across Europe more or less where we wanted our ocean to be located. The eastern seaboard of Iapetus was divided into two—a simple explanation of the differences between contemporary trilobites in the respective halves. Also, we knew that the Gondwana continent, with Avalon at its edge, was in cool waters, yet Baltica was at temperate latitudes. A physical separation provided a simple way of solving this conundrum. It took a while, but eventually palaeomagnetic evidence came to support the existence of Tornquist's Sea.* A buried mountain belt largely covered by younger rocks runs through Europe along the line of Tornquist's Sea. This line marked the trace of its closure: a closure that happened long before that of Iapetus. In a nutshell, a two-continent closure model had transformed to a three-continent model.

More complexity still: it was soon realized that Avalonia, too, was more interesting than merely being a promontory extending from Gondwana. It broke away from the main Gondwana continent and travelled under its own steam (carried on the back of the appropriate plate) across the ancient oceans, rather in the manner of the volcanic arcs that had preceded it. Both fossils and palaeomagnetic "fixes" support this interpretation. I have spent a substantial amount of British taxpayers' money—employing expensive computer systems and a post-doctoral assistant, David Lees—in plotting the movement of this chunk of crust—which, to recap, comprised England, Wales, southern Ireland, and a slice of eastern Canada, including eastern Newfoundland. In the early part of the Ordovician, about 480 million years ago, Avalonia was close to Gondwana—hence those similar trilobites on Bell Island and in France and Morocco (our old friend *Neseuretus* included). Later on in the Ordovician the similarity to Gondwana progressively declined, while the similarity to Baltica increased. Thus, Avalonia was moving across Tornquist's Sea towards Baltica; as the Ordovician continued it moved northwards. It may have had a "soft docking" with Baltica near the end of the Ordovician. Only then did the two, united together,

* I have told the full story of the battle to get the existence of Tornquist's Sea accepted in more detail in *Trilobite! Eyewitness to Evolution*—it took more than a decade.

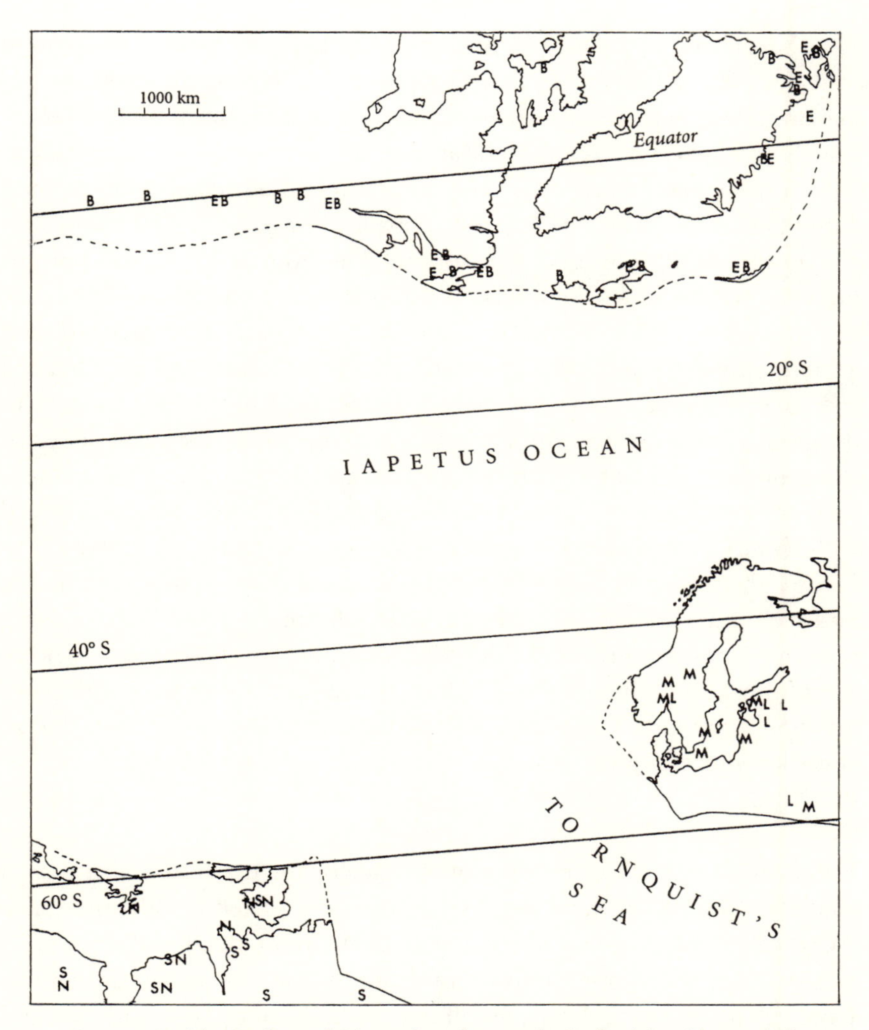

In 1982, Robin Cocks and I introduced an early Ordovician Tornquist's Sea between Avalonia and Scandinavia, 475 million years ago. The letters (BEMLSN) stand for typical fossils on the ancient continents. N, for example, stands for Nereuretus, *which we have met in the text. From the original paper in the* Journal of the Geological Society of London.

proceed to complete the closing of Iapetus. Avalonia was understood to be a "fossil" example of what is now termed a microcontinent: a comparatively small piece of continental crust with an independent history. It may have had something of the proportions of the microcontinental island of Madagascar: if so, it was pretty substantial—after all, "micro" is only a relative term. It was certainly large enough to push up mountains.

There is much evidence for Avalonia's journey preserved on the ground in Wales. The mountains of Snowdonia are rich in volcanic rocks. Avalonia's grandest peak, and its wild bare landscape, was once the site of eruption after eruption. You can see the ancient ash flows as ribs in the hillside as you climb upwards: rare ferns and ice-age relic flowers cling on to their damp flanks. These are the remains of explosive ashfalls, which prove that violent eruptions once darkened the same skies from which rain now plummets and trickles down your neck. Welsh volcanoes were the subject of classical studies by geologists like O. T. Jones and W. J. Pugh, who mapped their ancient shorelines and identified the sticky rock that had plugged the "necks" of the vents, in the era before "continental drift" was accepted. With the hindsight of plate tectonics, it is clear that these were the kinds of explosive eruptions associated with subduction, and modern chemical analyses on the composition of the lavas confirm this. Continental crust was mixed with basaltic magma of oceanic origin to give a characteristic "mixed" signature. There was, at least for a while, a *southward*-dipping subduction zone under the Cambrian mountains. The Welsh Basin—often called a geosyncline in the past because of its thick fill of sedimentary and volcanic rocks—was nothing more than the "sag" behind the series of Ordovician volcanoes that now command the heights around Cardigan Bay: deader than dodos, Snowdon supreme among them.

The story of our understanding of the face of the earth has been one of increased freedom of movement. First, there was a fixed world, immutable except for flood, plague, and elemental mayhem intended to demonstrate the power of the Almighty to point moral lessons for craven humanity. Time still had a human scale, for all that we knew our place in nature, and the likelihood of personal obscurity:

O time that cut'st down all!
And scarce leav'st here
Memoriall
Of any men that were.

So the seventeenth-century poet Robert Herrick described our obliteration. Later, the appreciation of the span of geological time rendered much egocentricity on behalf of our own species redundant. There was time enough to free the continents from their moorings. The Atlantic Ocean was allowed to open, to split apart Pangaea. The earth moved. We were free to imagine vanished worlds and vanished lives. It was a romantic vision, but also disconcerting, so to lose our bearings on historical certainty. Next, it was discovered that Pangaea itself was but a phase in earth's history. The supercontinent provided no deeper reality; it, too, was only in passing. Nothing is permanent except change. The Atlantic Ocean opened—but now it was discovered that its predecessor had already closed long before, at the time when Iapetus had been destroyed. Even the continents bordering Iapetus were not those we recognize. Avalonia and Baltica were not obvious from the map of Europe and America. They could only be revealed by understanding the deeper geological reality. That reality still influences what crops can be grown, and what towns can be built from local stone; it has outlasted the present configuration of continents. The geological Unconscious cannot be denied, for it still guides the way we use the land, and rules the plough. We are all in thrall to the underworld.

There is yet one more, final, mechanism of freedom to introduce. Recall the long faults trending south-west to north-east—like the Great Glen Fault in Scotland or those bounding Hank Williams' tectonic zones in Newfoundland. These are great fractures, cutting deep into the crust. They serve to chop the great range into segments. The individual segments often have their own features—unique rock successions not found in their neighbours, for example. It was soon realized that the faults bounding these segments might mark the site of extensive movement: two adjacent segments may have originally been far from where they are found today if they had slid past one another over a long period of time. There is a precedent for this kind of movement on the Pacific coast of the Americas—the most familar example being the San Andreas Fault in California, where one slab of crust is sliding past its neighbours. The units between major faults are known as "terranes," or, where there is doubt about their original disposition, "suspect terranes." The mountain belt came together as an accretion of terranes. Only when terranes were finally docked together would they share rocks with their neighbours, as a single blanket may be spread over sleeping partners. So now there is not only ocean closure to contend with, not to mention microcontinents and oceanic islands, but also a patchwork of shuffling slices as well. Once the moorings had been untied, what a kaleidoscope of moving fragments had been shaken up . . . Forget everything you think you know, and you can remake the world.

There are continuing arguments about the original location of major crustal slices. There are those who believe that Avalonia was originally situated off western Africa in the Ordovician. It was only brought to its present location by a later movement in the Caledonides. For once, the fossils don't help very much because they are similar over the whole of that part of Gondwana. How can we prove Avalonia's previous position one way or the other? What is needed is a kind of geological "fingerprint" that matches the slice with its nearest neighbour at its "starting gate." Current research is trying to identify such critical tests.

One promising line of investigation is provided by zircon crystals. Zircon, zirconium silicate, is a very stable mineral that is a common minor component of granite. It is so tough that when the granite weathers away, the zircon remains behind, to be washed down streams, eventually to reach the sea. Tiny crystals of zircon often finish up as a component of sandstones—one grain of zircon among ten thousand of common quartz. However, methods have been developed to rescue the precious crystals from the common herd. Zircons contain minute amounts of radioactive lead, and thus contain a "clock" that gives the geological age of the formation of the crystal within its parent granite. Extraordinarily accurate measurements of minuscule amounts of lead isotopes, however, are needed to give reliable results. Sam Bowring's laboratory at the Massachusetts Institute of Technology and Scott Samson's at Syracuse University in New York, are the most up-to-date centres of research in this area. I visited the Syracuse laboratory, and what most impressed me were the plastic bottles. These have to be cleaned continuously for more than two years to remove every last contaminating atom of lead: everything is purified over and over. At the end of it all they extract dates from within single zircon crystals—so small that they would fit onto the head of a pin—that are accurate to within a million years, a margin of error that applies even with dates of up to a billion years, which is an astounding technical achievement. Of course, this means that geological dates in general have reached a new level of precision. We now know that the base of the Ordovician is 489 million years plus or minus 500,000 years. But precise dates on zircons also have a use for terrane location. If Avalonia once lay off Africa, then it should have received sediments from that continent—including zircons. Dandruff shaken from the giant came to rest all around. It turns out that there is a close link in age between Avalonian zircons and those of the Precambrian orogenies that affected the African continent. These have a different age spread from those that happened in Brittany or Normandy, off which Avalonia now resides. The new evidence does suggest that Avalonia did indeed travel from off Africa to its present location. Will other evidence confirm its peripatetic history? We shall have to wait and see.

Now let us look again at the whole world. The Appalachian–Caledonian chain is a scar marking an ancient ocean, one that has disappeared. Pangaea evidently *came together* as a supercontinent from a collage of earlier, dispersed continents. The Ordovician world was more like our world than the Permian one of annealed land masses: it had scattered continents. Plate tectonics did not begin at the time of Pangaea; rather, the slow stroll of continents around the globe has been continuous. The mechanics of the earth have not changed for a far, far longer time than the breakup of the last supercontinent. To be sure, this vision embraces more transformation than Charles Lyell would have ever contemplated. But the application of his principles—to look at "living" fold belts and subduction zones to interpret fossil ones—has only confirmed the vigour of his scientific method. The reasoning that began in the "Temple of Serapis" rebuilt the world; and rebuilt it repeatedly.

If we proceed logically, *wherever* there is an ancient scar of a fold-belt running across the surface of the globe, we should expect evidence of a lost ocean, of past subduction. There will be volcanic rocks on the ground, and the thick successions of sedimentary rocks associated with them will be of the type that would once have been termed geosynclines. An assiduous researcher might well discover ophiolites tucked somewhere into the former range. It is what an advertising executive would call a "package."

Some of these ancient fold belts are obvious. Any topographic map of Russia shows the Urals almost bisecting it, like a persistent ruck in a carpet. The mountains run from the Arctic Island of Novaya Zemlya southwards, dividing the endless wastes of the Siberian *taiga* and the steppes from the Russian platform in the west. This line marked the eastern edge of the ancient continent of Baltica, which was separated from its Siberian neighbour by another vanished ocean. When that ocean closed in the Devonian period, another component of the supercontinent Pangaea was assembled. The Siberian continent was already recognized as an important piece of the earth by Eduard Suess: he called it Angara. Some of the largest copper mines in the world are related to ophiolites along the line of closure between Angara and Baltica. Similarly, across the middle of China there is a range called the Tsinling Mountains. Geologists have come to see this line as demonstrating that the most uniform civilization in the world, that of the Han people, comprises two separate geological foundations: another ocean consumed. Other ancient fold belts are less obvious, although they have been recognized for a long time. The Variscan chain—what Eduard Suess called the Hercynides—we have briefly met before. The fold belt passes west to east through Europe, from the rugged cliffs and intimate

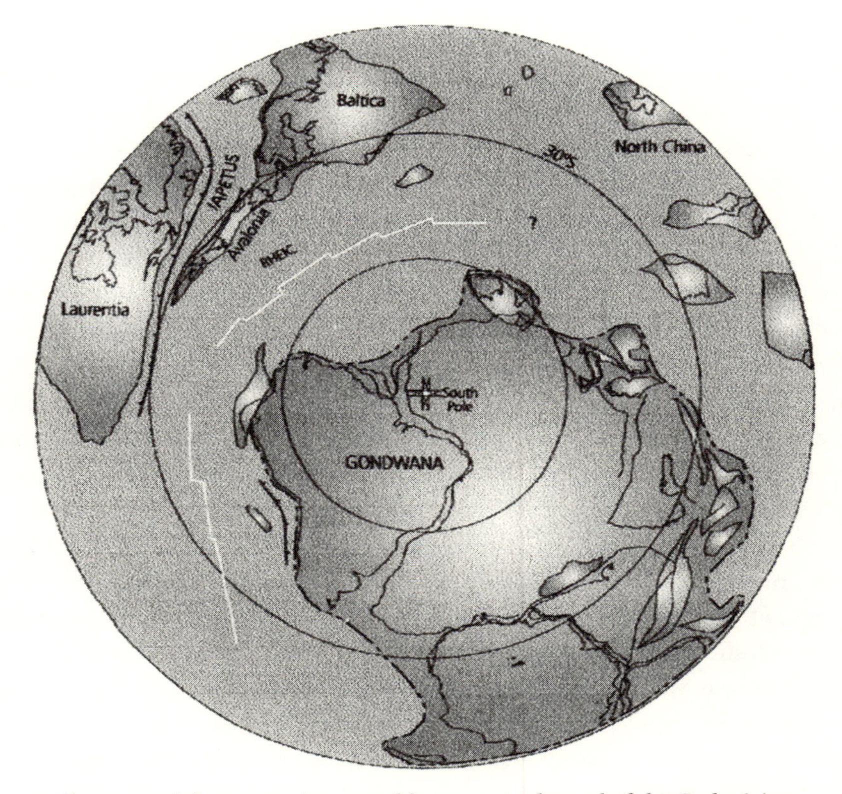

Change and change again: a world map near the end of the Ordovician. Tornquist's Sea has vanished since the earlier Ordovician, and a new Rheic Ocean widened between Gondwana and Baltica.

coves of Cornwall in the west of England all the way through central Europe north of the Alps. Another vanished ocean: one that had once divided Europe until its closure in the Carboniferous. This one has a name, too—the Rheic Ocean. For each one of these ancient oceans I might have undertaken a geological traverse from one side to the other, as I did for Iapetus in Newfoundland. You have to visualize a thousand geologists crawling over the ground, interpreting these ancient mountain chains in the light of plate tectonics: geophysicists probing ancient sutures with new equipment, trying to read the structure of the depths; geochemists looking at the distribution of rare elements. This is the research programme that continues unabated today, whether pursued in the wooded hills of central Europe, or the treacherous marshes of the Arctic. As with the Appalachian story, every question answered spawns a dozen more that need to be investigated. There are arguments about terranes; there are disputes about timing; there are disagreements about the role of oceanic crust in generating volcanoes. But without the plate tectonic model nobody would have known what questions to ask of the uncomplaining rocks.

The course of some ancient oceans opened upon older lines of weakness, as we have seen in the case of the Atlantic Ocean. Time reopened old wounds. We can now see the earth as a crude mosaic, stitched together in various ways. Scars of past tectonic crises cross continents. Some ranges are shadows of their former grandeur; the Urals, for example, have an average height of only 1000–1300 metres. Other old fold belts still carry glaciers on their heights, although none approach the mighty splendour of the Himalayas, which are the consequence of the latest round of continental movements. But these great mountains, too, will be ground down in millions of years. Where lighter continental crust has been thickened in orogenic piles as continents collide, the whole mass rises inevitably into the fierce realm of the levelling elements, ice and wind. The Himalayas, too, will one day be yet another seam upon the face of the old earth, another wrinkle added to testify to its character.

New York is the quintessential city: glitteringly artificial, Manhattan is a shop-window for capitalism. A city this vibrant and varied is a tribute to what my father would have called "commerce": the growth of towers from trade, the vindication of the market. It seems as far removed from the geological underlay as it is possible to get. Nature is pushed into a few holes in the sidewalks. Tough trees like ginkgos thrive around a few older houses and churches, but then, these trees have survived from the Jurassic and can face out a few fumes.

When you emerge from the ranked stores and offices into Central Park it is always a surprise. What is this undeveloped prime real estate doing here, in the midst of all this money and endeavour? A winding path or two will take you to where you can view through real trees the peaks and pinnacles of the city, as you might a mountain range from the plain. It does not diminish the sense of wonder at human diligence; rather, it lends perspective. Then you notice the rocks. Cropping out in places under the trees are dark mounds of rock, emerging from the ground like some buried architecture of a former race, partly exhumed and then forgotten. New Yorkers might sit on a sloping slab of it to eat a sandwich in the sunshine without giving it a thought. The outcrop is as unremarked as a hot dog stand or a fire hydrant; just part of the scene. But look more closely and you will see it is shiny and slick and layered. The light may pick out the glint of mica. The rock is schist, a metamorphic rock produced in the fold belts that have been the subject of this chapter. That New York can be built so high and mighty is a consequence of its secure foundations on ancient rocks. It pays its dues to the geology. This is just a small part of one of those old seams that cross the earth. But the rocks prove that skyscrapers not of man's making once covered the same ground. These are Grenville rocks. We have met them once before, in Newfoundland, where they formed the core of the Long Range Mountains. They are older by far than the Appalachian–Caledonian history. They are the legacy of a still older phase of opening along the line of the Atlantic, another manifestation of the same scar 1000 million years ago, deep within Precambrian time. The Grenville rocks make another belt, broadly following the western edge of the Appalachians. *Autres temps, autres mers.* When these mountains formed where New York now stands, there were no trilobites, no shellfish. The sea hosted algae, and perhaps only the most distant ancestors of animals. Yet it seems that even then, plates still moved, and the earth still carried on its stately dance of splitting apart and reassembling. How far back can we go? Even here, right in the midst of the most artificial habitat on the planet, a place where every second counts, there are relics of a deeper time when millennia counted for nothing.

7

The Dollar

Real Money: a 1922 U.S. silver dollar.

The most ordinary thing in the world, a scrap of paper: a buck, a greenback. The dollar: flimsy fuel for the world's economy. Gross national products are reckoned in it; national debts are totted up in it; currencies fluctuate against it. When I visited Kazakhstan after its separation from the Soviet Union nobody was interested in pounds sterling, or French francs, or German marks, let alone the despised rouble. What people wanted was the mighty dollar. When anything was paid for, the dollars were scrutinized closely. Crumpled ones and dirty ones were unpopular: they were thrown back as if they were contaminated. No, what was acceptable was the pristine dollar bill, the essence of money. I had always been puzzled by the apparent uniformity of U.S. dollar notes, regardless of their denomination—fifty bucks looking so similar to a twenty, or a singleton, and apparently conspiring to aid confusion. Now I realized why this had to be: the dollar bill *is* money: it's what money should look like. Everything else is just numbers.

Yet not long ago there was a silver dollar. When I travelled to the gambling state of Nevada in the late 1970s, these coins were still currency. They were a good size, like the largest coins in the sets of gold-clad chocolate money that kids still get for Christmas. Once I received a 1922 silver dollar in change in a small supermarket. Later, in New York, I saw the same coin on sale for twenty-seven dollars, thus making me one of the few people ever to leave Reno with a profit. The old dollar was more than an idealized form of money. The silver in it was, when minted, worth a dollar. Age didn't turn silver dollars into tattered rags: they hung about for years. This kind of money had a durability that feels out of place in the age of the virtual transaction, when money as an entity becomes more and more notional. In the old days, money had substance. The dollars weighed down your pocket. The $ sign itself is considered by some to be a transmutation of "8" and related directly to the Spanish "pieces of eight." Once upon a time their exchange values were linked. As for the name—dollar—it has a geological history.

One can find the same word in Danish, *daler.* In due turn, this is a variant of *thaler.* The thaler was the standard silver coinage across Europe in the sixteenth century. Each thaler weighed an ounce. There was a portrait of a bigwig struck on one side, often one of the counts of Slik who profited from their manufacture. At a time when there was much jiggery-pokery and adulterization in coinage, the thaler had a reputation for reliability. The reward of virtue in this case was eventually to bequeath its name to the ultimate lingua franca: the dollar that talks to just about everyone. Further, the name thaler was itself a contraction from Joachimsthaler, referring to the mine at Joachimsthal (the valley of Joachim) from which the silver was obtained. This small town in the Czech Republic was once part of old Bohemia, lying on the southern rim of the Krusne Mountains. What remains of the town is quite picturesquely dotted about the wooded hillsides, but it has declined from its great days. In 1530 its population was 18,000, of whom 13,000 were miners. The town will be found on modern maps under the name of Jachymov. So the Joachimsthaler became the thaler, which became the dollar. Thus history covers its tracks.

The rich silver veins were discovered in Jachymov in 1516. A few years later there were 800 mines. The first coins were struck by Count Stepan Slik, sometimes Germanized to Schlick, with due permission from the king; the Count had a licence to mint money, and of course he became rich. The relevant document dangling the royal seal is still preserved in the National Archives in Prague. Between 1520 and 1528 more than 2 million thalers were struck from something like 60,000 kilograms of fine-quality silver. Much of the silver occurred "native," that is, as the pure metal, and some of the coins could almost be struck on the spot. This kind of silver forms fine strings and sheets rather than nuggets. The silver is laid out in veins that follow lines of weakness in the country rock, often running along faults. Like all mining discoveries in the old days, the initial bonanza occurred on the surface, where some sharp-eyed prospector discovered an outcrop of rock, or weathered material, showing the cherished minerals. Then the vein was chased downwards, as deep into the earth as available technology and profit would allow. In the Svornost Mine at Jachymov, some of the tunnels dug more than 400 years ago are still there, 150 metres below the surface. They are crudely walled with stone slabs knocked in edgewise to help safeguard against rockfalls. A visitor in 1980 found some well-preserved tunnels with rock-lined ceilings that he described as resembling small naves; and there was even an inscription to St. Barbara, patron saint of miners.

Where humans have wormed their way into the underworld in pursuit of dollars, the heave of the violated crust often squeezes back again to obliterate their impertinence. In this connection, I discovered that this area of middle Europe is apparently the only part of the world where the cockroach is viewed favourably: their insect senses are peculiarly alert to changes in pressure in the rock—like the pressure-sensitive instruments used in Hawai'i to detect movement in the magma chamber. When a rockfall threatens, the cockroaches head out of their hiding places and scramble away along the conduits and passages in order to escape being crushed. They are the silver miners' friends, like canaries in coal mines, and it is bad luck to crush one. Cockroaches have been around since the Carboniferous period and will doubtless still be here after the last miner has gone.

All mines eventually become exhausted, and the same was true of the silver at Jachymov—but it was not the end of the industry. Different vein minerals are often found nestled together, or at least in close proximity, and when one has been worked out it sometimes happens that another acquires a commercial life. To put it into miner's jargon: today's gangue may be tomorrow's ore. In the mineral underworld of Czechoslovakia there were also plentiful ores of the metallic element cobalt, the heavy element bismuth, and poisonous arsenic. These elements all have medicinal and industrial uses. In their own way they are as precious as silver, even though their route to the dollar is less blatant. Cobalt is what gives a wonderful rich blue glaze to ceramics and enamels, and it can impart the same colour to glass. "Cobalt blue" is one of the standard colours of the oil painter's palette. Factories able to exploit these new minerals grew up around Jachymov in the nineteenth century: so there was life after silver. An additional mineral, dark and greasy-looking as pitch, was despised by the silver miners because when it increased in abundance the silver disappeared, like darkness shutting out light. This mineral is called pitchblende and has another peculiarity that belies its ordinary appearance: it is very heavy. This is because it is largely composed of uranium oxide, and uranium is element number 92 in the periodic table, almost as heavy as you can go, and heavier than lead. Useless, dark, heavy stuff this pitchblende, to be thrown into the woods, out of the way.

How time changes perceptions of value! In 1852, a chemist called Albert Patera discovered that uranium-based chemical compounds produced unusual and gorgeous colours—greens and yellows especially—when added to glass. There was an immediate vogue for these other-worldly hues, and so mining at Jachymov enjoyed yet another incarnation. At a more recherché level, the mine was a source for beautiful mineral specimens of rare uranium and silver minerals that still adorn great collections. But the major use of pitchblende had to await detection for nearly half a century, when the mines of Jachymov were crucial to a discovery that has changed the course of history. Working in Paris in 1898, Marie and Pierre Curie processed thousands of pounds of pitchblende from the same glass-factory tailings in order to isolate two radioactive elements that had hitherto been unknown: radium and polonium. Uranium's radioactivity had been recognized for a couple of years prior to this, but this pitchblende was *too* active to be accounted for by uranium alone. The Curies deduced that another radioactive element had to be present to make up the balance; they reasoned the existence of highly active radium before extracting it. But it was present in the tiniest quantities: a gram in seven tons of ore. The labour involved in refining a tiny visible grain from the unforgiving black ore is difficult to imagine. But they convinced the scientific world with sufficient material evidence of the elusive element to be awarded the Nobel Prize in 1903. This was the first international recognition that a woman could have as outstanding a scientific intellect as any man, and Marie Curie's name is still associated with European science funding.

With tragic irony, Marie Curie died of the consequences of the radiation she helped to explain to the world: cancer had yet to be linked with damage to dividing cells. Radiation was to become at once a medical weapon and the agent of mass destruction. Luminous watch-hands painted with radium compounds were formerly the everyday demonstration of radioactive decay: as a child I had an alarm clock on my bedside table that ticked visibly in the dark. Then it was discovered that workers in the United States who decorated these novelty clock faces, and then licked their brushes, had unusually high rates of mouth cancer. The delicacy of the genome proved to be a discovery out of phase with that of glowing radium. However, the realization that one element could decay to another at a definite rate also proved central to the endeavour of dating rocks, for this decay provided natural "clocks" which ticked in sympathy with the antiquity of geological time. This is how the world acquired a birthdate.

Joachimsthal has still other connections to reveal. The town physician at the time of the silver boom of 1527–33 was one Georg Bauer. He is better known under his literary name of Georgius Agricola. He wrote the first books about mining, of which his *De Re Metallica* (1555) is the most famous. Agricola's works became known throughout Europe, where Latin was still the language of instruction. Agricola's writings can claim to be the first widely studied geological texts. I have examined a copy of *De Re Metallica* in the library of the Natural History Museum; carefully leather-bound and stored in an inner sanctum, it seems an improbable handbook. But that is what it is: an illustrated compendium of practical advice for the miners and metallurgists of the day, with advice on ventilation and smelting and the safe construction of drifts. It also includes the first attempt at the systematic classification of minerals, by their properties and products. This was clearly the work of a good observer, a no-nonsense, pragmatical kind of mind. This probably accounts for its longevity as a reference work for more than two centuries. Agricola was at pains to say that his work was based upon facts he himself had observed. He clearly had little time for hearsay and recycling of mouldering classical sources, and he said as much. In spirit, he was close to his much more famous successor Francis Bacon (born in 1561), who set out the scientific agenda explicitly for the first time. *De Re Metallica* was translated into English in 1912 by Herbert Hoover, surely the only president in U.S. history with sufficient Latin to have undertaken such a task. Agricola had a near contemporary in Jachymov,

Georgius Agricola (1490–1555), father of mineralogy, and (opposite) one of his practical illustrations of mining practices—taking ore to the shelter.

the Revd. Johannes Matthevius, who was sufficiently concerned at the conditions under which the miners laboured to deliver sermons on their rights and safety, and on what would now be called unionization. As was customary among elite clergy, these perorations were eventually published in 1562, under the title *Sarepta oder Bergpostilla.* I would be intrigued to know if there were an earlier example of concern for the rights of the industrial worker. Ever since these early days, mining, above all other commercial activities, has been capitalism laid bare, where grisly toil has been at the service of boom and bust, and where the role of the boss has frequently been tainted with dubious profiteering. Miners have been among the first workers to fight back for their rights. On the side of collective labour, the coal miners have been the most militant trade unionists in the twentieth century, deified by some, vilified by others as communists and troublemakers. It is sad to reflect that there are parts of the world today where mining conditions are probably worse than they were in Joachimsthal in the sixteenth century, under the brooding eye of the Revd. Matthevius. It has to be added that in the Soviet era the Jachymov mines were one of the

main sources for uranium concentrates, the basis of bombs and power in the Eastern Bloc, and that the miners were mostly political prisoners. *Tempora mutantur*, as President Hoover might have said.

It is remarkable how all these different historical threads weave together in one mining area on the Czech–German border: the dollar and the first geological textbook; Marie Curie and the dating of the earth; even perhaps industrial justice. Jachymov also embodies much of geology over a few hillsides now strewn with overgrown tailings. It all has a ground truth in plate tectonics. The mines of Jachymov are just part of a great swathe of mineral riches dotted through central Europe and that run out eventually into the famous tin-mines of Cornwall at the westerly tip of England. All these mines have complex histories, many are played out, and they yield different treasures. But they are all intimately connected with one of the ancient seams that divide the complex European continent—what Suess and Holmes called the Hercynian fold belt, though modern textbooks seem to prefer the term "Variscan." There was a range of mountains—real, high mountains—running along this track through Europe, a range that was the consequence of subduction of ocean we have previously encountered, the Rheic Ocean. The silver and the uranium were the last exhalations from the dying seas, a gift from the underworld sealed in cracks since the Carboniferous. In Cornwall that gift was tin, tourmaline, lead, zinc, and silver. The miners in Jachymov had their counterparts elsewhere, speaking Cornish or Old French.

Silver and gold are unusual in occurring in nature as the pure element. Few other elements other than gases occur in such an unadulterated condition. We have already met native sulphur in the yellow-stained exhalations around volcanoes. Copper also occurs in sheets and nuggets, often so distinctively shaped they seem to have been beaten out by an earth sprite with an eye for abstract sculpture. Most elements are prone to combine with one or several others to form a compound—this is a mineral. The "noblest" metal, gold, is reluctant to become part of a mineral; it will not combine with oxygen in the atmosphere, for example, which is why it does not tarnish, and why gold hoards still glow after years in the ground. Iron, by contrast, is corrupted by water and oxygen and soon becomes crusted and fragile. Like iron, most elements cleave unto others. A mineral has a particular chemical formula, a characteristic combination of elements. An ore is just a mineral, or a set of minerals found together, that is of commercial value. Whether or not a given claim yields a worthwhile ore can change, too, as the market fluctuates, so mining is tied into the flux of capital: there is nothing objective about ores. Minerals carry scientific names that identify them uniquely, "official names," in the same way as the Latin binomen for animals and plants. Unlike the system developed by Linnaeus for the biological world only one name is required to identify a mineral. Some are ancient names that have a certain romance—sphene, tourmaline, galena, hornblende, topaz, garnet—but the majority are just "ites." That is, they have a stem name to which "-ite" is added as a suffix to indicate their mineralogical nature. Many are not very euphonious creations. Leafing through a dictionary that I have in front of me, I find bismutotantalite, cuproskodlowskite, fluorapophyllite, guanajuatite, kaemmererite, monteregianite, oosterboschite, shattuckite, vandendriesscheite, yofortierite, and, last, the not-to-be-forgotten zinnwaldite. Each mineral name is a label of chemical composition first, but also has a typical crystal structure. As we have seen, atoms can arrange themselves in more than one way even if they have the same chemical formula, and a different arrangement can merit a different mineral name. Each mineral type is termed a "species," thereby adopting the same language as biology.

It is somewhat confusing to find that some rock types, too, have "-ite" as suffixes. Charnockite is a rock, not a mineral. So are granodiorite and lherzolite. The definition of a rock type is generally looser than that of a species of mineral. In practice, most rocks are recognized from the characteristic assemblage and proportions of their constituent minerals. A rock made of feldspar and quartz will have a different name from one made largely of hornblende. But this system still allows for a lot of variation in appearance of the rock-in-hand specimen, according to how it occurred in nature.

Some minerals have very simple chemical formulae, as easy to remember as those of water and salt, which is a mineral. Many have horribly complicated ones. Some minerals are very common and others are very rare. Some are so rare that they are known from a single locality, the mineralogical equivalent of the Mauritius one-cent postage stamp that haunts the dreams of philatelists. There is a good reason for this. Rare elements combining under rare conditions produce rare minerals. Conversely, common elements make proletarian marriages. The commonest elements, silicon and oxygen, combine to form the remarkably common mineral quartz: the sand on the beach, that rounded cobblestone in the ancient church wall, the grit that gets into your sandwiches in the park—all quartz. This does not mean that a common mineral cannot be uncommonly beautiful. Perfect quartz crystals are much less routine than mere sand and gravel: they are six-sided prisms topped with complex pyramids. Its more massive form, rock crystal, has a kind of vivid transparency that makes most glass look insipid. And then quartz gets tinted by traces of other elements to produce named varieties that are commonly sold in rock shops: amethyst coloured purple by manganese; carnelian made warmly red because of a dash of iron. Such striking colours are not rare in nature: the quartz minerals that produce them are usually described as "semi-precious," but they are in no sense semi-beautiful.

Gemstones are crystals of greater rarity, and are a more traditional barometer of wealth than mere accumulation of dollars—silver or otherwise. All gemstones are minerals, but not all minerals can yield gems. Often, a common form of the gem mineral exists as a routine ingredient in rocks, but only rarely is it pure-grown, clear and crystalline, in the precious form. The frequent mineral corundum is often matt and uninteresting. Chemically, it is a simple oxide of aluminium. But grown deep in the underworld under special conditions, with the right impurity added by nature to induce colour, it becomes the most regal of all jewels—ruby. It is odd that this stone almost alone is used as a Christian name. If it were a matter of vulgar display, then surely diamond or emerald should be common names, or if exoticism were the important factor, "Opal" should be heard on the streets of London or New York. Some of the best rubies are found near the village of Mogok, in Burma. Benvenuto Cellini (1500–1571), greatest of goldsmiths, reckoned that a fine ruby was worth eight times the same weight in diamond. The ultimate rubies are called "pigeon's-blood rubies," allegedly because their colour resembles that of the blood of a freshly slaughtered dove: they are a lustrous but absolutely clear red and are the most valuable of all gemstones. They are rarely found in place in their rocky source, a coarsely crystalline marble that cropped out in the wet hills of Burma. Instead, they have to be sifted from ancient gravels into which they were washed and concentrated from their original bed. The largest rubies of all are from what is now Tadjikistan. If another quirk of contamination happens, corundum becomes blue, rather than red, and then you have a sapphire.

Crystals do not usually grow into gem-quality stones. Like most things in life, minerals are commonly compromised with impurities, or grow crushed by others, so cheek by jowl that they do not develop their most beautiful colour and character. The spectacular specimens that are found in museum cabinets are uncommon. More abundant minerals often accompany the rarities, as when a cluster of quartz crystals will form a plinth from which arise elegant columns of green or pink tourmaline. Beautiful gem-quality minerals are particularly found in nature in those geological sites where the crystals were able grow very slowly to achieve their perfection of form. This often happens in profoundly deep fissures in the presence of volatile elements like chlorine and fluorine, and where the crystallizing chamber is bathed in hot, siliceous waters. When miners eventually hollow out tunnels into these underworld crystallization chambers, the crystal seams can be displayed to the visitor as glistening, bejewelled encrustations lining ceiling or walls. There are immediate romantic connotations to such subterranean jewel-boxes. Thus John Keats described Endymion's descent into "the sparry hollows of the world":

> *Dark, nor light,*
> *The region; nor bright, nor sombre wholly . . .*
> *A dusky empire and its diadems;*
> *One faint eternal eventide of gems.*

Imitations of such displays were reproduced in the once-fashionable artificial grottos that still adorn the stately parks of Palladian mansions. Unfortunately, their creators often mixed geologically inappropriate materials: geodes of quartz placed alongside calcite stalagmites and stalactites. Their man-made "dusky empire" was constructed with scant regard to the truths of the real underworld.

Though minerals are named as species they are far fewer in number than the prodigious profusion of species in the biological world. Species of flies alone outrank the number of naturally occurring minerals several times; so do species of mushrooms and daisies. There is a simple reason for this. Life is based upon virtually infinite variations in the organic carbon compounds, of which DNA—the basis of the genetic code—is merely the most well known. The permutations and combinations of these molecules are capable of endless elaboration. Sex and genes are subtler markers of species differences than atoms alone. In the sphere of the rocks, there is a finite number of elements, and a finite number of ways of marrying them: and only those that can crystallize under the conditions pertaining on earth are capable of becoming mineral species. As I write there are about 3700 different named mineral species. The Amazon rain forest holds many times this total of beetle species still waiting to be discovered. Approximately 100 new minerals are named each year. Some elements, like aloof gold and platinum, do not readily combine with others under natural conditions, so this reduces the possibilities still further. Other elements, such as mercury, are very particular about their partner, so that cinnabar is one of a very small number of mercury ores. Its rarity is in inverse proportion to its importance, however, because as the source of vermilion it was once king of the painter's palette. Mercury, which is the only metal that is liquid at room temperature, was one of the pivotal elements in the whole philosophy of alchemy; it is still important as an industrial catalyst. So there are severe limits to the possible combinations of the fivescore natural elements. Then

William Wollaston (1766–1828), chemist, philosopher, mathematician and polymath. The medal struck in palladium (which he discovered) is the senior medal of the Geological Society of London. The mineral Wollastonite is named after him.

there are many elements of such natural rarity that they do not form pure compounds—the rare earth elements are most notable among them. Only eight elements make up 90 percent of the earth's crust—oxygen, silicon, aluminium, iron, calcium, sodium, potassium, and magnesium, in decreasing abundance. So the real choice of combinations of elements that might make natural mineral species is much more limited than arithmetical permutation alone might suggest. Minerals are families of natural partners born to exist comfortably on this planet. Their cradle is often where plates collide, or where the tectonic motor of the earth squeezes out unusual fluxes from the stressed crust.

A new mineral name has to be accompanied by the mineral's chemical formula and an account of its crystal structure. They are sometimes named after the localities in which they were found: thus it is not difficult to deduce from which country brazilianite originates. Others are named after prominent geologists and mineralogists. This is one way to have your name and reputation frozen in stone, in perpetuity. What could be more durable than the stuff of earth itself? Wollastonite is named for William Wollaston (1766–1828), chemist, philosopher, mathematician, and polymath. Among a number of achievements he is probably remembered today for his discovery of the noble element palladium in 1803, named after the great wooden image of Pallas, goddess of wisdom and war, that was supposed to protect the citadel of Troy from mishap. The Wollaston Medal is the highest award of the oldest geological club in the world, the Geological Society of London, and is presented annually to one of the greatest research geologists. It is the only academic medal that is struck in palladium.

Because oxygen forms 46.6 percent by weight of the elements in the earth's crust, and silicon 27.7 percent, it is not surprising that silicate minerals that include these two elements are the most abundant materials in the construction of the earth above the core. All the most common igneous rocks are combinations of different silicate minerals. That they have this creative capacity is because of the peculiarities of silica, and to understand the earth it is essential to know something of the chemistry of silica. Silicon and oxygen combine in such a way that the silicon atom sits at the centre of a four-sided pyramid, with an oxygen atom at each apex. You can make the shape of this tetrahedron if you take a small plug of dough and squeeze it between the thumb and index fingers of both hands. Make that finger-and-thumb gesture that French chefs are supposed to use to indicate perfection, then bring together both hands with this gesture at right angles and squeeze the dough between fingertips—a little tetrahedron will result. At the atomic level, these tetrahedra can then link together by their tips as chains, or sheets, or as three-dimensional frameworks. Their apexes may point upwards or downwards. This linkage explains many things about the properties of minerals: their hardness, the way they break, how they behave in the plates of the crust. Quartz is very hard because its pure silica tetrahedra bond in all directions to make a 3-D framework. It does not have natural breakage planes, or cleavage, for the same reason: there is no direction of relative weakness. Its rock-crystal clarity is a way of seeing its atomic purity, the eye appreciating intuitively features that lie beyond even the vision of the electron microscope. By contrast, mica has its silica pyramids bonded into sheets, layer after layer and strongly linked from side to side, but weakly linked in the direction at right angles. It is almost like a stack of leaves of papers. This is why mica breaks readily into thin sheets: you can pick at and flake a mica crystal with your fingernail; in theory you could go on splitting it down and down until it is a mere molecule or two in thickness. In practice, it is used as refractory "glass" windows in boilers and furnaces: it resists temperatures that would melt ordinary window glass.

Almost any configuration of silica tetrahedra that can be devised theoretically can be found in nature. The main mineral "families" correspond to different designs. The silica tetrahedra can be single, or linked together in pairs, or form rings, chains, sheets, or complex three-dimensional frames. The principal mineral varieties are determined by major differences in construction, much as you might define the main styles of architecture by aspects of proportion and design.

A few examples will illustrate how this works. The frequently beautiful mineral beryl is one of a few minerals that contain the second lightest of all metallic elements, beryllium. It is also one of a number of minerals in which the silica tetrahedra are connected together in rings, six linking one to another to form hexagons. Atoms of beryllium and aluminium make the joins between the silica rings. It is helpful to compare the whole structure with a honeycomb: the silica hexagons form the cells, which are sealed and bonded together with the metallic elements. The result is a distinctive mineral *species*. Because the structure is so well linked, beryl is hard—it can scratch glass. When beryl crystals grow under special conditions they may be transparent, and then beryl becomes a gemstone called aquamarine. If the colour happens to be yellow, it is known as heliodor; if green, it is emerald. These are more romantic names than beryllium aluminium silicate, the scientific description of beryl; but they are no more than the names of varieties, just as Rhode Island reds and leghorns are all chickens under the feathers. As usual, the colours are associated with "impurities" of other elements: a trace of chromium is what makes emeralds green.

Crystals of beryl are also often hexagonal in shape; hence the specimen in your hand closely reflects what happens at the atomic scale. The architectural analogy is inappropriate in this respect, for in minerals the finished shape of the construction arises directly from the shape of the bricks—not from the will of the architect. Beryl is a rather easy example to understand because of the distinctiveness of both its atomic and crystal structure—hexagons are familiar enough from close-ups of a fly's eye, or even from grandmother's quilting. Curiously enough, one of the most important structures in organic chemistry is also a six-sided ring: this is the benzene ring, which is a characteristic part of a vast range of naturally occurring carbon compounds. Plants and animals are full of them. It seems that anything silica can do, carbon can do better. It is interesting to speculate, as science fiction has sometimes done, on the possibility of silica-based life, given the similarities between the complexity of silica frameworks and chains and those based on carbon. There is no doubt that silicates are the closest rivals for carbon compounds in their complexity. But there are important differences: for example, carbon–hydrogen bonds are ubiquitous in organic molecules, but do not readily occur with silicon. The relatively tiny number of mineral species compared with carbon-based ones proves that silicon is just not so versatile. At high pressure and temperature, however, silicon bonds are less constrained, more capable of transforming from one shape to another. It is an intriguing thought that silicon life might have to grow deep *within* an alien world to have a chance of reproduction, where heat and pressure free up the rigour of its chemical bonds.

Mineralogy has moved a long way since the time of Agricola. X-rays can spy inside crystals to discover the "bones" of the atomic structure. Tiny amounts of a mineral can be punched out with a laser beam, and its constituent elements counted almost atom by atom. Electron probes tot up the proportions of the elements within a crystal by their atomic weight as mechanically as a coin sorter might file nickels from dimes, quarters from silver thalers. You sit in front of a computer screen and watch the piles of atoms grow as the analysis proceeds. In a mass spectrometer, the isotopes of single elements are sorted by their different atomic weights. Atoms in "vaporized" samples are deflected differentially by a magnetic field, like different-sized pigeons lured home to separate roosts, and it is now comparatively easy to measure abundance of elements in parts per billion. The rarest bird can be spotted alongside the pigeon. An atom or two of praseodymium is no problem, nor is it so difficult to sort out your yttrium from your ytterbium.

So the silicate minerals can be investigated in detail, another kind of intimacy in the history of the earth. We have to go back to Hawai'i to see why this matters. Recall the mineral olivine that formed green sand on one of the beaches. Olivine is a simple silicate of iron and/or magnesium. Olivine crystallized out very early in the igneous magma chamber. These first solid crystals would then tend to sink, like lentils settling to the bottom of a stewpot. In so doing they effectively remove magnesium, and other elements, from the remainder of the siliceous melt, which accordingly subtly alters in its composition. Other minerals can then crystallize out in their turn. The magma, in other words, *evolves.* Elements that are reluctant to shuffle into a silicate mineral early on will become commoner and commoner as this natural alchemy proceeds. You can now understand one of the ways in which elements that are very rare in nature can become concentrated. It can happen either by leaving early, or by hanging on late. Heavy minerals will often segregate together precociously. Most of the greatest deposits of ore for chromium were produced by early settling of the weighty mineral chromite within the magma chamber. Black, dense bands of chromite form dull layers of booty; the chromium miner is like the naughty boy scraping all the meatiest bits from the bottom of the stew. In the opposite fashion, after the magma has almost completely crystallized, a fluid residue remains, containing many of the most volatile elements, chlorides and fluorides, and metallic elements in solution. These seek out cracks and fissures in the country rocks surrounding the solidified magma, there to lay down their burden of minerals, the source of silver, copper, and zinc, and the wealth of industry. Hence Joachimsthal.

Feldspars are the most important single component of igneous and metamorphic rocks. They can be found almost everywhere: in granites, gneisses, and schists; they are even important in moon rocks. So it is important to know something about how they are constructed if we are to understand how much of the world is made. Probably the simplest way to see feldspars in the field is to walk to the nearest bank. Financial institutions seem to prefer to face their buildings with polished granite. There is something reassuring about granite—the fact that most modern buildings only have a thin "skin" of rock as a cosmetic is neither here nor there. The appropriate rock will have an overall pinkish or greyish hue. The feldspars are white or pink blades; they are the largest and most obvious crystals. They are frequently bounded by clearly defined crystal edges, because feldspars crystallize early from magmatic melt and nudge their way ahead of other minerals towards perfection of form. In some ornamental granites the feldspars are huge, making phenocrysts—from a distance they look almost as if they had been glued on to the rock surface like small white posters. Some rocks contain more than one kind of feldspar, pink and white together. Because they crystallize early, feldspars in particular are responsible for altering the chemical composition of the siliceous "stew" that remains behind; they become an important factor in the evolution of magmas. Since the generation of magma in its turn is the outcome of plate tectonic processes, feldspars are woven into the warp and weft of the fabric of the earth. Their chemistry really does matter.

Consider what we have seen on Hawai'i. The lavas there changed from fluid *pahoehoe,* constructing shield volcanoes layer by layer, to a sticky lava building sputtering cones, which still stick to the sides of the huge shields like Mauna Loa as if they were so many tumid pimples. The primitive lava, freshly minted from partially melted oceanic rock, crystallizes out first with calcium-rich feldspar and olivine. The magma held in its deep chamber evolves in composition as these early crystals are removed from the remaining magma. Later lavas are richer in silica, which makes them altogether more gluey; and the feldspars are pushed towards the alkali end of their range as the ingredients of the fecund stew gradually change. The stiff magmas cannot flow as benignly as the *pahoehoe.* Instead, vents build up cones surrounding them composed of pumice and bombs, and occasionally a destructive blast hurls material high into the air. Ultimately, the topography of the Hawaiian Islands is the responsibility of the atomic bonds of minerals.

The individual prospector of popular legend, discovering valuable mineral deposits through a combination of instinct, experience, and luck, is becoming a rare species. Remote sensing equipment can now detect buried minerals from a "fly past," using the most sensitive gravimeters to detect changes in gravity induced by buried bodies of heavy metallic ores. Rather than laboriously trudging the hills, inch by inch, samples can be taken from streams draining a prospect area and precise assays of the elements then taken—a good source will leave traces in the water. It is almost like smelling a potential deposit, with chemical sensors taking the place of the prospector's fabled nose. Large companies operate huge pits with sophisticated methods of recovery—silver, for instance, can be recovered as a by-product of lead and zinc extraction. It is all a matter of profit in a volatile marketplace, and the accountant has become as important as the geologist. Still, striking it rich is part of the mythology of geological exploration, and sometimes, rarely, it still happens.

Diamonds, as the saying goes, are forever. Unfortunately, this is not true. They are not really stable under the conditions found in the average jeweller's shop, for their natural home is deep in the underworld. They exist at the surface on sufferance, having been wrested from their profound origins. They are the nearest that carbon comes to imitate the structure of silica, that ubiquitous component of terra firma: it is no coincidence that these sparkling jewels are known as "rocks." Diamonds are the same material as coal, but transformed wondrously. Under sufficient pressure, the carbon atom is forced to make a three-dimensional framework analogous to that described for silica, except that diamond is pure carbon alone and each atom is strongly bonded to another four in a kind of unbreakable armlock. This is why diamonds are so hard. The largest diamond of all time was the South African Cullinan diamond, 3106 carats, and Thomas Cullinan was one of the few who did, indeed, strike it rich. This diamond, presented by the Transvaal government to the British King Edward VII, was cut into five huge gems, and ninety-six smaller stones to boot. In 1902 Cullinan had bought the ground on which the Premier Diamond Mining Company operates, since when more than 300 million tons of the country rock known as "blue ground" have been removed to yield something like 90 million carats of diamonds. Every year, a few stones the size of pigeon's eggs are discovered. In the Kimberley area—and many other places where they occur in situ—diamonds are found in "pipes"; these are huge tubular structures, punched deep through part of the ancient African shield in Cullinan's claim. Blue ground is the weathered relic of a volcanic rock known as kimberlite. The pipes are believed to be the product of a violent gas-charged magmatic explosion that dragged up diamonds from deep within the earth. It has been estimated that the temperature and pressure needed to turn black carbon to transparent diamond are 1200° Kelvin and 3500,000,000 pascals, respectively. Diamonds can be synthesized by reproducing these conditions. This suggests that natural diamonds originated about 150 kilometres beneath the continents. Most mineralogists now believe that these conditions are satisfied only in the upper mantle. Diamonds are a kind of offering from the gods of the underworld.

So precious minerals can be crystallized from magma, or distilled from its latest breath, or dragged from the depths by volcanic blasts. They will naturally be concentrated where the earth is in transit: where plates have converged along the seams of mountain ranges, or where those ranges once traversed the earth but have now been rendered low by millions of years of weathering and erosion. For where crust has been thickened and heated, there, too, is a crucible in which refinement and concentration can occur. Like almost everything geological on this mutable earth, minerals are genetically related to plate tectonics. The sulphurous breath of igneous activity engenders metal sulphide minerals. Some of these have been the source of riches, but they are usually accompanied by shiny yellow-metallic cubes of common-or-garden pyrites, iron sulphide—"fool's gold" for the credulous. Tin, copper, zinc, and lead make up the precious cargoes carried by sulphide seams hidden within the greater seams that stitch together the plates of the earth. Lead sulphide is called galena, and it, too, forms cubes that are dark as the metal they contain, and as heavy in the hand as you would expect. Where granites are melted in the depths of orogenesis, their subsequent cooling refines the rarer and aristocratic elements, or sequesters those that enter into pacts with volatile elements like chlorine. Vaporous waters carry cargoes of metals and gemstones into the surrounding rocks, fleeing the lumpen mass to hide in cracks and fissures. The silver of Joachimsthal was the product of such a refinement. In the limestone hills of Derbyshire, England, there are buried seams of fluorspar coloured green or mauve, yellow or blue, in which fluorine has married quotidian calcium. Beware testing this with sulphuric acid, for, if you do, hydrogen fluoride is generated, a gas that can dissolve glass, and burn holes in flesh, turning black what remains. Fluorine is the most voracious element, eager to combine with others; small quantities are found in many minerals, but fluorspar is its apotheosis.

And then there is gold, the aloof Vanderbilt of the periodic table. Gold is usually found "native," linked to no other elements. You will have wondered how an element so determinedly celibate can have found its way into veins, let alone gathered into nuggets. In most rocks gold is present in only the minutest traces, as scattered atoms or tiny flecks invisible to a prospector. To become concentrated it somehow needs to congregate. The answers are only just being teased out, particularly by Professor Seward and his colleagues at the Technical Institute in Zürich.* It seems that gold is not so aloof as it appears. Under conditions of high pressure and elevated temperature, and in the presence of water, it makes temporary alliances with hydrogen sulphide, to form hydrosulphide complexes that can migrate through the crust. The gold is deposited again in due turn as its temporary partner deserts. Certain other elements, such as arsenic, when in very fine, or colloidal form, can "scavenge" gold, making it adhere to the surface of the grains. So gold is swept up and captured. When the golden rock is eventually weathered at the surface, heavy and incorruptible gold remains behind and can be concentrated into gravels in river beds. This is where

* The same team have also shown that other metallic elements, such as lead, cadmium, and thallium, form somewhat comparable partnerships with chlorine in hot hydrothermal situations. This is one of the ways in which the emplacement of metals in mineral veins is becoming understood: they, as it were, hitch a ride on the back of a chloride complex. The study of such processes requires the latest equipment, such as X-ray absorption spectroscopy.

Gold sluicing on the Klondike in the summer of 1900 on Number 2 Claim, Anvil Creek, Nome, Alaska.

prospectors traditionally "pan for gold" by swishing away the lighter waste of sand with copious water until the gleaming flakes are concentrated in the bottom of the shallow pan.

Gold was formerly as synonymous with money as the dollar is today. The golden age is still the best of times; the golden rule is still the one that must be observed. Wisdom, like gold, comes in nuggets. Leonardo da Vinci described gold as "not the meanest of Nature's products; but the most excellent . . . which is begotten in the sun, in as much as it has more resemblance to it than to anything else that is, and no created thing is more enduring than gold." Gold was extracted early in history because its natural purity required no smelting. Its use in artefacts accompanies the appearance of the first civilizations. There were mines in the Sudan from which Pharaonic Egypt obtained plentiful supplies of gold: there is a record of a load which needed 150 men to carry it. The inner coffin of Tutankhamun was made of purest gold and weighs more than 100 kilograms. Like Leonardo, the early religions of both Egypt and South America associated gold with the sun, and with their greatest gods. The promise of gold was what led the conquistadors to the New World, and they were not disappointed. Sixteenth-century goldsmiths suddenly had a plentiful supply of the magnificent metal and produced incomparable work, although we might now regret the melting of the Aztec originals that fed it. Gold mingles happily with silver as an alloy, and the most durable gold-work is not 100 percent pure but has up to a fifth of silver in it. The silver puts some strength into a metal that is definitively malleable and ductile, and which can be beaten into gilding sheets thinner than rice paper. When properly worked, gold has unrivalled delicacy. Only gold is as good as gold.

Gold discoveries have always led to gold rushes, a mad scramble for wealth. The great California Gold Rush of 1849 may have inspired Mark Twain's definition of a mine as "a hole in the ground owned by a liar." New and fantastical claims led to hysterical occupation of gold-bearing ground by dreamers: scarcely one in twenty made any money. It was the same in the Klondike fifty years later. Conditions were appalling and the ground worked by the prospectors was permanently frozen. On the other side of the world, it was as hot in Australia as the Klondike was cold. The gold rushes to Bendigo in Victoria in the 1850s, and to Kalgoorlie, Western Australia, in the 1890s were parched affairs. The diggers were often desperate, excavating square miles of holes in favoured areas. Writing in 1852, the sculptor and poet Thomas Woolner described the devastation thus: "I never saw anything more desolate than the first sight of Mount Alexander was to me: it was what one would suppose the earth would appear after the day of judgement has emptied all the graves." No matter where you were, you ran the risk of being robbed of your wealth by a highwayman. In California it might be Black Bart, who performed twenty-nine hold-ups disguised by a headgear of floursack with crude eyeholes. After his depradations he left behind verses on morsels of paper:

I've laboured long and hard for bread,
For honour and for riches.
But on my toes too long you've tred,
You fine haired sons of bitches.

Which proves that Black Bart was a more effective robber than he was a poet. He was caught in 1884. In Victoria, the traveller would have had to keep his eyes skinned for Black Douglas, who might have attacked him in the Black Forest. Clearly, the opposite of gold is black.

There are fairy tales which exploit the naive lad who hears that the streets of London are paved with gold. To which he departs with only his cat, or his spotted handkerchief for company. Barnaby Rudge, the simpleton hero of one of Dickens' more esoteric novels, was probably the last literary manifestion of this fantasy. But there *has* been one time when the streets have truly been paved with gold. In the rush of 1893, Kalgoorlie prospectors threw away an excess of a bulky pyritous material as they delved deeper for pure gold, in the same way as the miners at Joachimsthal had once jettisoned pitchblende. They recognized the waste as being rich in yellowish "fool's gold," iron sulphide, and used it for hardcore, or to fill in ruts, or to bulk up crude sidewalks. Despite what was said above about the nobility of gold, there is, in fact, a stable compound that gold forms in nature with another rare element, tellurium. The resulting mineral is gold telluride, or calaverite, to give it its mineralogical name. The "waste" material used to pave the roads in Kalgoorlie in fact turned out to be this very uncommon mineral, but nearly three years passed before anyone realized that their streets were paved with the proverbial gold. It was not until 29 May 1896 that the results of an elemental analysis of the strange mineral leaked out to the mining community, resulting in a second gold rush—this time to the dumps where the blocks of waste had been disposed of. Dr. Malcolm Maclaren, writing an account of the affair in the *Mining Magazine* in 1912, described how "blocks of ore, assaying at 500 oz. gold per ton, had been utilized to build a rough hearth and chimney in a miner's hut." Forget the hills. There was gold in them thar bricks.

8

Hot Rocks

Driving in India reinforces your belief in a protective God. The dry and dusty roads north of Pune (formerly Poona) bend this way and that as they climb into the foothills of the Western Ghats. It is the dry season, which lasts for six months or so. The landscape is reduced to a uniform pale tan colour. It is hot enough to make the roadside farmhouses, with their small courtyards draped in leaves, look restful and inviting. The roads are just about wide enough to take two vehicles abreast. Disreputable trucks covered in dents and dust hurtle along the road with no regard for bends or goats or barefoot children. Worse, they overtake other, older trucks even dustier and more dented, whether or not they happen to be climbing up a hill or going round a blind bend at the time. Our own driver is more circumspect, possibly influenced by my drawn face and clenched fists. But it still does no good, because statistical laws beyond the reach of caution inevitably dictated that sooner or later there is an oncoming truck in mid-overtake on our side of the road. I close my eyes. A few milliseconds later there is a tremendous crash as our vehicle veers off the road and into a field. The statistic has been defied, thanks to a lack of roadside fencing, and providence. As we come to a halt our driver exhales between his teeth and says: "Oh my! We nearly had our chip!" One of the things I like about India is that they still know how to use English understatement.

As we drive northwards changes in the proportions of the landscape become obvious. For a few miles the road runs comparatively straight, then we climb for a while before another level stretch is reached. Any hills have flat tops, and in the distance more terraces rise one after the other, so that the hillsides look like a series of steps stretching away as far as the eye can see. We are in the Deccan Traps, one of the greatest volcanic features on earth, 1.5 million square kilometres of volcanic rock. Each flat level was formed by the top of a single lava flow, which evidently extended for many miles. So as we climb, we ascend through flow after flow. The top of a more resistant flow might make an extensive plain, and then its successor would comprise the next flat area, perhaps half as wide. Each terrace supports a farming population, with many smallholdings carrying humble, single-storey dwellings. There is no corner of the flat area capable of cultivation that had not been dug or tilled. Occasionally, there might be a small square of bright green where some local water source permitted salad plants to prosper, but it is clear that the crops were over for this year. Nothing to do except wait for the next rainy season. We pass only one factory run by the ubiquitous Tata corporation; the dependence on agriculture of this part of Maharastra Province is unlikely to change in the future.

The word "traps" is derived from a Swedish word (*trapp*) meaning "stairs" or "steps," and is certainly an appropriate way to describe the landscape. The Deccan is the apotheosis of the lava flow. The difference from the Hawaiian shield volcanoes is the sheer extent of the flows, which stretch for thousands of square kilometres. Rather than cones, they build up plateaux; for this reason the underlying rocks are sometimes known as plateau basalts. As you drive over the terraces for hour after hour, you begin to appreciate the scale of the traps. Single flows of highly liquid lava must have flooded out on a truly enormous scale, leaving flows the thickness of a house covering more than 100,000 square kilometres. Viewed from the moon at the time of eruption, this area of north-western India would have looked as if a great pot of black paint had been spilled out on the surface of the earth. The other difference from Hawai'i, of course, is that the plateau of the Deccan Traps is built upon continental, rather than oceanic, crust. It makes for an area of the world where dark, oceanic-style rocks have been plastered mightily on top of the continents. This happens on a scale that makes the ophiolites considered in the previous chapters seem insignificant.

Because the lavas are porous, water drains away through them. There are cracks and joints and holes running through the stack of layers, for all that they appear solid and impregnable as you pass them in a road-cut. Water sees the rocks differently: as a mass of tiny channels through which it can pass to lower layers, there to seek out other subtle passages still deeper. It is spirited away further and further from the surface. When the rainy season finally arrives, the landscape bursts into fecund productivity, but it is a time of plenty running against the clock. Even as leaves unfurl into the moist air, water is dribbling away unseen through the flows beneath the ground. I notice empty pits and scrapes all over the Deccan hills. These are temporary reservoirs that try to slow the draining. Roughly caulked, they are a temporary way of making ponds for irrigation. They only stay the inevitable drought for a while. As the sun beats down harder, evaporation takes its share, too. The whole region hunkers down for the mean season. You begin to understand how men are in thrall to geology even though there are many places where it does not blatantly announce itself; it rules unseen.

Native vegetation understands the bidding of climate and rock. Plants have developed all manner of techniques for surviving periods of hardship: leathery leaves, deep roots, deciduous habit. There is not much old woodland left in the parts of Maharastra Province I visited: too many hungry mouths need to be fed. That any remains at all is thanks to "sacred groves," which are small but inviolate areas associated with gods and legends. They retain a very rich natural flora, with many unique species. They, too, pay homage to the peculiarities of geology. Professor Varkar in Pune maintains a huge personal herbarium for conserving these species that threatens to edge him out of his apartment. Currently, there is concern about keeping some of these endemic species from disappearing altogether, a concern partly fuelled by pragmatic arguments about the loss of uninvestigated plant chemicals that may prove to be useful in areas such as cancer treatment. The case of the critically rare Madagascar rose periwinkle, which has yielded one of the best anti-cancer drugs, is rehearsed again. But the stronger argument is a moral one. Our single species has no right to extirpate what geology and climate have created over millions of years: a whole habitat. Unlike the Hawaiian chain, we cannot even say that these basalts are doomed to immersion. The only sea that swamps these inland terraces is a sea of humankind.

The road climbs more steeply after the town of Aurangabad, named after the last of the Great Moguls, Aurangzheb (1659–1707). Looking down from the hills beyond, you can see the magnificent, but now decayed, fortress walls that had been built around the town by previous Mogul emperors for its protection. In places, trees now sprout from the battlements. On a cricket pitch, albeit a dusty one, at least three teams of middle-sized boys dash about, ignoring the heat. "I am telling you, India is cricket mad!" the driver informs me, unnecessarily. Cricket is the legacy of the British Raj, and possibly one of the few enduring contributions of my own race to India, and certainly the most harmless. The other contribution is our useful and flexible language, which has helped India to take a prominent part in the international communications revolution. The sway of Moguls and British has passed away, but the tumult of the street carries on as it always did, only now enlivened with the cry of "Howzat?"

The Ellora temples and caves lie thirty kilometres on from Aurangabad. The hills in which they lie comprise another stack of horizontal lava flows. The monuments were constructed as rock temples and monasteries between the fifth and ninth centuries, starting more than a thousand years before the British sniffed Indian profits. The Deccan Traps host the most direct link between geology and culture that I know. The work grows from the rock, which provides both inspiration and material. There are three religions celebrated at Ellora; a guidebook tells me that sixteen "caves" are Hindu, twelve are Buddhist, including the earliest ones, and five are dedicated to that most peaceful of all religions, the Jain faith. The Jains are strictly vegetarian and will not contemplate harm to any living thing. The purist Jain will even sweep the road in front of his advancing feet lest he take the life of an ant. The caves lie in a series running north–south for about two kilometres, and apparently follow the "seam" of one particular lava flow that is more suited for excavation. Over some parts of the outcrop a more resistant flow overlies the caves; the small lake known as Sita's Bath is fed by a waterfall that plunges over this capping flow. Cave 16—Kailasa, Shiva's mountain abode—is the one that the tourists flock to first, and you can see exactly why. It is an astonishing construction. I should perhaps rather call it a deconstruction, a piece of architecture in reverse. For this edifice has been fabricated by *taking away* rock, rather than by building up with bricks or blocks. The temple was made by removing masses of lava to leave behind monumental buildings and reliefs. Effectively a gigantic sculpture, hewn from the rock with hammer and chisel, it was made during the eighth and early ninth centuries under the orders of King Krishna I. It is as if the temple lay nascent within the lava flow, until released by a century of chipping and digging by Hindu artisans. Something like 3 million cubic feet of volcanic rock had to be removed. The main pit dug from the cliff top was approximately 30 metres (100 feet) deep, 84 metres long and 46 metres wide (276 feet by 150 feet); at its centre a huge block was left behind, which was carved over decades into the huge, two-storey temple of Shiva. Inside, rooms were excavated. On the outside, rectangular columns frame sculptures that combine delicacy and extravagance: the richness of decoration is reminiscent of the Victorian embellishments of Pugin, for all that the subject matter could scarcely be more different. This has something to do with a sense of fecundity, an ebullience of ornament. The friezes of the temple are carved into life-sized elephants, stiff and grand. Standing separately from the temple, there is a victory pillar as splendid as any monolith, resembling a geological outlier, with its lines of detailed carving mimicking a kind of geological stratification, a case of art unwittingly imitating nature.

In several parts of the monument there are sculptures depicting the marriage of Lord Krishna with Parvati. The female figure is voluptuous: she has spherical breasts as plump as small grapefruits and a tiny waist. Krishna is manly, but also tender; his hands reach towards his partner with a decorousness that belies the nudity of the lovers. There is an intimacy between the couple that is affecting and tender. Even their token garments have an understated elegance: they conceal more than they expose. For all that they were gods, these figures were also made in human dimensions. The bodies are different from the Greek archetypes that adorned Pompeii. Here they have more of flesh and blood, even though some may sport extra pairs of arms, or hybridize in interesting ways with animals.

Looking at the rocks close to, it was clear that many of the lavas at Ellora were of the kind described as porphyritic: they show large, flat feldspar crystals "floating" in a matrix of darker, fine-grained basalt. It was as if bold white brushstrokes had been added to the faces of the sculptures. Such prominent, prismatic feldspars had crystallized out at depth, where they had had time enough to grow large, before the main lava eruption spilled out over the ground in the thick flow that would one day give birth to a temple. The crystals were carried along with the fluid lava and then trapped in its solidity like nuts in brownies. The crystals serve to leaven the relentlessness of the black volcanic rock.

An even grander monument lies seventy-five kilometres onwards from Ellora. The Ajanta Caves are carved into a vertical scarp along a horseshoe-shaped ravine. They, too, seem to follow a few flows; horizontal "ribs" above the caves show the boundaries between the flows that were left unworked by human hand. The caves were lost for a thousand years. After Buddhism declined at the end of the seventh century, they were simply forgotten; scrub covered their entrances, and bats were left to keep their secrets. A British horseman accidently stumbled upon the site in 1819; it must have been like the moment when the tomb of Tutankhamun was opened. A little road takes you to the base of the cliff, ending at an open space where there are dozens of stalls selling souvenirs. Mineral stores are as numerous as traders selling Thumbs-Up, the subcontinental equivalent of Coca-Cola. When the bus stops, tiny urchins mob you with handfuls of pretty minerals for sale. The minerals are, indeed, beautiful, and they are on sale here for a fraction of the price asked in western "rock shops." Pink or green crystals catch the eye; some specimens show gems arranged in delicate rosettes, or there are spiky balls made of elongate prisms. The crystals are another product of the ancient volcanic activity; and some of them belong to rare species, like apophyllite. When the lava cooled there were cavities, or vugs, left within it—many of these were originally air bubbles, part of the "froth" of eruption. At a late stage, fluids moved through the lava pile and minerals were deposited within the holes. Many of them are of a class called zeolites, which include several rare varieties sought after by mineral collectors. These beautiful crystals appeal to that queasy part of human nature that lies between aesthetics and avarice. I am as vulnerable as anyone to the appeal of rarity dressed up in crystal form. When the Pune–Bombay railway was cut through the Deccan Traps, glittering subterranean galleries lined with crystals were discovered, and these furnished some of the finest specimens now displayed in museums around the world. But even the scruffiest street vendor might have had a lucky day in the hills, and you can make a substantial difference to his family economy by buying an example for a few dollars.

No chance of turning a penny is left unexplored. There is a steep flight of stairs up to the Ajanta Caves. When I finally escaped the mineral vendors, two strapping young men offered to save me the walk by lifting me up the flight on a kind of palanquin. This I huffily refused, although I had second thoughts halfway up the stairs. The thirty caves are reached by way of an undulating path that has been scraped along the scrap. Originally, individual caves were reached along steps climbing from the stream below. These have largely collapsed. So now you go from one cave to another as you might from gallery to gallery in the Louvre. The comparison is appropriate, because almost every cave is an artistic treasure-house. Some have entrances supported by round columns of lava, but these were left behind as the excavation of the cave proceeded, rather than built up block by block in the classical fashion, so they are seamless. One cave has a roof made of twisting ropy lava belonging to the overlying flow. In a number of caves, Buddhist wallpaintings have survived in a remarkable state of preservation. They are by turns grand and intimate. Scenes from the life of the Buddha are complemented by episodes from everyday life; some are undoubted masterpieces. The torsos of the figures are unclothed, but all (except, on occasion, the Buddha) wear necklaces and bangles, garters and bejewelled headgear. It would be an understatement to call the latter "hats," for they come in every imaginable shape, from sparkling piles to elegant curlicues. The emotions of the figures are easily read by a modern observer; here the same tenderness, there the same sadness, the same humour. The subjects portrayed are those that have always interested artists: mother and child, death and redemption. On the back wall of the ante-chamber in Cave 17 there is a scene depicting the return of the Buddha to his palace as a mendicant, carrying his begging bowl. His son Rahula, whom he has not seen for seven years, in turn begs of the Buddha his recognition as a son. The poet Laurence Binyon (1869–1943), who did much to make the quality of Ajanta's art known to the world, said of this painting: "No picture anywhere is more profoundly impressive in grandeur and tenderness." I feel an affinity with this particular visitor to Ajanta because, like me, he spent much of his working life as an employee of the British Museum. There was another detail that moved me for a reason I struggled to understand. In Cave 2 there is a fresco that includes a half-naked girl on a rope swing, set against a rather dark and stylized field of flowers; it is an image of fun, but also curiously solemn. It has an exact counterpart in a painting called *The Swing*, by the pre-Revolutionary French artist Jean-Honoré Fragonard, in which an aristocratic girl, dressed as a notional shepherdess, soars carefree into the sky. Two girls from different millennia and from remote traditions enjoying an identical game, and captured in mid-swing by the hand of an artist. I take this as an emblem of the deep similarity of human responses, and, especially, of the importance of the artist in proving it.

Ajanta, too, was rooted in the geology, and not just because the caves were hollowed out from the appropriate flow. Clay derived from weathered strata was one ingredient of the plaster used to prepare the walls for decoration (the others being cow dung and rice husks), with lime being used as the "finish." The pigments were mostly of local origin, like ochre and sienna, the weathered products of iron-rich volcanic rocks. These earthy washes provide a tone to the paintings that makes them feel as if they have grown from the rocks themselves. It is an interesting question where the brilliant colours came from: surely the bright blue used to paint in some of the jewels must have been produced from a cobalt mineral. Other caves are full of sculpture hewn directly from the lava. Rock and man engaging directly via the chisel. Cave 1 has a huge hall nearly twenty metres square, with an aisle flanked by a colonnade. The crude cells for the monks behind would certainly have encouraged a decidedly ascetic life. My favourite sculpted images are the flying couples that adorn the columns: united in ecstasy, these *Gandharvas* loosely embrace as they fly to the heavens, eyes closed. In one cave after another there is a parade of larger-than-life Buddhas that display his various embodiments by means of exquisite hand gestures. Cave 26 is veritably crammed with sculpture, which has a profusion about it that is more organic than the Hindu work at Ellora. A carved tableau tells the tale of how the Enlightened One resisted even the charms of the daughters of his relentless tempter, Mara. At the centre of it all, the giant reclining figure of the Buddha (Mahapari Nirvana), head resting on one hand, exudes peace. Playing about his lips is that distinctive expression that is not quite a smile, but does undoubtedly convey bliss. Basalt, the commonest of rocks, has never been employed for a more rarefied purpose.

I have avoided reflections upon the vastness of geological time in this book. Its extent imbues every page, but I have passed over with hardly a second glance the millions, or even billions, of years that radioactive clocks tell us have elapsed since one geological event or another. If you are not careful, referring again and again to the huge span of time available for earth's evolution becomes something of a mantra, rendered subliminal through its constant repetition. Familiarity breeds unawareness. But at this point it might be worthwhile to think about what has happened over the historical span of the Deccan Caves. Images of three religions have appeared, carved in the rocks; Buddhism faded sufficiently for the Ajanta Caves to be lost for a thousand years. Neither Islam nor Christianity left its mark there, as they did elsewhere. The Moguls came and went, as did the British Raj. All this happened within a couple of thousand years, during which time the face of the cliffs altered hardly at all. The changes that mankind has wrought upon the landscape are much more profound: an ecosystem can be destroyed in a couple of human generations. What geology and climate married over millions of years, man can put asunder in a global heartbeat. All our human mess of history hardly registers on the scale of geological processes. It will be clear by now how much the world has changed to this infinitely slower rhythm. Lyell was wrong about the constancy of the earth's processes, but the method he gave us has been crucial to understanding the nature of the fluctuations that have occurred. Nowhere on earth today is a new Deccan erupting. This does not mean that it is beyond study, because its peculiarities are readily capable of being interpreted in the light of processes we understand. You cannot describe the world as "uniform" when such events occur without modern equivalents. However, there will be levels of history that we shall never reach; in "deep time" there will be millennia beyond recall, simply because the level of dating precision we need to "see" time is not available to us that far back. Some things are lost forever. Our own species emerged from Africa perhaps 100,000 years ago. This is still nothing, in geological terms. Yet all human biographies are crammed into those few milliseconds of geological time. I wonder if the adherence to the biblical timescale insisted upon by creationists is not partly motivated by a desire to *hang on* to history. Let time go into the millions, and beyond, and the insignificance of our own sector becomes patent.

Time again: the eruption of the Deccan Traps happened over a very short period. The pulse of history quickened. Increased accuracy in radiometric dating techniques has shown that the eruptions were bracketed by an interval as short as 1.5 million years.* It could have been even shorter: longer than the life of our species, to be sure, but still a blink in the gaze of time. Considering the volume of basalt erupted—an estimated 2.5 million cubic kilometres—and the vast area over which it extended, this was an extraordinary outpouring of magmatic floods.

The underworld bubbled over in this part of India. Many of the flows are twenty to thirty metres high and they maintain their thickness over hundreds of kilometres. Charged with gas, and highly fluid, they must have spread at remarkable speed, one after another. The lava emerged from a source that was plumbed into the mantle: the trace elements present in the basalts are consistent with such a deep source. The flows must have discharged from long fissures to account for their wide spread. These would leave their relics today as vertical feeder dykes cutting through the geological underlay, all the way to the depths. It has been estimated that the lava must have gushed out at a rate of hundreds of metres per second, which can be compared with the discharge from some of the world's largest rivers. What a sight it must have been at night! Having resisted the "rivers of fire" cliché when writing of Hawai'i, this is surely the place to apply it without embarrassment. Floods of flaming fury; a Ganges of geological gaudiness; an Indus of inflagration; it is scarcely possible to go over the top.

Deeper time: the radiometric dates prove that the flood basalts of the Deccan Traps were erupted 66 million years ago. This is on the

* A Web report on recent research suggests that different parts of the traps may have erupted over a total of 4 million years.

Cretaceous–Tertiary boundary. The same period saw the extinction of the dinosaurs, as well as of other kinds of animals that lived in the sea, such as ammonites. It is one of five mass extinctions that have punctuated the history of life, and probably the second greatest of them all. Not surprisingly, the two events have been linked: eruption and extinction. Quantities of sulphur dioxide and fine dust released into the atmosphere during the plateau basalt eruptions may have triggered adverse climate changes sufficient to kill vegetation on land and poison the seas. Herbivorous dinosaurs starved, while their predators briefly gorged upon their carcasses, and then followed them to oblivion. The marine plankton was decimated at the same time. Currently, rather more evidence favours the impact of a huge meteorite in the Yucatan Peninsula, Mexico, as the instrument of the dramatic crisis that reset the evolutionary clock. Its direct effects have been detected more widely in sedimentary rock successions around the world than in features that can be definitely attributed to the eruption of the Deccan Traps. Still, it is a remarkable temporal coincidence, and the eruption scenario is a hypothesis waiting in the wings.

It will come as no surprise to discover that the Deccan eruptions are yet another consequence of plate movements. Think back to the northward-"drifting" Indian peninsula and recall that it was not always where it is today. Remember that the world has been re-made, so to understand what happened 66 million years ago we have to reset the continental stage and shuffle the props. At the end of the Cretaceous the migrating Indian subcontinent passed over a "hot spot." As we saw in the Hawaiian chain, the heat source was stationary. The continent passed over it as a hand might pass over a blow-lamp. When it did so, the mantle plume injected its tholeiitic magma directly onto the continent with redoubled fury. This is one place where ocean and continent meet head on: piled up on the lighter continent, dark flows usually associated with the abyss transformed a part of the face of the earth, like a huge black blister. So much heat was dissipated that it has been suggested that there must have been a "super-plume" to give birth to the traps, a torch from the mantle bigger by far than the Hawaiian example. But then the Indian plate continued on its way northwards towards its rendezvous with Asia, and left the plume, super or not, behind. The eruptions finished as suddenly as they had begun.

A final twist to the story is that there are some scientists who believe that the huge outpourings of lava, and hence heat energy, in the Deccan could not have been accomplished by normal earth processes alone. They ask: where did the extra energy come from to account for a "super-plume"?

One answer invokes those meteorite impacts. The energy imparted by such an extra-terrestrial interloper, it is suggested, may have provided just the addition needed to induce the enormous lava outpourings; and since meteorites tend to arrive in clusters, the time equivalence between events in Yucatan and those in north-western India is no coincidence. The earth winced before a double blow. One meteorite left a crater in Mexico; the other provoked floods of basalt that gave the Hindu sculptors their stone and the Buddhist artists their canvas. Perhaps the stars painted on the ceilings of some of the Ajanta Caves were strangely prescient.

The Deccan Traps are not unique, although the richness of their human associations is without parallel. There are comparable massive outpourings of plateau basalts in several other places, forming untamed tracts of country. The Triassic-age Siberian Traps are, if anything, even more extensive than those in India, but are remote and inaccessible. They may have erupted over only half a million years. The Columbia River Plateau in the north-western United States was erupted 17 million years ago and occupies 130,000 square kilometres. The flows filled in an earlier mountainous topography, flooding and levelling former valleys with relentless masses of volcanic darkness. Something like 1500 metres of lava smothered this old landscape, with the effectiveness of a builder pouring in new concrete foundations over an old archaeological site. There may have been as many as 100 separate flows. Present-day rivers are cutting back into the pile to spectacular effect along deep valleys; one day, the old landscape will once again be exhumed. The Snake River basalts in southern Idaho (50,000 square kilometres) are even younger, of Quaternary age. There are also extensive plateau basalts in Brazil, and even Antarctica. There are ocean-floor flood basalts, too, where "super-plumes" have poured out their effusions far from the eyes of mammal or dinosaur—like the Ontong Java Plateau in the west Pacific, 120 million years old, or those to the south of Madagascar. These rocks are a major feature on the face of the earth.

As to their antiquity, there are indications that these extrusive plateaux may have besmirched the face of the earth as long as there were plates to move. Many researchers are convinced that "swarms" of basalt dykes dating to about 2.5 billion years ago, at the end of the Archaean, may have been feeders for vast, and now completely eroded, flood basalts. In the deep Precambrian there would have been no forest to clothe them, nor any kind of vegetable cosmetic to conceal their darkness. They would have been rawly black, explicit stains from the depths smeared over the landscape. The old earth must have been blackened many times.

. . .

Granite is an unforgiving and ungenerous stone, hard to break, slow to decay. Granite does not flinch under the onslaught of the elements. It lends its name to things that are dauntless and immovable. The Granite Redoubt were the grenadiers of the Consular Guard at the Battle of Marengo, 14 June 1800, whose unyielding square formation stopped the Austrian advance and helped win northern Italy for Napoleon Bonaparte. Granite does not give way: granite features belong to the face that reveals no secrets. It lends authority to public buildings. Even those who know nothing about geology can recognize its solidity. It means business. Granite constructions seem to say, "Here we stand, and here we stay." It is oddly consoling to know that the bank holding your life savings is made of granite. On the other hand, you do not expect granite to underlie gentle countryside. Seamus Heaney knew its properties:

> Granite is jagged, salty, punitive
> and exacting. *Come to me,* it says,
> *all you who labour and are burdened, I*
> *will not refresh you.* And it adds, *Seize*
> *the day,* And, *You can take me or leave me.*

Granite is abundant over the surface of the earth: moors and mountains, vast and inaccessible peaks in the Himalayas or the Andes, including some of the ultimate challenges for alpinists, along with monadnocks and inselbergs in Africa, and barrens and tors in Europe and America, all are the surface manifestations of this, the most obdurate geological underlay. After the Deccan Traps, one of the greatest of all basalt bodies, the granites of south-western England, form one of the least spectacular intrusions. Describing them is not perversity on my part. The granites of Devon and Cornwall are, as the tourist brochures relentlessly proclaim, "steeped in history." In the West Country you can understand how granite connects with scenery and tectonics—not to mention human character, history, and literature—better than almost anywhere. Dartmoor is often just another destination on the Heritage Trail, a brief stop on an inventory of historic Britain—the kind of tour from which most of the drama has been drained and the history condensed to a handful of brochures. But this does little justice to a wilderness that once struck such fear into the traveller that he would make his will before setting out to cross its dismal wastes. Even today, you have only to walk out of the sight of the nearest road, and hear

An ancient granite cross in a churchyard near St. Michael's Mount, Cornwall. The roughened surface testifies to the weathering of centuries.

the wind sweeping through the endless heather, punctuated by the chinking cry of the meadow pipit, while bold clouds scud across an enormous sky, to know that wildness endures here, for all the domestication of the beaten track. The granite moor still says: *You can take me or leave me.*

When you approach Dartmoor from the east its appearance is hardly dramatic. Off the new arterial roads there are narrow, beech-lined, sunken lanes, wedged between high banks. These were the ancient tracks into Dartmoor. In the spring, the hedge-banks are miniature botanical reserves, decked with bluebells and red campion, early purple orchids and greater stitchwort. It's like a patriotic flower show: red, white, and blue. Then you catch a view on the skyline of a low swell, with wide and unencumbered contours. This is the moorland, which is higher than it looks, and further away. It seems to take far longer than it should to reach the open skies. The twisting and turning lanes always seem to dodge off in a different direction as if reluctant to confront the bleakness ahead. Then you notice that the walls have changed. They now consist of pale, piled-up blocks: large and rectangular near the base, smaller near the top, and all fitted together in a kind of ad hoc jigsaw puzzle. Foxgloves sprout from the interstices, their pink spires always abuzz with bumble-bees when the sun shines; round leaves of pennywort slink in the cracks. These are walls constructed of granite blocks. Some of the walls might be a thousand years old.

After you cross the cattle grid you are on to the open moor. This is where diminutive tousled Dartmoor ponies acquired their hardiness. In the spring, the slopes on the moors can look almost black—even cheerless and oppressive. You could be forgiven for wondering if you were not, after all, looking at some kind of basalt flow, but it is the fine new growth of the ling that lends spurious volcanic heaviness to the moorland. Ling shoots are dark, with tiny leaves packed together against the harsh climate. Later in the year the same plant will wash the hillsides with millions of purple flowers. The heather family has learned the trick of thriving on poor soils almost anywhere in the world: we have already met a relative of ling on the raw volcanoes of Hawai'i, and the thin soil above the granite here is scarcely richer. Sheep pick delicately at fine-leaved grasses among the ling, white puffs of wooliness against the dark hillsides. Granite is used for almost everything out here. As well as the walls, vertical granite slabs are used as gateposts, whilst granite blocks have been employed for constructing dwellings since Neolithic times. Spinster's Rock, not far from Edwin Lutyens' grand granitic house, Castle Drogo, is a chambered tomb dating from about 3000 B.C., with three uprights supporting a great horizontal slab. The whole granite intrusion is peppered with several thousand Bronze Age "hut circles" that are often marked by stones (there may have been wooden huts, too). In the 1970s the low field boundaries, known as reaves, were recognized for what they were. These unremarkable lines of stones, often capped by gorse bushes, are still used as farm and parish boundaries: a tough farmer at the edge of the moors today might still follow a line laid out by his Bronze Age predecessor 3000 years ago. The farming habit endures, much like granite.

Stone circles and avenues are numerous over the moor, and isolated standing stones waiting like sentinels in the middle of a field, or by the roadside, are so common that you soon cease to remark on them. Here is a domesticated ancient history, lacking the grandeur of Stonehenge, perhaps, but giving a better sense of the density of occupation of land during the Bronze Age: this was no pioneer community, but a busy, thriving society. The people probably grew oats and beans and scoured the streams for tin. Grimspound is as impressive as Stonehenge in its own way, a stone structure with piled-up walls so thick it has been estimated they would have taken thirty-five man-years to build. Dartmoor seems to have been more or less abandoned in Roman times, a process that continued into the Dark Ages. It has been suggested that the eruption of Krakatoa in A.D. 540 so badly affected the climate that many of the inhabitants of Cornwall emigrated to Brittany in order to avoid starvation. Whatever the cause, the moor was left to itself for almost a millennium. Medieval farmers returned after about A.D. 950. The long-houses that they built still survive in a few places, for granite walls do not readily tumble down. The population was denser then. Near Hound Tor, there is evidence of an abandoned village, only the low remnants of house walls remaining. In the few surviving longhouses granite slabs make up the lintels and the door jambs, and the walls are thickset against the winter winds. Granite slabs were also used for crossing rivers, which can be intimidating when in full flood. Clapper bridges are simple structures comprising a few piers that support large, flattish granite slabs, over which pedestrians and packhorses could pass. In medieval times there were many tracks across the moor where footpads preyed on lonely travellers. Where these ways crossed one another a stone cross was often erected, and they endure today, even if the tracks have disappeared, so you might come across one standing alone and miles from anywhere, mysterious and somewhat forlorn. More often, you will find one tucked in the corner of a churchyard or on a village green. The cross may be crudely carved, but its import is unmistakeable. Before churches were built, the holy site may just have been an enclosure containing a granite cross. These monuments are peculiarly moving: faith itself, they imply, will endure like granite.

The underlying rock reaches the surface at the high points of the moors: the tors. From afar, tors can look like crumbled castles or ruined pyramids. Close to, you discover extraordinary arrangements of piled-up blocks, sometimes almost teetering on the edge of collapse. It is almost plausible that they were thrown up by some neolithic Henry Moore, for, like good sculptures, tors have the interesting property of looking quite different according to the angle from which they are viewed: from this side an obelisk, perhaps, from the other a medieval keep. You do not have to be very imaginative to see a twelve-metre figure at Bowerman's Nose gazing out over the greener fields below, like a protective god. Combestone Tor rather resembles piled bales or coffins, some tipped over as if they had been dropped from the sky by a careless giant. Vixen's Tor was the legendary home of a witch. All the shapes were the product of natural weathering; they are what remains when the granite around has been removed by hundreds of thousands of years of wind and water. The cracks that define the blocks are joints in the granite, natural cracks. Only the most resistant blocks remain behind. All else is lost to the great cycles of earth history, whittled away by time. The shapes, however fantastical, are no more than the legacy of the whims of the elements.

When you get really close to a tor you realize that you are not seeing the rock at all. The pale grey colour that tints the surface of the granite is a lick of paint. The surface has been decorated with lichens. This poorest of pasture can still support a patina of growth. Lichens grow in patches; they run into one another so that you cannot tell where one patch ends and another begins. The rain that falls so freely in the west of England is all they need to sustain their growth, with a meagre mineral supplement from the rock that provides their home. Many are so tightly bound to the surface that you cannot even scrape them off with your fingernail: they grow partly within the rock. Most are pale in colour, which is why the granite looks white and smooth from a distance. Occasionally, an orange patch reveals a different species. I have seen granites decorated with lichens at 80 degrees north, and I don't doubt that lichens can survive in Antarctica. A symbiotic collaboration between fungus and algae, lichens are tougher than either partner. Look closely and you may see the little pads or cups that are their reproductive structures. Many lichen species grow more slowly than the creep of tectonic plates.

If you want to see the native rock, you must go to a quarry. There are plenty of disused quarries around Dartmoor. Granite makes good road metal, and it underlies the sleepers along many of the great railroads constructed in the nineteenth century. The tough felons locked up in Princetown Gaol were once obliged to break such ungrateful rocks as part of their punishment. The prison is still there, even if the rock-breaking has ceased. It is a steep-sided block looking like an old cotton mill, or a factory from the Industrial Revolution, rather than a gaol—a factory for the manufacture of hardened criminals, perhaps. Merrivale Quarry nearby now lies silent, but the plant that used to lift the granite blocks remains behind, rusting, resembling a collapsed, gargantuan spider. The hole from which blocks were extracted is now full of water, but you can see a cross-section through the granite in the face of the quarry. The whole mass is divided into great slabs by subtle cracks running parallel to the surface of the ground above and by others plunging vertically. The granite was originally a nearly homogeneous mass, with hydrostatic pressure equally distributed in all directions through it. As the aeons of erosion slowly stripped the top of the granite away, the balance was lost and the consequent stresses caused joints to develop. Thus you could say that the shape of the tor was already anticipated deep within the ground. The rusting hulk once moved blocks that had been freed by blasting; then they were cut with diamond saws. To make a facing stone, the cut surface was then mechanically polished to a shine.

Bits and pieces of polished stone lie all over the huge waste tips that surround the quarry. There are other, more exotic granites besides, so the place obviously acted as agent for buffing several types of rock. I recognized the reddish Scandinavian orbicular granite, which is easily identified by its large, circular feldspars, each surrounded by a green rim of altered mineral. The feldspars reacted with the magmatic liquor with which they were surrounded to grow these neat fringes of crystals, so now they look like plums in a pudding. The commonest stone on the waste heaps is the local granite, however, mostly whitish and speckled, some blocks pale pinkish. The dominant colour follows that of the orthoclase feldspar that makes up the bulk of the rock. Sections through these big, white prismatic crystals are well displayed on the slabs and show apparently random orientation. It was recognized from the early days of scientific geology that the large crystals proved that the granite magma cooled slowly, at depth within the earth's crust, to allow such a coarse fabric to develop. The largest crystals may be up to seven inches long. There are also crystals of plagioclase feldspar. According to the British Geological Survey, the Dartmoor granite was emplaced as an intrusion, having risen from a depth of 17.5 kilometres (10–12 miles). If the plateau basalts were instant scabs on the face of the earth, granites were slow oozings from its deep contusions. The slabs also show clear patches between the feldspars, without clear crystal shape; this is quartz—silica—which fills the gaps between the larger, milky feldspar crystals. As for the black specks, they are probably dark biotite mica. This is a softer mineral than the others in granite's igneous jigsaw; they often pull away at the surface of the polished slabs, making tiny pockmarks in the otherwise flawless complexions of the elegant facing stones that are used as "skins" to clad buildings constructed of concrete and steel. Even granite can be a cosmetic.

If you want to see how slowly granite succumbs to the elements, look at one of the ancient crosses. It has been standing for 1000 years, yet still retains its shape. Run your fingers over the surfaces, and they are rough—not the crumbly texture of lichen, but something altogether coarser. Irregular lumps protrude from the surface. These are quartz crystals. Centuries of weathering have left these crystals standing proud as the other minerals around them have preferentially worn away. This is the extent of weathering over the time since the medieval serf system faded, the moor was abandoned, and machines that could cut granite to a polish were invented—just a roughening to the touch. Look in the little rills that drain the moors and you will see something similar. Tiny bits of granite, but mostly quartz, form a crunchy stream bed. This is the waste of the moors. The feldspar will rot. The quartz will edge its way seawards over millennia, tumbled

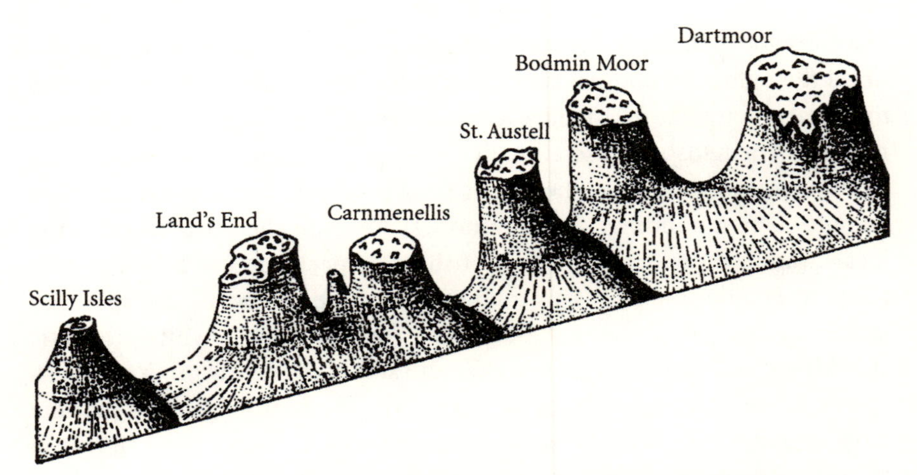

Even bigger than it seems: Arthur Holmes' portrayal of the subterranean continuity of the Variscan batholith in south-west England. The batholith "breaks the surface" periodically; in another 100 million years it may be exhumed entirely.

downhill during times of flood, but it will endure. The sandy beaches near Padstow, so popular with surfers, will gather the debris. At some time in the future a sedimentary sandstone will incorporate all the silica grains, and then it, too, will be elevated above the sea and eroded. A trillion grains or two might survive again, only to enter into yet another geological cycle. This is endurance. The longevity of quartz is reminiscent of the words of the hymn:

A thousand ages in Thy sight
Are but an evening gone . . .

Dartmoor is part of a still greater batholith. The same granite mass extends westwards to Bodmin Moor and St. Austell, and thence to Land's End, where it forms a mighty bulwark against the still mightier Atlantic Ocean. It extends beyond the mainland to the Isles of Scilly, where the early daffodils grow. Radiometric dates tell us this had happened by 290 million years ago. Arthur Holmes portrayed the whole mass as if it were a vast plutonic whale whose back broke the surface in a few places. Its shape is rather longer and thinner than that of a typical whale; but you can, if you wish, view the folded sedimentary rocks that surround and overlie the mass as born on the back of a hidden granitic leviathan.

What should have been one of Britain's greatest wild places at Land's End has been turned into a theme park. The drama of land and sea, with waves heaving and sucking at the obdurate cliffs, is played out only a few yards away from a snack bar. Zennor, on the north coast a few miles away, is a much better place to observe the confrontation between erosion and granite. Zennor is a medieval hamlet and has changed little: it still comprises church and farm and few houses. A bowl (granite, of course) is displayed there in which vinegar was placed during the plague years to sterilize money that had to pass between the infected village and the outside world. The pub is called the Tinner's Arms, and there are the relics of mines nearby. The road shies away from the sea along this coast, so you have to walk along an ancient walled path to see it. When I was there the hillsides bore pools of bluebells. The granite plunges sheer to the sea, or else the cliffs are buffered by a zone of huge fallen boulders. Running along the boundary between cliffs and sea there is a blackened zone fifteen metres high—another kind of lichen painting the rock—marking where the winter storms batter the land. The sea is clear enough on a sunny day to see the sandy sea floor, the long legacy of erosion. Jackdaws and stonechats busy themselves among the gorse that covers the steep slopes. Higher still, the granite emerges from the springy sea-turf and yellow-flowered gorse in pallid or greenish crags. Off the coast, outriders of granite form steep-sided sea-stacks that provide a kind of advance guard against the onslaught of the elements. Even on a calm day the waves smash foaming against the columns, greedily seeking out the joints, the weaknesses that lie hidden in the fabric of the rock. The sea seeks to make low everything that has had the temerity to rise above it: it is the great leveller.

Where hot, light granite magma rose deep within the ancient mountain belt that once traversed south-west England, it baked and altered the sedimentary rocks surrounding it. The magma also bodily engulfed chunks of these folded rocks. The zone of thermal alteration around a granite is known as a metamorphic aureole: in Devon and Cornwall this can be as much as four miles wide. You can expect a metamorphic effect upon all the slates of Devonian or Carboniferous age that surround the granite masses. St. Michael's Mount is a little boss of granite off the south coast of Cornwall not far from Penzance. It is spectacular when seen from afar: a conical island with a castle perched on its craggy apex. It is almost too romantic to be true. At high tide it is cut off from the mainland and a ferryman takes you across, to disembark in its own granite-walled harbour. At low tide a causeway is exposed, enabling you, rather more prosaically, to walk back to the mainland. The whole island, with its cobbled harbour backed by pretty houses, has been taken over by the National Trust. The dominating granite castle, seat of the St. Aubyn family, is everything you expect such a building to be, with thick walls, battlements, an old chapel, and a dizzying view out to sea. It seems to grow out of the rock, and indeed it does, for the building had to accommodate itself to the profile of its granite foundations. Life was much tougher here in the past. Where tended and wooded gardens now flank the heights, a dairy herd formerly provided essential milk and cheese and were pastured on bleak hillsides. The castle started as a Benedictine monastery; after becoming a fortified manor it was sacked and retaken several times. Bleakness can be clothed in trees, but granite cannot be modified so readily. Towards the top of intrusions, however, their cusps may fracture. Just below the castle entrance you can see swarms of parallel veins of quartz cutting through the granite, filling such fractures. They make white stripes running over the surface of the rocky steps, worn smooth by hundreds of years of tramping boots. And a few feet from the granite itself, behind the old dairy, there is an exposure of the metamorphosed country rock, the rock into which the granite was intruded—a brown, crumbly outcrop ("hornfels") that could not contrast more obviously with the granite above you on the hillside. If the tide is low you can examine other altered rocks on either side of the causeway. Similar rocks are better displayed around St. Ives, the impossibly picturesque resort and artist's colony on the north-western edge of the Land's End granite. These rocks are spotted with cordierite, a mineral that grew within the slates as they were heated by the granite intrusion, a transformation in situ, an alchemy of the underworld. Twisty quartz veins are abundant, writhing through the rock as pale wisps. Pebbles of the same material can be found in rock pools, white as pigeon's eggs.

All around Cornwall there are old chimneys. They arise improbably out of the sides of gorse-clad river valleys or stand alone on the moors. These mark the sites of mines—"wheals" in Cornish. Many of them were formerly rich in tin, and some were rich in copper as well. Several other metals—silver, tungsten, antimony—have been mined at one time or another. Most scholars believe that the classical Greek name Cassiterides—the "tin islands"—refers to the Scilly Islands and Cornwall: cassiterite, tin oxide, from the same linguistic root, is the common ore of this valuable metal. The Veneti are said to have bartered for tin and sold it on to the Phoenicians. Bronze, of course, is an alloy of copper and tin. Nearly 4000 years ago the Bronze Age inhabitants of Dartmoor eagerly extracted the tin ore from the beds of streams that concentrated it. The smelting of tin ore was high technology at the time. Nor did the practice of looking for reworked tin ore in stream beds die out. The eighteenth-century "streamers" followed the course of former stream beds,* often now buried beneath younger river terraces. But even by Roman times miners had learned to pursue the booty nearer to its source: veins—or lodes—that traversed the country rock surrounding the granite intrusions, or within the upper parts of the intrusion itself. The charmingly named Ding Dong mine in the Land's End Peninsula has been claimed as a Roman survivor. There are dozens of such lodes, and many of them follow faults or other fractures in the country rock, which provided preferential passage for the hot fluids that laid down the valuable minerals. The fluids were derived from the latest stages of the intrusion; the lodes within the granite itself are often found along joints, so it must have already solidified. The majority of lodes are described as "hydrothermal" deposits, a technical term that is unusually self-explanatory ("hot water"). Often they are layered with different mineral products, like a vertical club sandwich. There are almost as many chimney-stacks as there are lodes.

Eighteenth- and early-nineteenth-century wheals have a melancholy, romantic feel in ruin that they doubtless lacked when they were working. Wheal Betsy on the western edge of the Dartmoor granite sits on a hillside on one side of a valley, a steeply rectangular building with gaping windows and a chimney adjacent. It marks an ancient tin site that was redeveloped in 1806 and ran successfully for seventy years. It was originally worked by water power, but in 1868 the present building was erected to house the famous Trevithick Cornish beam engine, which did all the necessary pumping and grinding and crushing. Such engines were the work-horses of the Industrial Revolution. The building somehow looks much older. It is built of the local Culm—slabs of slaty material that make up the country rock in those parts. As always, the top of the chimney-stack is finished off with bricks; I imagine the stone could not be trimmed with sufficient precision. You can still pick up bits of quartz rock containing glinting black ore from the edges of the grassy waste heaps. On the northern Cornish coast, Wheal Coates, near St. Agnes, teeters close to the cliff edge, surrounded by gorse

* These were probably the beds of rivers that received the products of erosion of the granite—and concentrated the tin ore—during the Pleistocene ice age. Dartmoor escaped direct ice cover, but a permafrost climate and exaggerated weathering probably served particularly to enrich these ancient stream beds. A one-ounce gold nugget was once discovered in the Carnon Valley.

and sea-turf. Ridges in the vegetation mark the courses of medieval open-cast workings. It is a challenge to imagine the wheal in 1881, when there were 138 people employed here; the clanking of machinery must have drowned out the sound of the surf breaking against the cliffs far below. The Towanroath pumping station drained a 600-foot shaft—the tin ore was actually chased out under the sea. It is difficult to conceive of worse working conditions: the cramped tunnels, the weight of rock and sea above, looming but unconsidered. There were accidents to contend with, and silicosis. The consolation of religion was of the bleak variety signalled by mean, grey chapels, which can be found on the edges of many villages. The last working tin mine in Cornwall (indeed in Europe), at South Crofty, closed in 1998, so ending a tradition that had endured for two millennia. Who knows if the products of the hot fluids linked with granite will ever be sought again? With the closure of the mines went names that would once have been as familiar to the miners as those of their household pets. Near Basset, there were Theaker's Lode and Paddon's Lode, Doctor's Shaft and Marriot's Shaft, and the Great Flat Lode: all forgotten now, covered by gorse, hardly visible even to the buzzards wheeling above the abandoned workings.

Granite intrusions are found in mountain belts. They, too, follow the bidding of the plates. They appear at the centre of those most dramatic upheavals of the crust, where one plate collides with another: they are the sternest legacy of the tectonic cycle on the continents. The Cornish granite is but a ground-down stump from the great Variscan chain that stretched eastwards through Europe at the end of the Palaeozoic era. From a beginning in the Permian, we know it took nearly 150 million years for the granite to be deeply excavated by erosion. Some of its most characteristic minerals appear in sediments of Cretaceous age lying to the east of Dartmoor. This English pluton is now among the least spectacular of granites, for all its historical resonance. Greater granites lie elsewhere in the British Isles: the mass of the Donegal intrusion in Ireland; or the Cairngorms in the Highlands of Scotland, a granitic heart within the Caledonides. Carved by the glaciers of the Pleistocene ice age, the Cairngorms are still rawly gouged: it will take millions of years to reduce these mountains to domestic proportions. Glaciated granites make fearsome vertical rock faces: Mont Blanc in the Alps, or the sheerest cliffs of El Capitan in the Yosemite National Park, California. Climbers delight in the meagre cracks offered by such rock faces; those who suffer from vertigo can only wince at the mere thought of tackling them. Doubtless, new challenges still await brave or foolhardy alpinists on granite peaks in the Himalayas. There are granites in the heart of the Appalachians, and we have met them already in their northward continuation, the "barrens" of Newfoundland, whose name so well describes the scenery they produce. The Tatra Mountains of Slovakia are "barrens" in the heart of Europe. The greatest series of intrusions of them all forms the core of the Andes. Massive granites extend discontinuously along the heights of the mountain chain all the way from Tierra del Fuego almost to Panama; they cover something like 465,000 square kilometres. The batholiths of western North America are hardly less considerable: in Alaska, British Columbia, Idaho, the mighty Sierra Nevada, the Peninsular Ranges of California, granites comprise several millions of cubic kilometres of the most intransigent rock, here of dominantly Cretaceous age. These granite ranges parallel the confrontation of the Pacific Plate with the western continental Americas. On the opposite side of South America the Sugarloaf at Rio de Janiero, Brazil, rises like a massive, stuck-up thumb from the eroded gneisses that surround it: the mountain looks almost organic, as if it had grown upwards like a monstrous termite mound, yet it, too, is the product of granite's obstinacy in the face of the elements. Go to the other side of the world, and Mount Kinabalu, on Borneo, is a bald eminence rising from lush and dangerous jungle; it is flanked by cliffs 1000 metres high, broken only by the occasional gulley. At its summit it carries a few jagged peaks and weird, lobed, weathered protuberances, like malignant growths. Kinabalu is a surreal island of bareness amidst a green sea of tropical profusion. The fiercest, strangest, least tractable features on the face of the earth belong to granite.

But even these proud peaks will be humbled in time. There are ancient granites in the oldest parts of the earth that have been brought low. In central Africa, they form low swells, exfoliated layer by layer as if the granite were the sloughed skin of the earth itself. Tropical weathering peels off the outer layers of granite blocks as if they were onion skins. The granite is reduced to heaps of reddish rounded boulders. These often provide the only feature to relieve the endless savannah, the last remnants of mountains fifty times older than the Alps.

The durability of granite is a gift that has served to preserve the monuments of ancient human cultures. Wherever it has been used as a building stone, the structures survive—whether in the architecture of the Near East, Egypt, and South Africa, or in the pyramids of the Mayan civilization in Mexico, or the temples of Nepal. Steles made of it endure. Granite has outlasted the pomp of despots and recorded the faces of gods whose names have been forgotten. Its hardness and coarseness impose limits on its tractability for sculpture. The "rose syenite"* of Syene—the Greek name for Aswan—is famous for its use in the sculptures of Pharaonic Egypt at Luxor and was also used in the construction of the obelisks at Karnak. The material itself imposes a certain gentleness of contour and limits the detail that can be portrayed. In some respects this is a virtue, for it obliges the artist to consider essentials of form. The grotesque visages of Mayan deities are the more terrifying because the sculptor has been compelled to simplify and formalize. Thus sculpture is constrained by geology. Much more detail can be imparted to the limestones and sandstones used by medieval sculptors of Romanesque churches and cathedrals for gargoyles and leaf tracery, but time has served most of them badly. Noses crumble, roses rot. The Sphinx is proof of the same deficiency in the Nile Valley. Marble—metamorphosed limestone—serves the virtuoso chiseller best, especially the white marble of Carrara, which can record every fold of a garment or the delicate swell of the least muscle on a torso. Greek and Roman sculptors sought marble out. How much, one wonders, would Michelangelo have achieved had he been obliged to use only granite?

Abraham Gottlob Werner, Professor at the Mining Academy at Freiberg in Germany at the end of the eighteenth century, taught that granites were part of the *Urgebirge*, the first deposits precipitated from a primeval sea that once covered the earth. Younger rocks, he proclaimed, including all the ones we would now recognize as sedimentary, overlay them. Granites played a part in disproving such theories, not least because they could be seen to *cut through* sedimentary and metamorphic formations—and must, therefore, be younger than the country rock they intruded. Here is how the pioneer experimentalist James Hall of Edinburgh described intrusions in 1790, not many years after James Hutton had outlined his ideas about their fiery origin: "Wherever the junction of the granite with the schists was visible, veins of the former . . . were to be seen running into the latter and pervading it in all directions, so as to put it beyond all doubt, that the granite in these veins, and consequently of the great body itself . . . must have flowed in a soft or liquid state into its present position." By the time Charles Lyell made his

* There are many varieties of granite, with different names according to the detailed mineralogy of the rock. Syenite is another coarse-grained igneous rock that lacks the abundant free quartz found in the typical granite. Confusingly, the "rose syenite" is a granite, despite the name.

journey around the Bay of Naples the deep igneous origin of granites was established. But questions remained: why were they associated with mountain belts, ancient and modern? And where did the magma from which they crystallized originate? How did they arrive in their present position?

Eduard Suess appreciated the significance of granites in mountain belts: he had seen splendid examples in Switzerland, where he had learned his geological trade. He recognized that they were not, as it were, bottomless, but were intruded as masses into the deep crust late in the mountain-building cycle. He wrote that "they lie imbedded in older stratified rocks; their form is of large irregular loaves or cakes." He was right: geophysical evidence can now be used to "map" the bases of intrusions. Suess also recognized that in places he could see how overlying strata were baked and metamorphosed by the granite, which is "therefore younger than the overlying strata." But he also went on to say: "It is absolutely necessary that the injection of a granitic mass, possessing so high a temperature as to be capable of altering the surrounding rocks, *should be preceded by the formation of a corresponding cavity.*" As usual, Suess' italics indicate that this was a particularly important point and show that he believed strongly that the granite was intruded into a kind of welcoming space that opened up before it.

It is now timely to return briefly to Newfoundland, to recall some of the rocks in the central Mobile Belt. This is part of an ancient mountain belt, the Appalachian chain, stripped down to its innards by millions of years of erosion. There were granites there, as there should be. But there were also rocks—migmatites—in which pods and folded veins of granite intermingled with banded gneisses, which are high-grade metamorphic rocks. It was confusing: some of the rocks cropping out among the stunted conifers seemed more granite than gneiss, with darker wisps alone betraying a kind of gneissic memory, but in other places veins of true granite cut through what was indubitably banded gneiss. There was a transformation happening in front of you. Gneiss seemed to "dissolve" before your eyes and merge into granite. Then, magma generated by this process in turn squeezed out into adjacent gneisses—or so it seemed. This area of central Newfoundland was evidently exposed to erosion so deep that it penetrated into the tectonic crucible where granite is born. Exactly how this birth happened proved to be controversial.

In both editions of Arthur Holmes' textbook there is reference to a process he termed "granitisation." This is the conversion of gneisses and other rocks into granite in the solid state and in situ by a kind of wave of transformation. He, too, recognized the importance of deep erosion into

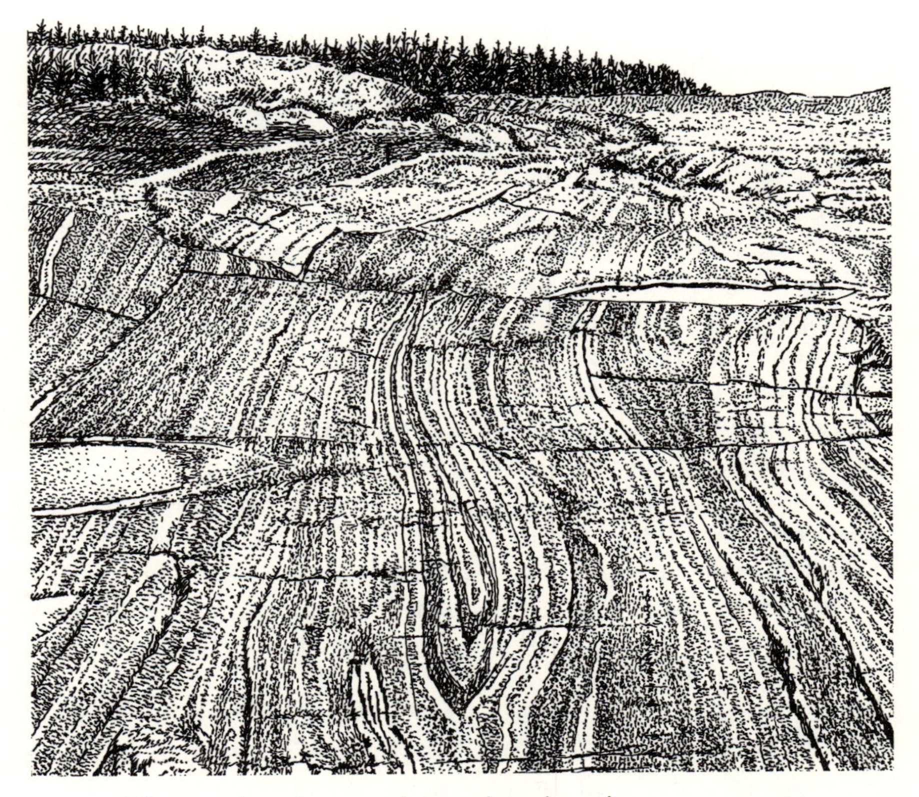

When granite veins merge into gneiss: migmatites on a wave cut a platform in Precambrian rocks on the Baltic Shield.

the interior of the crust, where "we can see the granitised rocks of former ages and the crystalline rocks associated with them, all arrested at the particular stages they happened to have reached when the processes of metamorphism and granitization ceased to operate"; or, to put it simply: rocks frozen in time and trauma. The transformation of other rocks into granite had been suggested by J. J. Sederholm, Director of the Geological Survey of Finland, who had studied appropriately ancient Precambrian rocks of the Baltic Shield. He had proposed that the changes happened "as if by magic," as he said, when a transforming agent he termed "ichor" circulated through the rocks. To invoke the name of the ethereal fluid that flowed like blood through the veins of the gods was a suitable way of signalling the mysterious properties of granite generation. Holmes opted for the less classical though equally insubstantial term "emanation" to describe what had been modified to "rock-transforming gaseous solutions" in his textbooks. Moving fronts of these "emanations" migrated through the rocks deep within mountain belts effecting their wondrous work, transforming gneisses and schists into what would eventually become the substance of unscalable cliffs and bleak uplands. The more extreme granitisers thought that all this could happen without the necessity of invoking magma at all.

Holmes adopted these transubstantiationalist views on the origin of granites because they were also those of his strong-willed second wife, Doris Reynolds. In 1931, while Holmes was Professor of Geology at Durham University, he met Doris on a geological field trip to the Scottish volcanic island of Ardnamurchan. Before two years had elapsed, Holmes had secured Doris a job in his own geology department, sitting across from him on the other side of the same large desk. An affair of this kind would scarcely be worth remarking in a modern university department, but in the 1930s it was unusual enough to be something of a scandal, the more remarkable because Arthur Holmes was, by nature, unusually levelheaded. Holmes and Reynolds were married after his first wife's premature death in 1938. Doris Reynolds had to be a tough-minded and determined woman to prosper in what was in those days an overwhelmingly male profession (which, regrettably, it still is). You disagreed with her at your peril. She became an ardent, even extreme, adherent of granitisation, and Holmes dutifully followed suit.

There was, however, an opposing school, headed by an outspoken Canadian petrologist, Norman Bowen. This group has been labelled the "magmatists," which is self-explanatory. They claimed that granite had indeed solidified from a liquid magma, and that magma had been generated by the partial melting of gneiss under sufficiently high temperatures at sufficient depth in the crust. Like the granitisers, they could point to field evidence to support their point of view—it was just different evidence. Places like Newfoundland could be used to support both views, depending on which bit of outcrop you examined. The debate between the two schools soon heated up to magmatic temperatures. One of the problems the granitisers had to account for was the loss of certain elements, especially calcium, iron, and magnesium, in the transformation of a typical gneiss into a typical granite. These are the elements characteristic of "basic" rocks (as opposed to "acidic" ones, like granite). The granitisers said that these elements migrated away from the freshly transmogrified granite in what they termed a "basic front"—a kind of fugitive chemical miasma spreading away from the growing granite behind. Bowen described this idea as, on the contrary, "a basic *af*front to the geologic fraternity"—not a great pun, admittedly, and with a nasty dig, intentional or not, at the sex of his principal opponent. Reynolds returned that Bowen was "accusing Nature of an intentional break of politeness." Bowen's friends in the United Kingdom included Professor H. H. Read, who became Reynolds' implacable opponent. He speculated that the "basic" elements might, after all, remain behind as a kind of residue as granite generation proceeded, and in 1951 he added punning insult to injury by writing, "I suggest for discussion that some basic fronts may be better interpreted as *Basic Behinds*—a somewhat indelicate term, I admit, but one which expresses the possibility that we may here be dealing with subtraction rocks [i.e. gneiss minus 'basics' = granite]." Doris Reynolds coped with this kind of antagonism in spirited fashion when the Granite Controversy, as it came to be known, was played out in the scientific meeting rooms of London. As she wrote, she bought herself "one of those high hats like a witch. I thought that if I kept that on at the meeting I could not be overlooked." Attitudes had hardened, like granite itself. As Seamus Heaney said, *You can take me or leave me.*

I cannot report that Reynolds triumphed in the battle between miasma and magma. The crucial work in settling the issue was the outcome of improvements in the design of experimental equipment, notably, O. F. Tuttle's cold seal pressure-vessel. The mineral ingredients of rocks could be "cooked" under pressure; it should then be possible to see under what conditions they melted—became a magma, if you like. Bowen teamed up with Tuttle to study how a "granitic mixture" of feldspars and quartz behaved under conditions of temperature and pressure like those in the deep crust. Previous experiments under low pressure had only produced a thick, sticky magma at excessively high temperatures. The crucial added ingredient was water. When the mixture was heated with the addition of the aqueous phase, the temperatures and pressures at which a granite melt resulted were lowered considerably, quite consistent with a presumed origin deep within mountain belts. It was shown that natural granite magma should be the first to "melt out"* under the ambient conditions at depth if there were

* The partial melting of the original rock is known as anatexis. There is actually a set of different pressure-temperature conditions under which melting will occur—for example higher pressure might require lower temperature. These stability fields are conventionally plotted onto pressure/temperature (P/T) graphs.

pore water present. Since water is well-nigh ubiquitous in the crust, this was not an unreasonable assumption. To put it the other way round, if rocks such as gneisses were progressively subjected to higher temperatures and pressures—and this might happen as the crust thickened where two continents collided—the first product that would "melt out" as magma would have the composition of a granite. Granite magma sweats out in the profound depths of the underworld. The swirling migmatites of Newfoundland, like those that Sederholm had studied in Scandinavia, were granites in the process of their birth, "frozen" just as Holmes had described it. Like many births, it was a messy business; hence the bewildering arrays of texture and detail—some rocks were already nearly granite, or they were gneisses that had been invaded by granite sweated out elsewhere; other rocks again were neither one thing nor the other. Deep flux flourished.

By 1965, the outlines of the process of granite magma generation were clear. Progressively refined experiments in partially melting rocks continue to this day. As we have seen before in this history of the earth, few issues of geological truth are ever clearly black and white. Doris Reynolds may have been wrong about her mysterious transforming emanations, but much recent research has been focused on the details of how elements move through rocks at the sub-microscopic scale. In a sense, the granitising spirit lives on. As for Arthur Holmes, his association with the discredited theory did his reputation no good at all in the latter phase of his career. This was hardly just, since he was a great geologist with many other achievements to his credit.

It should be added here that granite magma can also be generated from much more basic, oceanic-type lava. This can occur by a continuation of the process of evolution within an oceanic magma chamber, as described previously, whereby the heavier, basic (and darker) minerals progressively crystallize out and "push" the remaining liquor more and more towards granitic composition. It is rather like making applejack: fermented cider is buried in the freezing ground and as the ice crystallizes out little by little the remaining liquor becomes enriched in alcohol. This process of refinement and magma generation may occur in relation to the subduction of an oceanic slab. The final granite rock may look similar in the hand-specimen to its counterpart generated by the partial melting of gneissose, continental rocks, but its deeper chemistry betrays its origins. The isotope ratios of rare elements (specifically, strontium $^{87}Sr/^{86}Sr$) are not altered by the refinement process and give different signatures for the granites with continental, as opposed to oceanic, origin. This method proved that the large batholiths of California were the product of partial melting of the continental crust—a scientific point that had been in dispute in a place where continent and ocean approach one another so closely. The greatest granites are mostly of this continental kind.

What, now, of Suess' contention that granites must fill in cavities already present for the reception of magma? Clearly, granites like those in Devon and Cornwall had intruded as a mass—and one that was still hot enough to bake surrounding strata in the country rock and give rise to all those lodes filled with hydrothermal ores. The magma also engulfed lumps of the rocks through which they intruded. It seems evident that huge quantities of magma, once it had been sweated out at great depth, moved upwards through the crust. Detailed studies of large batholiths have shown that there were often several pulses of granitic magma. The reason for the upward movement turns out to be rather simple. The hot granitic magma is less dense than the other crustal rocks through which it passes: it rises through heat and sheer lack of weight. Far from entering ready-made cavities, it insinuates itself into the emerging mountain belt. The process resembles what happens when a lava lamp—that iconic 1960s objet d'art—is switched on. A heat source at the base of the lamp sufficiently alters the density of an oil layer to make it rise as a weird plume through the overlying liquor; all the materials are rather viscous at this stage, so the process takes some time. For the rise of the granite magma, millions of years must be allowed. Nonetheless, the shapes of such "diapirs" can be almost as remarkable as those cooked up in the lava lamp. Now you will understand the precipitous oddity of the Sugarloaf at Rio de Janiero.

Like the lava lamp, the generation of granite demands a source of energy. Granites do not rise up from the depths to disrupt the chalklands of England, the Paris Basin, or the Texas plains. These are all areas underlain—at least near the surface—by undisturbed sedimentary rocks. Granite eschews the quiet, uneventful regions of the earth. For the source of heat is subduction: granites are the sweat of earth movements. Whether along the Pacific coast of the Americas, or through the Alpine to Himalayan mountain chains, or in the ancient Caledonides, or in Suess' "Hercynides," granites rise to dissipate the energy that plates produce by their marginal friction. Particularly where continent collides with continent, the crust is thickened enormously. Nappe piles upon nappe. The concatenation of sediments and volcanic rocks yields a deformed and packed pile sandwiched between approaching continents. The pile may bulge downwards deeply towards the mantle to form what have been termed "roots" beneath the growing mountain chain, the image perhaps being more appropriate to dentistry than to a tree: mountainous molars have downward extensions into the tissues of the earth. While this happens the temperature gradient can be depressed for a long while, pushed downwards along with the folded sedimentary pile to a region of high pressure at depth. But, given time, the heat flow from the interior of the earth will be restored. The isotherms will bounce back (though "bounce" is probably too enthusiastic a word for the leisurely progress of geological time). Then, the deep alchemy can proceed further, as additional heat sweats out the granite magma from the already metamorphosed rocks, ready to mass up and rise as plutons through the developing mountain chain. Naturally, this rising body will cut across the earlier folding and distortion that accompanied the thickening of the tectonic pile. Granites are a consequence, a spin-off, a late expression of tectonism and subduction. Given the chemistry of rocks, the physics of their melting, and the temperatures and pressures involved in tectonism at plate boundaries, granites are almost a logical conclusion to a dialectic of the earth. They arise inevitably in mountain belts. Once emplaced, they may themselves be subject to late phases of orogenic activity: sliced or thrust, or even squeezed so hard as to be metamorphosed anew. Nothing, not even this proud rock, is exempt from the mill of the earth.

Both Suess and Holmes realized that the thickened, "rooted" mass of a mountain chain would ineluctably present itself to the blast of erosion. Such a concentration of light continental rocks as is found in mountain belts would be compelled to rise as a result of isostatic readjustment. As it did so, the resulting mountains have to enter the realm of ice and wind, which diligently do their elemental duty to render the mountains low again. The more the weather slaves away and succeeds in removing rock, the more further isostatic adjustment will successively deliver the depths into the maw of the storm. Deep in the underworld, this flexure of the lithosphere is accommodated by the creeping flow of the underlying plastic rocks of the mantle, which moves as a viscous fluid. I have tried various homely metaphors, like depressing a big crouton on a bowl of pea soup, or plastic ducks in hot baths, to try to make this process more transparent; but the conditions at depth are so different from those in saucepan and bathroom that a bald statement of what goes on will have to suffice. As erosion proceeds, the granite's turn for exposure will eventually arrive, its rounded bulk exhumed from its seat in the folded pile. This rock will certainly give erosion something to chew on, but frost is eventually able to shatter it, mercilessly prising it open along its joints. If the granite forms a

Deeply weathered feldspars in granite is the source of china clay. Cornish clay pit in the early years of the twentieth century.

high redoubt, glaciers will grow upon it. As the ice sheet creeps downhill it will use boulders of granite entrained within it to scour and gouge the surface beneath it, forcing the rock to conspire in its own destruction. In the end, the forces of erosion will win: they always do. Quartz sand will pass onward into the next cycle of the earth. The ground-down granite will end up as no more than a gentle slope topped by remarkable boulders, standing as witness to their strange journey from deep in the earth's crust.

Where granite has rotted deeply without eroding away, the feldspars may be converted to deposits of the white clay mineral known as kaolinite. This is the china clay used for making fine porcelain. All around the Cornish town of St. Austell there are what look like snowy terraces, or perhaps small white volcanoes. They look unnatural; from afar you might think they were made of piled salt. They are in fact the waste heaps from the china-clay pits. The feldspars in the granite of this area have been converted wholesale to white clay. A number of the great chasms from which the clay was extracted remain. One is now the site of the imaginative ecological experiment, the Eden Project, which has reproduced various natural ecosystems inside huge geodesic domes. There is something rather wonderful about growing a tropical jungle in an old clay pit: it is boldly traducing the poverty of granite soils. The clay, though, was always productive in other ways. The very best plates and vases were manufactured from this startlingly white clay. Josiah Spode developed the techniques for making bone china in the mid eighteenth century, but the fine goods were not made on the spot. Instead, the clay was transported all the way to the Potteries, the region around Stoke-on-Trent where the great names in ceramics had their factories. The trade was a stimulus for the construction of canals, which could move great loads, cheaply. Wedgwood, Spode, and the Mintons were enthusiastic experimenters, always coming up with new pottery recipes and new glazes. The delicate blue-and-white ware manufactured by Spode in the early 1800s show rural scenes, or illustrate fables, or reproduce an embellished Orient, with feathery trees and borders of frothy foliage. It is ironic that the exquisite height of refinement in ceramics was grounded ultimately in granite, that coarse and implacable rock, whose natural sculptural expression is the tough stone cross that faces out the centuries in the bleak wastes of the Great Moor.

9

Fault Lines

Some people love Los Angeles. They love its endlessness, the way one conurbation merges with another along freeways that promise more liberation than they ever deliver. They love the sidewalks lined so regularly with teetering palm trees like overgrown fly-whisks. They love the unbridled eclecticism of the prosperous villas, where Spanish haciendas combine with Corinthian columns, and maybe a soupçon of Elizabeth mock-Tudor, all flanked by implausibly verdant, perpetually irrigated lawns of coarse grass, upon which citrus trees seem always to be either in flower or in fruit. Downtown L.A. may be like any other towering downtown of glass and steel, but away from the commercial centre there are places that are unlike anywhere else. Venice Beach is to narcissism what Rome is to the priesthood. It's a place of impeccable pectorals and toned tans, but it is also a town with one of the best bookshops I have ever visited, where it is possible to sip a leisurely cappuccino and discover the meaning of "chill out." In older towns like Riverside there are still distinctive hotels, Art Deco villas, and orange groves that have escaped redevelopment, that allow you to understand how enticing this part of California must have been before everything became joined together, and the drive-in mall homogenized everyday commerce.

California has a reputation for being laid-back, and the warm climate does indeed produce a kind of dreaminess in the visitor. The smog-filled air backing up on the built-up coastal plain against the San Gabriel Mountains contributes to an unreal, impressionistic haze. A little more experience quickly proves that this is where a Mediterranean climate mingles with the North American work ethic, and the people that I know in L.A. are every bit as frenetic as New Yorkers, only in Hawaiian shirts. It seems entirely appropriate that the microchip revolution was nurtured here. I was slightly disappointed to find that Häagen-Dazs ice cream was manufactured in a huge factory in greater L.A. I had naïvely imagined that it was mixed by Flemish maidens in cowsheds. Movies, microchips, and chocolate fudge sundae are all made on a global scale here. When I paid a working visit to UCLA, on its incomparably well-furnished campus, it was humiliating to find that I was always the last person at the desk every morning. I may well have been the first to leave every evening. The art of being Californian, it seems, is to cultivate a loose-limbed insouciance while secretly working away like a frantic ant.

This curious combination of devil-may-care and industry may be a response to living in the shadow of doom. California is one of the least stable parts of the earth. This is the state with the shakes. All that beavering away in the pursuit of fame and the dollar could be regarded as diverting attention from tectonic truth. There is no S. Gennaro to intercede, as there is in Naples. There is only a kind of collective amnesia, produced by keeping your eyes down and your pockets full. Suddenly the ground might start to heave, and the porticoes of the most extravagant villas will come tumbling down, the sprinklers on the lawns will dry up, and automobiles on the freeway will be tossed about like dry beans on a sieve. This will happen when there is an earthquake; and an earthquake will happen when there is sudden movement on a geological fault.

If you had to ask the man in the street to name just one geological structure, he would probably say: the San Andreas Fault. Indeed, in the minds of some people, it is almost synonymous with geology. There are other great faults—like the Anatolian Fault that has wrought periodic havoc in Turkey—but it is the San Andreas that has somehow entered common consciousness as symbol of the power of the earth: the great shaker.

Faults are breaks in the earth's crust. Many of them are brittle fractures, which happen when rocks snap. Most rigid materials break if they are strained beyond endurance. Bones break if they are bent in the "wrong" direction or if they are forced to carry more than they can bear. Even steel girders will fracture given enough goading, or if some microscopic weakness lurks within them. Why should the earth itself be any exception? For common materials, the strain that has to accumulate before breakage can be measured by experiment: much modern architecture depends on precise knowledge of what a load-bearer can carry without fracturing. When a

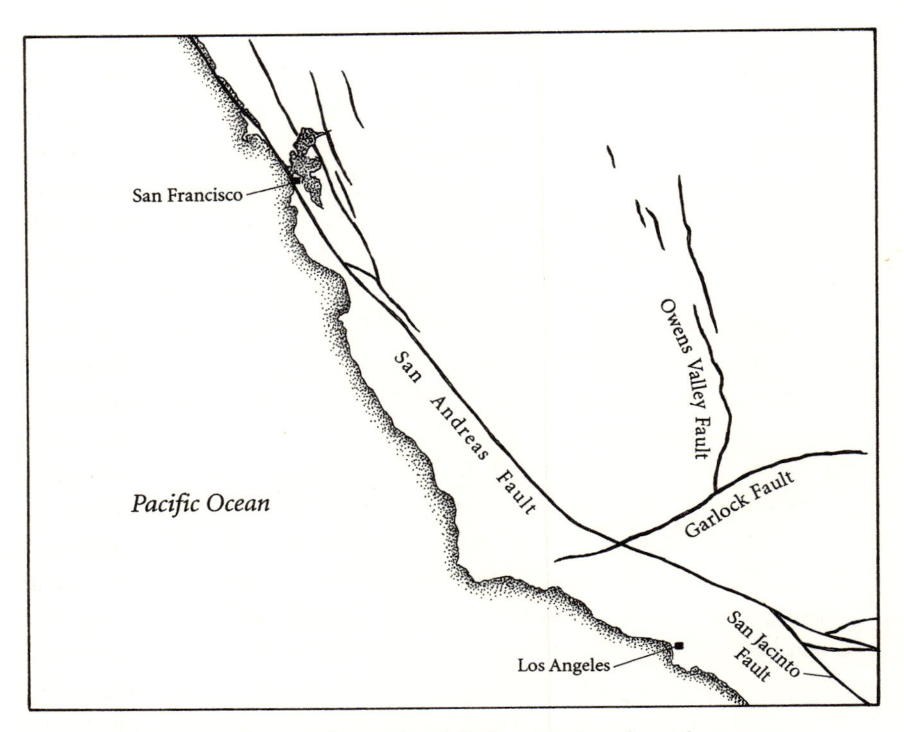

The San Andreas and associated faults running along the western seaboard of North America.

break occurs, the material is said to have "failed." So the language of tectonics has resonance with our own character, our own faults, our own failures. Faults run deep. "They say best men are moulded out of faults," as Shakespeare put it in *Measure for Measure:* every man his own planet, with weaknesses ingrained. The possibility of failure always lurks when pressure is exerted upon our nature, which is when we may be sure our faults will find us out. Perhaps more than any other geological process in this intimate history of the earth it is faulting that we understand at the most visceral level, for we all know of failure, and most of us acknowledge the flaws that cut through our own thin crust.

The San Andreas Fault is just one of several faults that make up a complex of potential catastrophes, the flagship of a fleet of faults that run close to the western edge of North America. Most of the faults trend more or less parallel to the coast. In places, maps of interweaving faults look more like a braided mesh than the single, deep cut of our imagination. Los

Angeles lies off the main fault, but is vulnerable to movement on others. The whole system is 1280 kilometres (800 miles) long—about the same length as the entire British Isles. It is one of the great fractures on our planet. In places, the fault zone is a mile wide. You can see it from space, for many features line up along it. From far above, this scarp lines up with that depression, and the whole picture is gradually revealed. Lakes tend to "pond back" along the fault zone, both because of its weakness and because different kinds of rocks are brought into juxtaposition on either side of the fracture. Bays and valleys follow it. Rarely is it displayed as an open crack, an explicit fissure, but it can be seen clearly, running straight as a die, as if scribed out by a titanic burin, in arid areas such as the Carrizo Plain in central California. The San Andreas Fault, *sensu stricto,* runs from northern California and San Francisco Bay, southwards to Cajon Pass, near San Bernadino. There, it splits into a number of branches, including the San Jacinto and Banning Faults.* Faults, like football teams, often take their names from towns, although the San Andreas itself is named after a lake of that title in the San Francisco area, along which the main fault runs. New and active side branches are still being discovered, usually when a sudden earthquake reveals what has until then been concealed beneath a cover of rocks or soil, quiet as you like. It is a messy business, this slippage of the earth, and liable to spring a surprise in a vulnerable place.

The rocks along the San Andreas move one way. On the seaward side of California they are sliding past the rocks on the coastal range side and moving northwards. Imagine if you were able to stand and watch the movement from the western side of the fault. You would see the rocks on the far side of the fault slipping away to your right—hence the fault is properly called dextral. If you average out all the movement in the long term, it amounts to only about thirty millimetres a year in some segments. You might find it rather boring watching the earth move, if it were continuous. Of course, the segments of crust do not slide past one another in a stately fashion. The reason that there are earthquakes at all is that the crust moves in jerks; it sticks, it judders, it jams, until it is compelled to give way. Faults are not smooth, well-oiled affairs, nor is movement distributed evenly along them all within a system. Strain builds up in a particular segment until failure occurs. Then this one segment may shift dramatically,

* The north-easternmost branch is still called the San Andreas in this area.

The aftermath of the 1906 San Francisco earthquake as fire engulfs the town.

letting out energy in destructive waves. The longer the period of quiescence on an active fault, the more worrying it is to seismologists. The prelude to seismic cacophony is a silence longer than it ought to be. When the fault finally "gives," the displacement direction becomes as obvious as if there were a signpost marking it: "➡ Crust This Way." Streams are

diverted. All along the fault those running off the California mountain-sides towards the sea show kinks in the course of their little valleys. After the 1906 San Francisco earthquake, a road in Tomales Bay was offset by nearly 6.4 metres (21 feet). The strain of a century or more was dissipated in a dramatic jump. The same event caused destruction in the city on a massive scale. Ironically, the earthquake generated data that helped scientific understanding of the very forces of destruction. It struck at 5:13 a.m. on 18 April. Many of the dignified Victorian houses were not constructed with earthquakes in mind and collapsed. To a geologist briefed in disaster, it is surprising that there are so many of the old houses still standing today. Mature trees were uprooted, and even tossed entirely root-over-crown in places close to the fault line. It has been estimated that some 375,000 square miles were directly affected by the seismicity, half of it on land. Hardly had the dust settled than human faults began to move the tectonics of character. There was plundering. Mayor Schwartz was forced to act. Police and army units enforced order. But the Mayor's problems were just beginning. At 8:14 p.m. there was an aftershock—the rocks were "settling" after the main event. But the shock was sufficient to bring down buildings already weakened earlier: thousands panicked. Then the real killer started: fire. Conflagration spread through the city. Over the next few days, magnificent hotels crumbled into ashes, and neighbourhoods were destroyed. At one point there was a mass rescue by sea that was probably the greatest such exercise before the evacuation of British troops at Dunkirk in the Second World War. At the final account, the death toll stood at some 3000 souls.

The San Andreas and associated faults will not calm down. The system has been operating at least since the Miocene, for some 20 million years. The fault was already old when the first primates came down from the trees to sample terrestrial life. As I write these words, I have checked the U.S. Geological Survey Web site: it reports that an earthquake of intensity 5.2 on the Richter scale struck Gilroy, California, on 14 May 2002. There were no casualties.

The intensity of earthquakes on the Richter scale is not like temperature measured on the Celsius scale. One degree up is not one degree quakier. It is a logarithmic scale, based upon the energy released during the event, which means that one notch up the scale represents a considerable increase—a magnitude 7 earthquake releases thirty times more energy than a magnitude 6, for example. Charles Richter was a geologist who developed his famous scale in 1935 based on the movements of the San Andreas Fault, so the language of seismology was born along the Pacific coast. The scale is calibrated precisely with modern instruments, but it has also entered the common vocabulary. The San Francisco event of 1906 has been estimated as 8.3 on the Richter scale.

I have a sneaking regard also for the Mercalli scale, which describes earthquake intensity in terms of what is felt on the ground. Arthur Holmes outlined a twelve-notch scale of this kind (using Roman numerals), which has appealing features. Here are some samples:

> Intensity II. *Feeble:* noticed only by sensitive people . . . Intensity V. *Rather Strong:* felt generally; sleepers are wakened and bells ring . . . Intensity VII. *Very Strong:* general alarm; walls crack; plaster falls . . . Intensity VIII. *Destructive:* car drivers seriously disturbed; masonry fissured; chimneys fall; poorly constructed buildings damaged . . . Intensity XII. *Catastrophic:* total destruction; objects thrown into air; ground rises and falls in waves.

Imagine the scene: a Californian ranch-style hacienda with Ionian portico; Myron (a seismologist) vaguely notices something. "Can it be an earthquake Intensity II?" he muses. "Must call Christabel, she's *so* sensitive . . ." The tingling increases. "Uh, oh, it's stronger than that . . . in fact, is that a bell I hear ringing? . . . it must be a V!" Almost immediately, an Ionic column collapses: "Oh my God! I feel generally alarmed and my plaster's falling! We have a VII!" A wild-eyed motorist staggers into the front yard . . . "Heavens to betsy! It's a seriously disturbed car driver! We've moved to an VIII" . . . "Aargh!" (Myron is thrown into the air . . . it was a XII).

I have seen roadcuts slicing through the fault zone. The rock that is thrown up is hard to characterize—just a tan-coloured mash of little pieces, with an occasional larger fragment. This is the debris of the grind of plate against plate, a kind of tectonic hamburger meat known as fault gouge. Geophysical evidence proves that the San Andreas Fault extends at least 16 kilometres (10 miles) down into the crust. It becomes easier to understand how the great fault might "stick" long enough to build up the kind of strain that would topple the Winchester Hotel in San Francisco when it was suddenly released.

In order to understand what is going on under the sunny coast of the western United States, we have to look globally. Plate meets plate at the edge of Asia on the opposite side of the Pacific Ocean, but here it is along oceanic subduction zones. California is very different. To appreciate what is happening, we have to look beyond the hedonistic shores of Venice Beach—out and under the sea, southwards to the Pacific sea floor adjacent to Central and South America. From quite early days in modern oceanographic exploration, a range-like, relatively shallow region was recognized here rising above the general level of the sea floor. This was named the East Pacific Rise. Later, this structure was identified as a spreading ridge, and was mapped as a more or less linear structure for several thousand kilometres. Its smooth course was offset by transform faults* that shuffle its crest step-wise into a dozen or so chunks. But it is not a typical mid-ocean ridge, not least because it is nowhere near the centre of the Pacific Ocean. Instead, its course heads towards, and under, North America, passing into the continent along that most curious of long, narrow seas, the Gulf of California, which separates the peninsula of Baja California from the Mexican mainland. The conclusion can be drawn that the East Pacific Rise has "disappeared" beneath the North American continent. A whole chunk of the eastern Pacific has been tucked away beneath the great mass of what is now the United States. To put it another way, the rate of production of new crust at the ridge centre was exceeded by the rate of consumption of oceanic crust at the western edge of the great continent. The result was that the ridge was "sucked" towards America, and then beneath it. In Oligocene times, 30 million years ago, there was still a continuous East Pacific Rise off the coast of California. Three million years later it had partly disappeared under North America. The subsequent shuffling of the continental and oceanic blocks set up the northward rotation of the western coast that still happens today. The San Andreas was born. Proof of this long-term movement was discovered in fossils of sea-shells occurring to the west of the San Andreas Fault. If this seaward segment had indeed been shuffling northwards for a sufficiently long time, it should carry with it a cargo of fossil clamshells and snails—the remains of animals that would have originally lived in the subtropics but which were transported tectonically all the way to cold northerly climes. These fossil "strangers" were indeed discovered by a palaeontologist applying his hammer to ancient marine sedimentary rocks. The beauty of such theories of the earth is that all branches of science have their part to play in unravelling the sequence of movements of the crust. Charles Lyell would have rejoiced in the solution.

The next place to go is downwards. The latest methods of analysis of earthquake waves create a three-dimensional picture of "slices" of the earth's crust in horizontal sections at various levels, much as CAT scans can image the interior of the body. This method is known as seismic tomogra-

* We met this kind of fault in Chapter 4, where blocks of oceanic crust slide past one another during ocean-floor spreading, an altogether gentler process than the kind of transcurrent faulting represented by the San Andreas, where plate grinds against plate.

phy. There is an ongoing project in the San Francisco Bay area to understand the deeper connections down to ten kilometres or so into the earth. It carries the happy acronym of BASIX (Bay Area Seismic Imaging Xperiment). The 3-D models developed show the plunging oceanic lithosphere underlying the northward-moving part of the coast. The different faults in the area show up as vertical planes cutting emphatically through the crust. But, more surprisingly, they seem to connect with one another by way of a deep, nearly horizontal zone of weakness—a "detachment zone"—where the North American Plate and Pacific Plate interact at depth. When Pacific push comes to American shove, all those diligent Californians are evidently dancing to a deeper tune.

All such advances in knowledge help towards improved prediction of when the next earthquake will strike. This was once regarded as an extraordinarily difficult business, as much art as science. Sudden drops in the level of water in wells might indicate movements along faults at depth, and, as in Mercalli's scale II, some animals might be sufficiently sensitive to notice early signs of which humans might be unaware. Today, early seismic signals that indicate when strain on a particular fault is reaching a critical level can be distinguished from all the other "noise" produced by our creaking earth. Predictions have improved so remarkably that the Gilroy earthquake of May 2002, briefly mentioned above, was anticipated in an article published in the *Proceedings of the National Academy of Sciences* by John Rundle and Kristy Tiampo of the University of Colorado's Cooperative Institute for Research in Environmental Sciences (CIRES). The article was published on 19 February—a clear example of being wise *before* the event.

Similarly, building technology has risen to the challenges of accommodating violent shaking without falling down. Shock absorbers have been designed that rest whole buildings on giant pads of rubber and steel. Webs and struts of steel reinforce concrete slabs to the same effect. Even so, the world was shocked by pictures of collapsed, though reinforced, concrete overpasses after the Kobe earthquake of 17 January 1995, in which more than 5000 people died. This devastating event in Japan was exactly one year after an earthquake in Northridge, California, one of the suburbs of greater Los Angeles, lying on a hitherto unremarkable branch of the San Andreas Fault. Roads collapsed there, also; sixty people died. If an earthquake is violent enough, human ingenuity counts for little. Nonetheless, engineers are confident enough to build high towers within earthquake zones. It is impossible to gainsay the confidence of the TransAmerica Pyramid in the middle of downtown San Francisco, an 853-foot needle piercing the skyline, peerless among its towering neighbours; it is earthquake-proof—or so it is said. Sadly, in China, Turkey, and Afghanistan the effects of plate grinding against plate have not been mitigated by the ingenuity of engineers; buildings still collapse like houses made of playing cards when the earth finds the strain too much.

Kobe and Northridge are both symptoms of destruction. Oceanic rocks are being destroyed beneath Japan—and tumbling freeways are a consequence. Equally, oceanic rocks are disappearing on the Californian side of the Pacific Ocean. This, the greatest of the world's oceans, is getting slowly smaller, eaten away at its edges. In 50 million years' time its trans-navigation will not be a challenge.

It must be admitted that the San Andreas Fault is not particularly impressive on the ground. Professor Peter Sadler, of the University of California, Riverside, has been mapping the San Andreas and its splays for years. He smiles somewhat lugubriously when we drive up into the foothills of the San Gabriel Mountains. "There it is," he says. "Where?" I respond. With a little work, you can make out a linear depression in the sparsely vegetated ground running northwards, and perhaps a modest bank on its eastern edge. It does not seem adequate to account for bucking freeways. Erosion in the soft fault gouge soon modifies the famous fault's discreteness. We pass up into the mountains, leaving behind the slipping margin of the most prosperous nation on earth. The mountains themselves are the legacy of earlier tectonic episodes—phases of terrane "docking" such as I described in the northern part of the Appalachian chain in Newfoundland. Much of mountainous western North America has been "plastered on" to the continent. Beyond the mountains lies the Mojave Desert.

There are probably as many mixed feelings about the desert as there are about Los Angeles. Many people drive through it as quickly as they can—usually heading straight for Vegas, where they need never see a salt pan. Most geologists I know are desert lovers. There is nothing like the clarity of light in a desert. There is nowhere better to see geology writ large. Shales that usually weather away to mush can be collected for their fossils. Structures that normally take months to work out are laid out for all to see. Geological maps are written out on the ground, explicit for once. The natural history is intriguing, and no observer capable of detecting an Intensity II earthquake can fail to wonder at the adaptability of organisms in the face of adversity. Some way over the Mojave Desert there is a particular region

ABOVE: *The Welsh part of the Caledonian chain: a distant view of Snowdon.*

RIGHT: *Ancient rocks come to the surface in the middle of New York City. Belvedere Castle in Central Park is built upon metamorphic rocks.*

Real money: an early-sixteenth-century silver thaler. Silver from Joachimsthal provided reliable currency across Europe.

Marie Curie (1867–1934) in her laboratory. Pitchblende from Joachimsthal supplied the ore from which the element radium was extracted.

Mineral as gemstone: an uncut ruby from Pakistan—a typical hexagonal prismatic crystal.

Wollastonite crystals. This mineral, named for William Wollaston, is a calcium silicate typical of metamorphosed limestones.

A stunning vase made from differently coloured fluorite from Castleton, Derbyshire, U.K. Compounds containing fluorine are often associated with hydrothermal deposits at times of mountain-building.

A gold nugget. The Wicklow Nugget, found in Ballin Stream, County Wicklow, Ireland, in 1795. It was owned by King George III and weighs 682 grams.

ABOVE LEFT: *A temple at Ellora: a case of negative architecture. Massive temples and ornaments carved out of a basalt lava flow.*

ABOVE: *Basalt allows for delicate and erotic sculptures at Ellora.*

LEFT: *Wonderfully delicate painted murals in the hidden caves at Ajanta, in the Deccan Traps.*

LEFT: *A granite remnant left by erosion, a tor on Dartmoor: Bowerman's Nose is either comical or sculptural according to the ambient light.*

BELOW: *A piece of polished Dartmoor granite, showing the large white feldspars, dark mica, and greyish-looking (actually transparent) quartz.*

OPPOSITE TOP: *The Deccan Traps, India, a plateau countryside made from basalt flow piled on flow, originally erupted very quickly. A view of the Ajanta outcrop with temples excavated into a single flow.*

Propped on granite: St. Michael's Mount viewed from Marazion, Cornwall.

BELOW RIGHT: *Tin mineralisation follows the Hercynian granites. The ruins left by a once-great industry have a certain romance: Wheal Coats, north Cornwall.*

BELOW LEFT: *One view of granite magma rising: the lava lamp model.*

ABOVE: *Granite batholith towers above the tropical jungle: Mount Kinabalu, Indonesia.*

BELOW: *An elevated road collapses as a result of the Kobe earthquake of 1995.*

Aerial photograph showing the trace of the San Andreas Fault across the Carrizo Plain.

where Joshua trees grow abundantly on gentle slopes. Their growing tips carry aloft a ball of sharp leaves, like an elephant's trunk bearing a porcupine. Their bare trunks branch sparingly below and are frequently bent fantastically, as if they had sagged once under the burden of hardship and perked up again in times of plenty to point skyward. Then there are yuccas, which may, if you are fortunate, sport a great spike of white, bell-like flowers arising from a spiky rosette of leaves. There are many other varieties of spiny or woody plants—nothing nourishing is given away to grazers in such a hard place. But little footprints impressed on the sand prove that there is plenty of animal life out and about after dusk. The arthropod specialist in me recognizes scorpion tracks.

Turn left at Baker and you can make a visit to Death Valley. Everyone should. Like almost all major features in the far West, the valley trends north-north-west–south-south-east, and is under structural control. Faults define the edges of things: the eastern edge of the mighty Sierra Nevada at the Owens Valley; the western edge of Death Valley. The high country of the Sierra is, to say the least, difficult of access. The snowy peaks are almost intimidating when viewed from the more modest White Inyo Mountains to the east. They are cut off as if by a knife at the edge of the green and domesticated Owens Valley—one of the greatest scenic contrasts in the world. The pioneering conservationist John Muir is said to have burst into tears at the sight. The exploration of the Sierra Nevada, as Eduard Suess said, "resounds to the lasting honour of Whitney." Josiah Dwight Whitney first mapped the geology of the area during 1864–70, and has the highest mountain in the Sierra Nevada as his memorial. He it was who recognized the major fault boundaries that are still defined on maps today. These are not faults that shuffle rocks huge distances sideways, in San Andreas fashion. Instead, they are faults that drop chunks of the crust. Gravity alone allows things to sink, so wherever there is crustal tension or extension there are faults of this kind. The rocks relax to a new level, with a fault marking the boundary. There is still movement along the older faults in eastern California—and this is nothing unusual. Like old soldiers, old faults never die, they simply fade away. I once felt a tremor in the agreeable old town of Bishop's Castle in the Welsh Borderland that derived from a fault several hundreds of millions of years old. Death Valley drops to eighty-five metres below sea level; this dramatic disparity is a result of movement ("downthrow") on the fault. It is so hot that sweat even drips off the tip of your nose. At Badwater, salty springs evaporate in the sunshine, leaving a white crust dusting a saline waste. Astonishingly, there are *things* living in the pools, and scrubby, halophilic plants that seem to eke out a living even on the barest ground. The depth of the valley below sea level means that it is hotter than you believe possible if you're broken down by the roadside. Air-conditioning systems simply cannot cope with the job of cooling the inside of a vehicle in the fierce heat; engines expire with the effort. The surface of this valley has just been let down too low: it overheats more than it should. It is the fault's fault.

Death Valley is not journey's end. Eastwards again, we cross the border into the state of Nevada, heading for the Great Basin. Unlike geological faults, state borders are arbitrary lines on the map, but sometimes they do flag cultural differences. The contrasts between California and Nevada are the most striking. First, Nevada is so empty. There seems to be nobody there. When you are driving along the paved highways you are often alone with the horizon. After a turnoff onto the dirt roads, you soon get used to going for an hour without seeing another vehicle. My colleague Mary Droser and I were accustomed to explore the lesser tracks off the greater tracks in search of fossils. All the *people* are in Las Vegas and Reno, which leaves the rest of this beautiful state to the enthusiasts.

On the slow climb into higher altitudes, you observe that Nevada is sage-brush country. The open valleys in the Great Basin are full of it, dominated by it—a tough little shrub that mostly grows thigh high, bearing thin woody stems and greyish leaves. It takes its name from the aromatic smell of the foliage. This sage-like odour is strongest after rain, when the whole atmosphere is infused with a distinctive fragrance, reminding the European visitor of walking through the Mediterranean *maquis*. Sagebrush has to be tough to endure climatic extremes: the winters are hard in the Great Basin, but the summers are long and dry, and can be very hot. There is little free-standing water in the summer: most of the lakes are salt pans that shimmer white in the sunlight. This is where the winter waters gather and evaporate, leaving behind another thin layer of salt each year. You can drive as fast as you like across the pans, for they are perfectly flat. Even in summer, there are occasional torrential squalls, and then you had better get out of the pan as fast as you can, for rainwater drains rapidly into it. Mary Droser and I once had to flee such a storm in our all-terrain vehicle, close to the Confusion Range. We saw lightning playing in the hills and felt no more than a hint of anxiety. Within minutes, everything turned dark. Soon it was deluging, and we could see the track over which we had driven washing away in front of our eyes: no choice but to head through the brush. It was startling to discover how quickly the draining flood reached the axles of our vehicle; we could all too easily imagine being carried away to early graves. Observing a semi-desert convert to a shallow sea in half an hour was truly terrifying. Reaching the hard-top was one of those moments that returns to you again in dreams.

The roads across the Great Basin run straight for a while and then climb and loop. This is the Basin and Range province. Straight courses cross the basins, which are long valleys that run approximately north–south for perhaps a hundred miles. Between successive basins are ranges of comparable length, and up to 3000 metres (10,000 feet) high, one after another. The sage-brush wilderness is used for cattle ranching, but it is poor pasture. You can go for miles without seeing a cow. When you do, it comes as a surprise to see animals that would not seem out of place in a lush European meadow. While you are speeding along the roads across the basins, there is a delicious openness about the world. These roads, you feel, could go on forever. The ranges appear as distant prospects, backed by white clouds, and then quite suddenly are upon you. Each one is a mystery, a surprise. Some are high enough to retain snow late into the year, so as you climb, you cool. The road has to buck and turn back on itself to cross each range. Naturally, the rocks that geologists seek crop out in the ranges, and when the road begins to ascend there are immediately crags and bluffs and road-cuts that reveal the character of the land. The exposures on the flanks of the ranges are superb. You can follow the rock-beds for mile upon mile: the only problem is reaching them, for to get close you rely on rocky old tracks made by hunters and backwoodsmen. The flat tyre becomes an occupational hazard, and walking back to the main road is not an appealing prospect in the heat.

Much of the rain in the Great Basin falls on the ranges. Hence the vegetation is utterly different from that of the valleys, dominated as it is by fragrant sage-brush and almost completely lacking trees. On the ranges, there are large numbers of conifers, often aged and twisted, which line the hill-sides without crowding them, so that you can easily amble between the trees in search of rocks. On hot days, the conifers give off an entirely different smell from that of sage-brush, a pleasant whiff of cedarwood and resin. Among the conifers is the piñon pine, the source of pine nuts, which were an important source of protein for indigenous Indian tribes like the Shoshone, whilst the bristle-cone pines are the oldest living things on earth. In the woods, you feel you are alone in the world. You even get to welcome the rare intrusion far overhead of jet aircraft from the Nevada Test Site. From the top of the range the view commands the adjacent basin, and thence to ranges beyond. All the names are evocative: the Monitor Range, Egan, Toquima, Hot Creek, Lost River, the Ruby Mountains.

Towns are small, and a long way apart. Some are ghosts, relics of silver discovered and worked out; others are picturesque in their near abandonment. Carson City used to be the state capital. Now it is an endearingly ramshackle collection of wooden houses scattered over the hillside. Eureka, Nevada, is billed as "the loneliest town on the loneliest road in North America" on a big hoarding outside the main street. The claim may even be true: Route 50 traverses a lot of sage-brush. There are historic buildings from the days of the great gold boom in the 1870s: a fine old theatre, all painted up, standing opposite the courthouse. I have visited Eureka sporadically over thirty years, and it has changed subtly; smartened up, if anything. It is surprising to hear an enigmatic language spoken in one of the bars: it is Basque. During the boom years, Basque shepherds established themselves in these parts. The miners drifted away when the good times were over, as miners always do, but the farmhands stayed on, as farmhands always do. So here in the loneliest town there is also the strangest tongue. Philologists puzzle over the origins of Basque in the way that geologists puzzle over the deep structure of faults.

Almost everything that I have described in the Basin and Range exists at the behest of the geology: the long mountain chains and the valleys between; the metals sequestered in the hills; the internal drainage that produces the *playas* or salt pans. Faults are the deep control. They chop the landscape into pieces. They direct the ranges north to south. They provide the general control over the most intimate details of the landscape.

The steep scarps of the ranges are defined by huge faults. These let down adjacent chunks of crust that form the basins, so the whole landscape is divided into a series of massive, elongate blocks. The main faults lie on the western sides of the ranges, and they run north–south, albeit with many local complications; hence the direction of the faults determines the grain of the landscape. The downthrow sides of the faults mostly lie to the west: these are cracks in the brittle crust, deep faults, but lacking the transcurrent motion of the San Andreas. The winding roads that require such intense concentration from the driver are, you might say, climbing up the sides of the fault blocks. The level swathes across the basins reflect instead what erosion has done to the mountains. For even as the basins have been let down, so they have filled up. Sediment has flooded in to raise what faults have made low: 1000 to 1500 kilometres (3000 to 5000 feet) of such sediments have filled in the basins. The flood that Mary and I experienced was just one moment in the great levelling, one incident among millions. The sage-brush knows the geology, with the unwavering intelligence of its precise adaptations. Nor would the pine trees lead their slow lives were it not for the gift of the mountains. In the comparatively recent past, when the climate was wetter, many of the basins were united in one vast lake, known as Lake Bonneville. We could see evidence of the ancient shorelines of the lake on the edges of the basins—a kind of bevel in the hillsides. The *playas* were the last remnants of the great lake. Further east, where the basin and range pass into Utah, a more considerable legacy is the Great Salt Lake, a huge expanse of whiteness, testimony to the evaporation of all the salts formerly leached out by rain from the eroding mountains, made solid by evaporation under the fierce sun in drier times. It is a weird place. From a highway passing alongside it you can see mirages, when the horizon suddenly goes liquid. Small buttes arising from the lake appear to float eerily above the ground. Everything shimmers.

The faults that define the edges of the basins dive downwards at a steep angle. Seismic reflections show that the faults actually flatten out at a depth of perhaps twenty kilometres. They are more like a series of scoops than simple vertical cracks dropping chunks of crust to make a step, like a failure in a concrete floor. So now we have what my American friends would call a "visual": the whole vast Basin and Range area stretching, and the crust accommodating to tension by extending and sliding down faults in parallel chunks; the greater part of several U.S. states shattered and adjusting. The process must, of course, be related to deep things, just as the Californian coast shakes at the bidding of the subterranean engine. Like that area, the evolution of what we see today has a history of 30 million years or so. We have to look far down beneath the sage-brush and the piñon pines to see what guides the character of our journey across the Great Basin.

Arthur Holmes was already familiar with one fact about the deep structure of the Great Basin: the crust is thinner there. It has an average thickness of twenty-five to thirty kilometres—something like fifteen kilometres less than under the Great Plains, which comprise the flat heart of America. At this point we have to recall that the deeper you go into the crust the more the rocks are likely to flow rather than fracture, as pressure and temperature increase; the twisted gneisses in the centre of Newfoundland were an example. Some of the great minds of tectonics, like Dan McKenzie, have pondered the Great Basin, and have used the contrast in behaviour of the brittle upper crust and the ductile lower crust and mantle lithosphere to explain what is seen at the surface, and to infer what happens far beneath it. Deep seismic reflection profiles have revealed that *below* the normal faults that define Basin and Range there is a low-angle "zone of detachment" flanked by a region of ductile deformation. It is a plane of weakness at great depth where slow slithering happens: one major piece of lithosphere over and past another. In one version of the theory, this shear zone is thought to cut obliquely downwards across the entire lithosphere, providing the deep control of the crustal extension over the Basin and Range—which in turn generates the faults. For example, the crust is thinned where the zone of detachment approaches the surface. Remember that the crust is less dense than the mantle beneath; hence, where it is thinned, this results in subsidence. This is the converse of the situation in mountain ranges that we have considered previously, where *thickening* of crust during continental collision results in its elevation. It is all a matter of checks and balances. Where the crust has to accommodate to such deeper controls, it shuffles in blocks in response, brittle-fracturing to obey the bidding from below. At the same time, extension in the crust may induce melting of its lower part, and lavas will erupt at the surface as a consequence. Mary and I saw evidence of these on our traverse across the Great Basin, where piled ashes formed terraces or slaggy piles. Some of them were the product of the kind of explosive aerial eruptions that generate extensive ash flows; these are preserved as ignibrites, like those we met around the Bay of Naples at the start of this book. And faults can be conduits for the hot fluids that distill the bounty of the earth, where minerals and metals escape upwards from the magma chamber. The miner, too, must nod to processes* deeper by far than his shaft can ever penetrate. Pines, sage-brush, salt pans, gold, twisting roads . . . it is a *Gestalt* united by tectonics.

There are other fault lines that traverse the earth. Cutting north–south through eastern Africa there are great valleys: a depression passes south-

* The genesis of the different ore deposits is much more complex than this implies and involves several separate mineralization events, as well as re-activation of older deposits.

wards from the Gulf of Aden through Ethiopia, then splits into two arms, passing either side of Lake Victoria. The eastern arm strikes southwards across Kenya, its course marked by lakes sitting in the valley base: Lake Turkana, Lake Baringo, Lake Magadi, Lake Natron. The last named is one of the great natural sources of soda, for it is another Great Salt Lake, where evaporation has concentrated soluble salts. Natron is an old name for hydrated soda and has left its legacy in the chemical symbol for the element sodium (Na). The lake supports a huge population of delicately pink flamingos, which use their extraordinary bills to filter out nutritious microorganisms from the warm, alkaline lake. For these birds, a soda fountain is a literal description of their source of nourishment—and they derive their colour from their food. The western arm is one of the most blatant lines on the face of the earth, an almost continuous ribbon of water running southwards from Lake Albert to Lake Edward, and thence to Lake Tanganyika and Lake Nyasa (Malawi): the Great Rift Valley. Eduard Suess would have used the term "graben" for this kind of structure. The total length of the system is some 3000 kilometres (1800 miles). The width of the valley is about 30 miles (50 kilometres) along its entire length. You do not need to be a naturalist to read this line, for political boundaries and, before them, tribal boundaries are drawn along it. The faults to either side of the Great Valley let down the stretch of crust between: slithering cracks delineating a sag. Earthquakes are concentrated along the rift, so much so that a map of earthquake epicentres provides a kind of pointillist map of it. Clearly, the surface of the earth is jittery along these lines. When seen from the air, the faults defining the edges of the rift are clearly visible and can be as straight as a ramrod. They are geological lines graven on the ground. There is often a series of faults at each edge, stepping down, as it were, to the valley floor, rather than a single, mighty fault. Near Lake Turkana the valley is filled with volcanics and sediments to a depth of 8000 metres; since the scarps are up to 2000 metres high, this implies no less than 10,000 metres of movement on the fault. Staggering though this total might seem, when divided by the 10 million years over which the movement has been happening, it is slippage averaging just one millimetre a year: lichens grow faster.

Volcanoes follow the rift line less slavishly than earthquakes (the greatest, Kilimanjaro, is outside it), but throughout Ethiopia and Kenya, and further south, there are many active or dormant volcanoes hard by the fault lines. Many of them erupt curious lavas rich in carbonate minerals—which we usually associate with sedimentary rocks—known, rather unsurprisingly, as carbonatites. Ol Doinyo Lengai (2891 metres, 9485 feet high) is a typical volcanic cone that has spewed out lavas that have helped to supply the white crusts of Lake Natron. The crater at the summit of the volcano is one of the most desolate places on earth. A crumpled and petrified sea of carbonatite is dotted with steep-sided conelets, the dark holes in their apexes revealing the volcanic chimneys below. Everything is white. It is a hellish inversion of all we have come to expect from a volcano, a bleached-out counterpart of the realm of Pele. The lava still pours out onto the surface of the crater from time to time, and when it does it is black, like any other lava. But as it cools it changes—first to a chestnut colour, and then, after some hours, it, too, becomes white and crusty. Erupting in the heart of Africa, the lavas have enriched themselves with juices acquired from the continental crust, thereby transformed utterly to a new and strange composition. Black becomes white.

Colourless volcanic gases can be more dangerous than the obvious sulphurous exhalations of more dramatic volcanoes. Carbon dioxide is heavier than air and collects in hollows (as Oliver Goldsmith described so graphically in the "cave of dogs" near Naples). Investigators in the rift must not be caught unawares in such a pocket. The most lethal carbon dioxide eruption was not in the Great Rift, but in an altogether less impressive volcanic area in the Cameroon Highlands in West Africa. On 26 August 1986, gas bubbled up from Lake Nyos and crept like an invisible stream downhill, a 150-foot thickness of lethal miasma. Twelve hundred people were suffocated in the town below, along with their beasts and the very birds perched on the bushes. Only cattle-herders in the hills escaped to tell the horrible story.

The Rift Valley had another role to play in history. Game has flourished there for millions of years and, in pursuit of it, our distant ancestors and relatives walked upright among the bushes and trees. If Africa was the cradle of human evolution, it was around the lakes and volcanoes in the great valley that the bones and artefacts of our ancestors stood the best chance of being preserved.

Expeditions to Lake Turkana in the Ethiopian Rift discovered fossils which showed that hominid history must now be reckoned in several millions of years. It was here that our history began to reveal its true complexity—that there were many different animals in the family of *Homo sapiens*, from which we, alone, eventually prospered. Just how many species were involved is the subject of endless debate among anthropologists. The details of human evolution are not part of our story, but I am obliged to mention that the flow of new discoveries has not abated—as this was written in 2002, a scientific paper in *Nature* has claimed a new fossil as marking the branch between the human line and that leading to our close relatives, the great apes and chimpanzees. It is no surprise, in such a disputatious area of science, to find that claim already challenged. What is not in dispute is that early hominid evolution involves a cluster of long-extinct species, some of which are not close to our line, and some of which are. Nor that the genesis of the line leading to modern humans was in Africa. The growing richness of fossil evidence is a measure of how much we are learning, but I do not doubt that current textbooks, like their predecessors, will eventually have to be rewritten. That there is such a fossil record, whatever its entanglements, is thanks to the geology. The volcanic rocks that periodically spread, or exploded, over the rift valleys both buried the precious remains and provided the means to date them, for the radioactive clocks which measure time are interred with the bones. And the fact that faults let down the rocks, step by step, preserved their precious cargo of history, which might so easily have been eroded away as if it had never been. What has been dropped down in the rift has survived. We owe our history to our faults.

When Eduard Suess was writing, the structure of the African rift valleys was already partially understood. With his characteristic confidence he extrapolated from what little pioneering geology had been undertaken in this part of Africa. Of Nyasa and Tanganyika he wrote: "I do not see in what way these two depressions, each of which, although of very trifling breadth, is prolonged through about five degrees of latitude, could have been produced except by trough faults, and in my opinion their origin is very similar to that of the Red Sea, [and] the Dead Sea." Suess had made a connection that we still observe today—and which we can explain through deep processes. The African rifts ("trough faults") pass northwards to the Horn of Africa. To the west, the Red Sea has rifted margins. So, too, does the great depression running northwards beyond the eastern edge of the Mediterranean Sea. The Dead Sea lies at its lowest point, below sea level, a sink for all the waters of the Jordan River. One of the first pictures I remember from my children's encyclopedia was of somebody lying on their back in the Dead Sea reading a newspaper. Salt is concentrated by evaporation there to the point where you cannot sink. Faults have let it down deeply: if there was another biblical Flood, the sea would rush in and inundate the whole area in the twinkling of a geological eye.

The Red Sea is a young ocean. Look at the map: it requires little imagination to close it up again. It is like a simple break in a platter. Imagine Africa and the Arabian Peninsula united as a single continent. An irresistible force parts them. First, a rift valley records the tension in the strained crust: it sags and weakens. As Suess noticed, the faults are preserved along the flanks of the Red Sea today. There are associated volcanic eruptions, many related to readier access to the underworld provided by faults along the rift. Then, the true ocean opens out as new, oceanic basalt rocks are emplaced within the nascent basin; a mid-ocean ridge grows. There are two "arms" to the ocean, one along the Red Sea, the other along the Gulf of Aden, separating the Somalian Plate from the Arabian Plate, and meeting at Afar, at the end of the Eastern Rift of Africa.* Inexorably, the African and Arabian coasts move apart. In its early phases, the new sea was cut off from the Indian Ocean from time to time and evaporated to dryness, leaving behind thick deposits of salt. Today, there is much active geology within the juvenile ocean: earthquakes are aligned where you would expect them along its mid-line, where oceanic lava can also be sampled; thirteen pools of hot brines have been identified on the ridge in localized "deeps." These are places where valuable metals like zinc and copper are concentrated. Originally of volcanic origin, they have been distilled into brines by circulation through the underlying rocks. The young ocean is a kind of chemical brewhouse where the ingredients of continents and mantle can interact and cook up new recipes. Many ancient ore deposits are the fruits of such encounters. In the Red Sea, perhaps, we may obtain a vision of the birth of other, greater oceans. We can imagine the early days of the Atlantic Ocean; it, too, transferred from "rift to drift" as Pangaea dismembered itself. We can envision how today's familiar continents first shuffled into their separate identities, just one late phase in the slow dance of the plates over the face of the earth.

If the Red Sea shows the birth of an ocean basin, the rift valleys of Africa are thought by many geologists to represent a still earlier phase in continental break-up: a new geography *in statu nascendi.* The Red Sea and

* There are thus three diverging systems meeting at Afar: the Red Sea, where the African and Arabian Plates separate; the Gulf of Aden, where the Somalian and Arabian Plates diverge; and the Eastern Rift, where the Somalian and African Plates are pulling apart. Afar is accordingly described as the site of a triple junction.

Gulf of Aden are but one part of Suess' interconnected system, which extends southwards into the great continent. Given enough time, perhaps the eastern part of Africa will separate from the western, and a new ocean will divide the Somalian Plate from the mother continent. Another movement of the dance will have begun.

As for the deep cause of the separation of the neighbouring plates, there is much evidence for mantle plumes impinging on the crust beneath the eastern part of Africa. The rising limbs may provide the deep root of the tension that causes the progressive rifting in the valleys: like huge, hot hands parting the brittle skin of the earth. The same volcanoes that buried the remains of early hominids are no more than a consequence of these profound motions, occurring where hot mantle partially melts deep beneath the continent to generate magma. The plume spreads out like a mushroom beneath the lithosphere. This explains why major volcanoes like Kilimanjaro lie beyond the compass of the rift itself. They arise like suppurations from hot sepsis under the skin of the earth. Much evidence from modern measurements along the Rift Valley testifies to the rising of hot mantle material beneath it. For example, the general heat flux is greater within the rift than on the stable continent to either side. Measurements of the minute changes in gravitational anomalies across the region are also consistent with a hot plume beneath the rifts. Current geophysical work seeks to model what is happening at depth. Some geologists believe that the rift system within Africa will develop no further, that it will, in effect, remain forever a "failed ocean." Given the slow march of geological change, it is unlikely that mankind will be here to confirm whether this is true or not. Who knows if the fossil remains of the last of our kind will lie entombed beneath some ash-fall within the rift—just one more species among the millions that have already become extinct?

The surface of the earth is criss-crossed by faults like a sheet of crazed glass. How could it be different after thousands of millions of years of the crust being subjected to stress? Uplift produces cracks; loading of the crust often produces the same effect; stretching of the crust again induces faults. Tectonic squeezing may thrust rocks pick-a-back in stacks as the crust shortens, wincing brittlely away from forces greater than it can accommodate. Only in deeply buried rocks will plastic deformation predominate. For the rest, cracking and evidence of movement is everywhere. On average, the longer rocks have endured, the more likely they are to have been affected by faulting at some stage in their history. A walk along almost any beach backed by cliffs of sedimentary rocks will discover places where the

Portrait of the artist as a young man . . . the author by the Pony Express Trail across the Great Basin, Nevada.

strata have been displaced by small faults; sometimes as neat and obvious as cutting a triple-decker sandwich and dropping one side—you can match the layers and tell just how much the fault has displaced the strata. In other places, a greater fault may juxtapose quite different sets of rocks. You may see crushed rock along the fault, or find milky quartz, or the sea may have picked out a cave along the fault plane. In quarries cut into the guts of ancient mountain belts you may inspect the polished planes of faults long since sunk into quietude: scratched and gouged in the direction of movement, what once might have rocked the earth is now no more than a passing stop on a geological field trip. The faults of old earth are beyond counting. Green fields may cover them over; or, as in the western United States, they may delimit mountain ranges. Fault lines cut the character into the face of the earth.

10

The Ancient of Days

Dressed in sensible boots and tweeds, in the summer of 1883, the Scottish geologist J. J. H. Teall strode down to the harbour in the little fishing village of Scourie. He looked back on a scattering of cottages, an inn, and a kirk tucked into the back end of a protected bay, not many miles from the far north-western tip of mainland Britain. Teall crossed by Scourie Lodge on his way and passed the time of day with its tenant, the factor of the Duke of Sutherland, Mr. Evander MacIver. There were not many educated men in those parts—and Mr. MacIver had attended no less than three universities—so he was naturally curious about the presence of a middle-class geologist upon his land. Then, too, as one who had prosecuted the Clearances, he may not have known many among the local population who cared to flatter his erudition. He may well have invited Teall to inspect the wonders of his stone-walled garden, constructed upon one of the few patches of fertile earth in this most barren part of Scotland. The Gulf Stream warmed it to a degree that defied its northerly latitude beyond 58 degrees; in this protected haven the astonished geologist might have seen exotic red-and-blue fuchsias, and even a thriving New Zealand cabbage-palm that had been grown from seed sent by one of the factor's relatives on the other side of the world. But the geologist had other interests today, and, with the factor's blessing, took off along the walled track behind the boat-house that led up over the hillside above the lodge. He approached the sea along the northern shore of the Bay of Scourie, where the coast was indented by a dozen small inlets backed by modest but rugged cliffs. The rock forming the underlay of the country announced itself everywhere: in lichen-covered crags, in stones lying about the grassy fields, in the crude but durable patchwork of the stone walls, in rounded boulders on the narrow beaches. The rock was gneiss, grey and heavy-looking, finely banded and speckled—mile upon mile of it, apparently. When Teall tramped onwards towards the small headland of Creag a'Mhail his eye caught something different: a steep, dark seam of rock running towards the headland. The sea had eaten through part of the seam, but now he could see that it continued to the east and struck inland. As a good Lyellian geologist, he recognized the seam for what it was—a dyke of igneous rock, cutting through the gneisses. It must, therefore, be younger than the rocks it violated. He clambered down to collect a piece of the "trap" (he probably thought it to be the volcanic rock called dolerite). A few weeks later, back home in the laboratory, he recognized something else: the igneous rock of the dyke had itself been metamorphosed, like the gneisses it intruded. Teall published his results in volume 41 of the *Quarterly Journal of the Geological Society of London:* "On the Metamorphosis of Dolerite into Horneblende Schist." The Scourie Dyke had been introduced to the world.

It has since become one of geology's holy places. Dozens of famous geologists have followed Teall's trail, and you somehow feel there ought to be a plaque somewhere about. There have been structural geologists, geochronologists, geochemists, and palaeomagneticists—the whole tribe of specialists spawned in the century since Eduard Suess wrote *Das Antlitz der Erde.* When I visited the site late in 2002, Mr. MacIver's cabbage-palm was still thriving inside its walled garden, 150 years old and perhaps fifteen feet high, with many heads that had been sprouted over the years. It was, I was told, the most northerly palm tree in the world, and as I dodged the hailstones I could well believe it. The track behind the boat-house was now boggy and clogged with rushes, and higher up was overgrown with prickly gorse bushes; but the tough gneisses were of course unchanged. It could have been these rocks that Henry M. Cadell of Grange, Bo'ness, had in mind when he wrote in *The Geology and Scenery of Sutherland* in 1896: "A hand specimen of gneiss resembles a closed book with chapters of varying lengths, each chapter consisting of pages of a different tint"; in other words, striped, but irregularly so. To the geologist, these rocks are known as the Lewisian gneiss. I brought back a beach cobble of the stuff: it was satisfying when clasped in the fist, like a goose's egg. I also found a smaller pebble derived from a dyke: black with subtle lighter speckling, like a piece of the map of deep space. Since Teall's discovery, the area around Scourie had been logged in detail by Dr. Clough of the Geological Survey, who had found not one but many dykes, all with a trend west-north-west–east-south-east. Dykes, like angry wasps, tend to occur in swarms; they are intruded in response to extension over a great area, a crustal cracking filled with magma. Their direction clearly had absolutely nothing to do with the familiar Caledonian trend running from north-east to south-west across Scotland—indeed, they lie almost at right angles to it. Looking at the map, the course indicated by the dykes seems to be rather common in the north-west Highlands, guiding the shape of lochs and the flow of rivers running towards the irregular coastline. What now controls the lie of the land is a legacy of something far, far older than the Caledonides, an antecedent world, a scrap of immemorial crust: the ancient of days.

What with the covering of lichens, the wind and the rain, and the sea splashing up, it is not as easy to distinguish gneiss from dyke as it should be. When a cold wind is stinging your eyes, one dark rock looks much like another; but eventually the famous dyke was identified. My mentor, Graham Park, who has spent much of his life on these rocks, recommended that we go to another dyke locality at Achmelvich Beach, a few miles to the south, where it was easier to clamber over the outcrop. This proved to be good advice. The single-track coast road to Achmelvich follows the contours of the land with precision; it is as if somebody had simply stuck a very narrow strip of tarmac over the switchback terrain. When a vehicle comes the other way, etiquette dictates that you must scuttle into a marked passing-place, while waving at the oncoming vehicle in comradely fashion. The coastline is wonderfully varied, indented with little sea lochs, and strung with half-hidden offshore islands of every size and shape, at low tide all marked out with a bright fringe of yellow bladder-wrack plastered to the shore-rock. The Lewisian gneiss emerges from the ground in numerous grey, rounded hillocks, rank upon rank of them. They look almost like cumulus clouds, and towards dusk they can merge uncannily with the sky. The soil produced from the weathering of old gneiss is thin and poor, as if the passage of billions of years had somehow leached all the goodness out of it. There are boggy pools to either side of the road, each with a meagre fringe of rushes and cotton-grass, and hollows filled with sphagnum moss and carnivorous plants, equipped with sticky glands, that supplement their hard rations with what insects they can trap. The late autumnal bracken and dry grasses paint the slopes with a shade that falls somewhere between rust and russet, laid warm against the pallid outcrop. Somehow, the landscape *feels* ancient, worn down, as if it had seen it all before, and could only now manage a hilly shudder when once it might have drawn itself up into great mountains.

Crofters' cottages dot the tiny road to Achmelvich; each one has a few trees of downy birch on the hillside behind it, which is just now carrying lemon-coloured autumn foliage, and perhaps a mountain ash by the garden wall, heavy with orange berries. The older croft cottages are stone built, with one end of the building reserved for winter quarters for animals. Rough lumps of gneiss make for solid but unrefined buildings: they seem almost to grow out of the land. I saw several examples of plain but comfortable-looking newer houses alongside the original croft buildings, which in some cases had become tumbledown. The tenancy of a croft is rigidly administered by the Crofter's Commission, and crofters enjoy a security of tenure, and of inheritance of that tenure, that many lessees might envy. Each croft has an inviolable parcel of land assigned to it, and rights to run perhaps a dozen sheep and cut peat, all designed to keep a family above subsistence level—the Commission was partly set up to ensure that the commotion of the Clearances did not recur. A modest rent is paid to the landlord, in this part of Scotland usually a landowner like the Duke of Sutherland. There is simply no opportunity for speculative development in such remote communities, but you *can* replace your old croft house with a spanking new bungalow. Of course, modern materials imported from outside are quicker to build with, and cheaper to maintain, so that the umbilical connection to the local geology is lost in most of the newer erections. Something—a little character, a morsel of regional idiosyncrasy—is lost with the transformation; it is a small step towards Sutherland becoming like everywhere else. Graham Park told me that there is a waiting list in the prettier areas to get a croft, so the appeal of life close to the elements in the least populous part of Great Britain evidently endures, although only a minority now follow the farming option.

Achmelvich proved to be a perfect little bay backed by white sands—shell sands, surprisingly, like something you might find on Hawai'i. The shells have been washed in from the Atlantic Ocean, smashed to pieces in storms and blown into marram-covered drifts. On the north side of the bay the old gneiss runs into the sea. You can get a more comfortable look at the geology here than at Scourie. The speckled appearance of the gneiss is more apparent where the sea has washed it clean. These are rocks that have, to borrow a phrase from Cadell of Grange, been "crumpled by the long-continued ill usage to which they have been subjected." They have, in fact, been subjected to almost the greatest extremes of pressure and temperature that rocks can endure without completely melting—"ill usage," with a vengeance. The even-sized grains look almost like some kind of granite: it is as if they had been squeezed so hard that their every component was forced into having similar dimensions. Geologists refer to this high pressure and temperature regime as the "granulite facies" in recognition of the distinctive texture of the rocks. The strong banding, or foliation, betrays the fact that deformation was the *fons et origo* of the rock's characteristic fabric. Laboratory experiments have shown that the minerals typical of these rocks only co-exist under conditions close to those that pertain near the base of the crust. The gneisses that are now just a few feet away from white, shell sands blown from the Atlantic Ocean have been on a journey deep into the earth, from which they have re-emerged to become encrusted with lichens and moistened by sea spray.

I once went in search of the type locality of another granulite facies rock, known as charnockite. The locality was the gravestone of Job Charnock, founder of Calcutta (d. 1693). It was originally recognized from the great man's memorial, and so, naturally enough, took his name: ancient rocks with origins as deep as the Lewisian gneiss evidently also occur on peninsular India. I struggled through the fug of Calcutta to get to the Anglican Cathedral. Were it not for the presence of that irrepressible city on every side, and the relentless heat, the building might have been in Tunbridge Wells or Ashby de la Zouch; it seemed so *frightfully* English. Inside there were various elaborate memorials to army officers and civil servants, but I never found Job Charnock, nor his charnockite gravestone. Some day I may return to try and find the correct church.

The chemical composition of the Lewisian gneiss on the foreshore at Achmelvich indicates that it was originally an igneous rock known as tonalite. Metamorphism at depth changed it to gneiss. The dingy colour of the feldspars imparts greyness to the whole region; I am used to clear pink or white feldspars, like those in Cornwall. Under powerful modern probes, the Lewisian feldspars can be shown to contain myriads of minute inclusions—comprising elements such as titanium that are "soluble" in the feldspars at very high pressures—and these, like the chemical smog that fuzzes up the streets of Calcutta, impart a kind of smokiness to the commonest mineral in the gneiss. The world, after all, is painted at the molecular level. On closer inspection, the gneisses can also be seen to contain irregular balls of very black rock—some larger than a football—that look as if they had been caught up in the original igneous intrusion. They are reminiscent of the xenoliths, those messengers from the underworld we first met in the lavas of Hawai'i. They are believed to represent fragments of the original ocean floor from which the tonalites were derived by partial melting when this ancient crust was first formed. The mere fact that we can use such comparisons shows that the processes that operated so long ago can still be interpreted in the light of processes we know today. And then there are the dykes, much more visible than in Scourie: dark walls decked with contrasting white lichen patches like paint splashes, each dyke bounded by narrow sheared edges. If you look hard you can see remnants of the original igneous texture, mere ghosts of crystals of feldspar long since changed into a granular mosaic by transfiguration in the deep earth. We are looking so far back in time that ghosts and chemistry are our best informants.

But how far back in time, and how do we know?

To unravel the argument we have to follow the footsteps of the early investigators a few miles to the east—to the Moine Thrust. Recall that the north-western margin of Scotland is, in its deeper geological reality, part of the old Laurentian continent. Scourie and its environs are but a piece of the old heart of North America, stranded by the subsequent opening of the present Atlantic Ocean on the "wrong" side of the water. Mr. Evander MacIver would no doubt have had to light a second cheroot to take it all in, but the idea is familiar to us by now. To the east of Scourie, forming the highest of the Highlands today, lies the eroded remnant of the great Caledonian mountain chain. The Lewisian territory forms part of what Eduard Suess would have called its foreland—and we have to imagine that foreland stretching far westwards over into what is now the core of Canada. The final heave of the Caledonides towards the foreland happened along the Moine Thrust. The geology is shown at its simplest at the famous locality at Knockan Cliff, a few miles north of Ullapool, close to the A835, the main road northwards. The interpretation of the geology of this unremarkable grassy slope topped by a small crag sparked one of the great rows in the history of science.

When I first visited Knockan Cliff more than twenty years ago there was not much to mark out the place—a small monument, no more. Nowadays, it has a full-fledged visitor centre topped by a turf roof and a marked trail, set with plaques, leading upwards to the cliff. I noticed that the construction of the centre had been sponsored by no less than eight different organizations, which reflects how difficult it is adequately to mark sites of scientific pilgrimage. The marked trail was excellent. You are led up through a succession of rocks of Cambrian (about 540 million years) age piled one on top of the other. Near the base is the "Pipe Rock," riddled with the tubes made by worms in the sands of an ancient seashore: from the top, the exit holes of the tubes look as if someone had stubbed out a whole series of cigarettes in the sand; then comes a rock bed including small fossils of the mysterious tube *Salterella;* above again lie shales yielding the beautiful early trilobite *Olenellus,* the size of a crab, which proves as clearly as if the outcrop had been date-stamped that these rocks are not merely Cambrian in age, but *Lower* Cambrian. This was the time period when marine animal life carrying shells apparently suddenly burst into variety and complexity. A Lower Cambrian sedimentary pile, then, and above it, along the outcrop, a thickness of early Ordovician limestones. Then you are led onwards and upwards to the famous thrust. A dribbling waterfall marks the spot. At the top, forming the little cliff, are some completely different looking rocks. For a start, they have been strongly metamorphosed, whereas the Cambrian rocks beneath them on the slope show no such alteration. These uppermost rocks are known as the Moine schists, a dreary formation that covers a huge area of mountains and heather-clad moors to the east. At the base of the cliff is a rock bed—perhaps two—of a creamy yellow colour, forming a kind of boundary layer between the Moines above and what lies below.

The nineteenth-century argument centred on the interpretation of the rock succession I have just described: Was it a normal sequence, in which case the Moine rocks would have been younger than the rocks that evidently lay below them? Or had the upper part of the sequence been thrust bodily over the underlying sedimentary rocks, and might therefore be much older than the rocks on which it came to rest? In many respects, it is the same problem already examined in detail in the Glarus Nappe in Switzerland, and the arguments do not need to be rehearsed again. It is sufficient to say that reputations were made and lost in defending one view or the other. The name that is in use today embodies the eventual solution to the problem—it was, indeed, a thrust. The yellowish rocks beneath the Moines were mylonites, the ground-up rock paste upon which the huge mass above rode. Look at them closely and you can see evidence of the intense deformation to which they have been subjected: dense pleating and banding. Because it squeezed *over* early Ordovician rocks, the thrust event must have happened after that time, which marks it out as part of the Caledonian deformation. Elsewhere along the long outcrop of the Moine Thrust, which runs from Durness in the north to the Isle of Skye in the south, there are true nappes interposed along the boundary, so the Swiss analogy becomes even closer. The Highlands Controversy, as it came to be known, took more than thirty years to play itself out, during which time tempers inflated to almost orogenic proportions. The visitor centre portrays the heroes of the controversy as the two Geological Survey geologists, Ben Peach and John Horne. Peach and Horne's 1907 *Memoir* is certainly one of the great works of geology, and it mapped out the Moine Thrust in incomparable detail. But the intellectual leap towards the thrust solution was made more than a decade earlier by the great geologist Charles Lapworth, in *opposition* to the then Director of the Geological Survey, one of the most famous scientists of his day. Lapworth himself suffered a nervous collapse as a result of his "anti-establishment" role in the Highlands Controversy. He dreamed of being crushed beneath the mighty thrust. Peach and Horne were, in a sense, righting a wrong that had once been sustained by their own paymasters. It is all done and dusted now, a controversy subsided, and as I stand in the cold sleet on this bleak slope it is hard to believe that the rocks before me once generated so much heat. In the visitor centre an animation shows how this part of Scotland—in early days a piece of Laurentia—has changed its latitude over the last billion years, playing its part in the stately progress of the drifting continents. It began near the South Pole; by the time that the *Olenellus* trilobites were grovelling around in the Cambrian mud it had reached the equator; thence it moved at an irregular pace towards its current position at high latitudes. This is another lesson in how we must unmake the world as we go back in time—lose all familiar geographical outlines, rearrange everything. The deeper we go, the stranger everything is, the less recognizable.

As we look westwards from Knockan Cliff, we are also looking further back into geological time, to the ages before the Cambrian. Before us lies the mountain of Cul Mor, 981 metres high, presenting a curiously rounded, rather than jagged, outline and rising with almost improbable steepness from the surrounding undulating plain. A dusting of snow has picked out the higher strata. It is as if white, parallel, horizontal lines had been drawn upon it, lending it the appearance of a layered cake, sprinkled with icing sugar. These are bedded, sedimentary strata, neither folded nor sheared. Yet they are also the rocks that lie *below* the Cambrian succession—there are several sites nearby where you can see the sandstones that form the base of the Cambrian rock succession lying upon this earlier "package" of sediments. There is also evidence that there was a break in time, an unconformity, below the base of the Cambrian rocks, which records the invasion of the sea over a rocky surface that had previously been weathered down to close to sea level. So the rocks that make up Cul Mor were, by definition, deposited before the Cambrian began—they are Precambrian in age. All along this part of the west coast of Scotland there are wonderful mountains composed of the same sedimentary rocks: Suilven, Stac Pollaidh, Canisp, Quinaig. Almost wherever you are along the coast on the Lewisian gneiss, one of these mountains seems to peek down at you from beyond one of the low hills in your immediate vicinity. None of them belongs to the select group of Scotland's "Munros"—the 284 mountains catalogued by Sir Hugh Munro that rise to more than 914 metres (3000 feet) high. But they seem higher than they really are because of their sheer sides—in some cases there is only one way up for the climber without pitons. Uncountable years of erosion have left these peaks standing proud, and their intractable, uniform hardness has resulted in their curious, monument-like topography. You feel an urge to bound to the top, to feel that you have conquered your own piece of deep time. The rocks take their name from Loch Torridon, thirty miles to the south-west—they are Torridonian in age. Careful mapping in the last thirty years has shown that there is an older and a younger series of rocks within the Torridonian as a whole: a further division of Precambrian time. Perhaps we should take a closer look.

It has to be said that this wonderful countryside has one drawback: rain. On my previous visit during the summer twenty years ago, I concluded that I might just as well jump into a lochan early in the morning to save myself the trouble of trying to keep dry later in the day. There was scarcely such an option in late October, with snow on the hills, and even the sheep looking a little depressed. My visit coincided with the heaviest rain in eastern Scotland for years, and so I inspected the Torridonian through a haze of water. Graham Park optimistically insisted that, in western Scotland, the weather is always liable to take a turn for the better. "It's getting lighter in the west!" he would exclaim, pointing towards a slightly less dense bit of strato-cumulus. We examined the Torridonian along the roadside between squalls. Here was a place where you could see that the rocks were mostly reddish sandstones. Years of rain had weathered out the sedimentary bedding into a kind of naturally sculpted mural, which preserved the traces of the kinds of channels that are produced by vigorous flash floods today. In a broken rockface, crystals of feldspar, still quite fresh and pink, indicated that these ancient deposits were laid down under arid conditions: in moister environments feldspar soon decays to clay. So the Torridonian was indeed torrid—a curious, but entirely coincidental marriage of name and historical fact. A few miles to the south, by Loch Maree, the Torridonian makes an impressive waterfall, the Victoria Falls (Queen Victoria herself had inspected them when staying at the Loch Maree Hotel). The falls are reached by walking up through a majestic conifer forest, underlain by moss and ferns, and no scrub at all. In sheltered places, fir trees make stately specimens in western Scotland, and their branches whisper and sigh. The falls plunge over a rock bed made of massive sandstone. The rock beds underlying it are much shalier, and very much softer, so that the falling water has excavated an impressive overhang, like a miniature Niagara. Little by little, the Torridonian rocks are being worn away, even after having survived for so long. Later that day, Graham Park was proved right: the sun came out. We were in the very eye of the storm.

From the southern shore of Loch Maree we can see the whole story of the rocks laid out on the cliffs on the opposite side. Loch Maree is one of the most beautiful lochs in Scotland. In the seventh century, an Irish monk lived on one of its islands, and converted the local people; his holy well was credited with curative properties, though no one I spoke to could tell me its location. On the flanks of the loch, patches of the old Caledonian forest linger on, with groves of Scots pines with branches that make such interesting crooked-elbow shapes compared with other conifers. Scraps of wild forest also survive on islets in the loch, where seedlings are protected from the depradations of nibbling sheep. The long loch follows the old grain of the land from north-west to south-east, tracing out a fault more than a billion years old. The northern shore is so straight that it might have been ruled out upon the ground. Fundamental faults never entirely go to sleep, and there is evidence that antique faults like the one underlying Loch Maree came alive again when the Atlantic Ocean opened, in the latest of the many tectonic cycles to affect the Highlands. Sir Walter Scott, who virtually invented the Romantic view of Scotland, supplied us with an unexpectedly accurate description of the durability of faults (in *The Lord of the Isles*, 1815, canto iii):

Seems that primeval earthquake's sway
Hath rent a strange and shatter'd way
Through the rude bosom of the hill,
And that each naked precipice,
Sable ravine, and dark abyss,
Tells of the outrage still.

The original lakeside track ran along the straight northern shore, but now the main road runs from Kinlochewe—where Peach and Horne lodged—

to Gairloch along the southern shore. From this road, under the shadow of Ben Eighe, we can see the structures written out in the cliffs on the far side of the loch. It is golden eagle country, with open crags and mountainsides. To the east, the Moine Thrust is repeated, but here complicated by one of the nappes, the Kinlochewe Nappe, which actually turns the whole sequence upside down. But we can clearly see the Cambrian sandstones underlying the thrust forming a steep slope, and beneath them again the thick-bedded, pinkish Torridonian rocks, of which we had already seen so much on the hillsides around the Victoria Falls. But here there is something else displayed: the relationship between the Torridonian rocks and the Lewisian, with which this chapter began. Above the shore opposite to us, around Glen Bianasdail, the low, greyish, mound-like hills have a familiar cast. They recall the low hills around Scourie—and they are, indeed, made of Lewisian rocks. We can now see that the Torridonian sedimentary rocks lie on top of the Lewisian. Canisp, Suilven, and the rest of the western mountains rise above, and lie upon, an undulating basement of Lewisian gneiss; as Cadell of Grange put it in his colourful way: "These mountains stand like giant sentinels round the margin of a heaving sea of gneiss." Not only that, on Loch Maree we observe that the Torridonian beds fill in valleys in the original Lewisian surface; they are spread over an older landscape, as a builder might level irregular ground before commencing construction. The ancient surface is being exhumed as erosion slowly, very slowly, strips away the Torridonian rocks above. It reveals a hidden landscape. When we look over the Lewisian terrain we are, literally, looking back nearly a billion years into the past. Strip off the thin veneer of vegetation—it is no more than a flimsy film—and you can imagine walking through late Precambrian territory, while the rocks you are strolling upon provide an inkling of still earlier times. The Torridonian rocks have *not* been metamorphosed; equally, we know that the Lewisian rocks have been altered deep within the earth's crust. So it follows that the Lewisian had already been through its entire "life cycle" of formation, deep burial, metamorphism, uplift, and profound erosion, before the Torridonian rocks were even laid upon it. And the massive thickness of Torridonian must itself have been derived from the erosion of yet another mountain range* lying to the west on the ancient Laurentian continent—a range now banished to the far side of the Atlantic Ocean. An awesome vista of geological

* This is proved, for example, by the selection of pebbles found in the conglomerates in the Torridonian, including volcanic rocks, most of which cannot be matched by anything in the Lewisian gneiss.

time suddenly opens up before us, a vision of age after age of mountain-building, of continents remaking themselves, stretching far back into the distant reaches of the Precambrian. It should provoke a sense of our own insignificance, but it also stimulates a sense of wonder that we, alone among organisms, have been privileged to see these vanished worlds, and challenged to understand the immensity of time. A sudden squall makes us shiver, and turn up our collars.

The Precambrian is divided into two huge sections: the Archaean, greater than 2500 million years; and the Proterozoic, 2500–542 million years ago, underlying Cambrian and younger strata. Since the earth was formed 4550 million years ago, any rocks that might be discovered dating between then and 2500 million years will be Archaean, by definition. The oldest history of the Lewisian gneiss is Archaean; the Torridonian belongs with the later part of the Proterozoic. Our intimations on the shores of Loch Maree concerning the vastness of geological time were indeed correct: more than 2 billion years was laid out before us there. Most of earth's history lies in the time before shelly fossils of animals became common in Cambrian and overlying rocks. However, life was present on earth during most of Precambrian time, but it did not include large organisms with shells. Small fossils—thin threads and rods for the most part—miraculously survive from these distant eons. They endure because not all early rocks have been ground through the mill of metamorphism. In a few places, special rocks, mostly cherts, survive that are privileged time capsules carrying traces of these pioneering organisms. The earliest fossils of bacteria date to about 3500 million years ago, deep within the Archaean. The authenticity of these remains has recently been debated, but even if the earliest ones are chimeras, there is geochemical evidence that life has been present on earth for at least 3600 million years. The names used to divide the Precambrian are therefore inappropriate in the light of modern knowledge: there was proto-life before the Proterozoic, and the Archaean was not as barren as was once thought. The story of life is not my story here, but it cannot be excluded from the story of the plates because life and earth are meshed together, as we shall discover. Living organisms have been more than mere passengers on the shifting tectonic jigsaw.

There are many places in the world where Precambrian rocks are more extensively exposed than they are in north-west Scotland, and other places where there are rocks older than the Lewisian. However, the Highlands have a pivotal role in unscrambling Precambrian time. Although the Lewisian takes its name from the Isle of Lewis in the Outer Hebrides, where high-grade gneisses are abundant, it was on mainland Scotland that the main intellectual battles were fought. We have focused our attention there for this reason. When Eduard Suess wrote *Das Antlitz der Erde* little progress had been made in subdividing Precambrian time. The earliest rocks tended to be lumped together as "basement," or described as "fundamental," and then swept under the conceptual carpet. "Basement" still seems rather appropriate for something underlying the whole edifice of geological time. Even in Suess' day, although most serious scientists knew that the world was very old, the calculations designed to find out exactly *how* old all proved to be gross underestimates. They were based on false assumptions—for example, assuming an erroneous rate of cooling of the earth. More objective methods were to change everything. Arthur Holmes was a crucial figure in the development of radiometric dating techniques, especially Uranium-Lead methods, which improved upon the earlier estimates. He spent hour upon hour hand-cranking calculators, performing tasks that a modern computer would take a millisecond or two to complete. No matter, he and his colleagues established—admittedly with huge margins of error—that the earth had to be several billion, rather than hundreds of million, years old. The Precambrian consequently grew both in extent and importance. Between the two editions of "Holmes" (1944 and 1964) the earth roughly doubled in age. To understand more about the Precambrian became a research priority. Could you still invoke the same tectonic processes for deep time that were applicable to the last few hundred million years? And could such apparently intractable rocks as "fundamental" gneisses be further divided in the field? This is where the Scourie Dykes come back into the story.

We must return to the coast from our tour along Loch Maree. The little seaside town of Gairloch nearby is strung out along a broad bay. Every cottage is rendered for protection against the winter blast, and then whitewashed; the overall effect is pleasantly harmonious. The rain was still coming down in torrents when we arrived, and the place was empty, except for some bewildered but game Dutch tourists. The ground behind the town displayed what had by now become a familiar topography, the grey rounded hills of Lewisian country. The main road heads eastwards out of town through a stretch of boggy terrain, and after a couple of miles passes a small freshwater lake, Loch Tollaidh (Tollie), which, in the curious brooding light of a stormy day, manages to look both silvery and leaden at the same time. Like many lochs, it has a fish farm located within it, the long-term ecological consequences of which are coming under increasing scrutiny. The gneiss in the region, is, if anything, more strongly striped than it was at Scourie, blobby pink and grey seams picking out the strong foliation. Graham Park pointed out to me the outline of the Scourie Dykes intruding the gneiss on the hillside across the far side of the loch. A century ago the skilful Geological Survey geologist C. T. Clough mapped out the course of these dykes across country. It became clear that the dykes could be used to trace out broad "folds"* and similar structures within the Lewisian confusion: they provided a key to unlocking another phase of the remote past.

It may be helpful to recap the sequence of events. First, the Archaean crust was formed from tonalite; then this crust was buried and deeply metamorphosed to the granulite facies, implying temperatures of 1000° C and pressures of 10 kilobars. This tectonic trauma generated the banded gneiss we saw exposed in undulating, grey masses along the Scourie coast. Then came the intrusion of the dykes, which cut through the earlier rocks. All this history is laid out in the strata around Scourie Lodge, not far from the most northerly palm tree in the world. The ancient events are therefore referred to as Scourian, the oldest rocks in the British Isles. But, in the vicinity of Gairloch, later events have remoulded the earlier Scourian rocks. Insult has been added to injury, Pelion has been heaped upon Ossa. To the uninitiated, one gneiss, to be frank, looks rather like another. But that careful mapping around Loch Tollie proved that earlier structures, *including* the dykes, had been *re*folded by a subsequent tectonic event. This later event was termed Laxfordian.

Structural geologists love such things. They read the flexures in the rocks with the certainty of a blind man reading braille. They can easily imagine what it is for a fold system to be caught up in another, subsequent phase of contortion, fold imposed on fold. I admire this capacity to think in three dimensions—four, if you include time—more than I can say. In 1951, midway between the two editions of "Holmes," a paper by Janet Watson and her husband-colleague John Sutton outlined the history of the Lewisian in essentially its modern form. All the rocks in the Gairloch area had been re-metamorphosed—heated and baked again—and reset to a lower metamorphic degree than the granulites, known as the amphibolite facies. Less deep, less extreme. A mineralogist reads this change by logging

* Folding of foliation in metamorphic rocks is not the same as simple folding in sedimentary formations. Because metamorphic rocks are so altered and contorted, it is sometimes not possible to tell their original orientation. So structural geologists call a downfold a synform rather than syncline, and an upfold an antiform, not an anticline.

transformations that have occurred to the minerals within the gneisses—technically, the minerals are said to retrogress. Minerals never lie. They tell us exactly the conditions of temperature and pressure to which rocks have been subjected, betraying another phase in their "long continued ill usage." This is still a form of cookery. It is just that, in metamorphic geology, cookery can go backwards as well as forwards.

Now we just have to add dates. This is not quite as simple as it sounds. It is not like looking up a date on an old calendar. Dating relies on setting the radioactive "clock." This might happen when a mineral crystallizes out from magma; year zero in this case is truly a birth, followed by its own trajectory of aging as the radioactive clock winds down. But the clock can also be "reset" by a subsequent metamorphic event. Even the same zircon crystal can respond to one event after another—the inner part of a single crystal can be much older than the outer shell. Zircon is a kind of Methuselah among natural substances and seems to survive five times longer than almost any other mineral, which is why it contains such a useful "clock." Modern techniques can obtain separate dates even from within that single zircon crystal. Even the most enthusiastic geochronologist will tell you that all dates still require intelligent interpretation, but radiometric ages have transformed our understanding of the Precambrian, and they are getting better all the time. So here are the dates for north-west Scotland. The crust-forming event at Scourie (the oldest Lewisian if you like) happened 2960 million years ago—within the Archaean. The main metamorphism to granulites happened almost on the boundary between the Archaean and the Proterozoic, around 2500 million years ago. Graham Park and his colleagues have recognized a third phase, termed Inverian, following the Scourian, but predating the Laxfordian, which produces extensive shearing of the Scourie gneisses in several regions. As for the Scourie Dykes themselves, a sample from Teall's original locality has provided what is currently the best date for their intrusion at 2400 million years ago. Mr. MacIver would doubtless have raised both his eyebrows in astonishment had he realized the antiquity of the ground on which the lodge was built.

Then there is the subsequent phase in the long history of the Lewisian, when the dykes were folded—as traced out in the hills around Gairloch—and when the Laxfordian metamorphism remodelled the ancient gneiss to a whiter shade of grey. The best date at the moment from the Laxford area is 1740 million years ago. To put it in perspective: the events frozen in these ancient gneisses span more than twice the time that lies between the human race and the first trilobite.

As for the Torridonian sandstones overlying the Lewisian, those pink rocks that we saw by the Victoria Falls, as rainwater dribbled down our necks, prove to be some 800 million years old; the oldest sandstone group around the village of Stoer may be as much as 990 million years old. So, for a billion years or more, the area occupied by Lewisian gneiss was eroded, ground down grain by grain, until time had breached a window deep into the ancient crust. Then it was blanketed by the sediments that now form Suilven and Stac Pollaidh. The undulating grey Lewisian landscape that we admire today would have looked much the same long before the first trilobite scurried in the mud, or the first snail crawled upon the sea floor. And it will endure long after our species has vanished from the world. "Time, that agèd nurse / Rock'd me to patience."

This is not yet the end of the story. Around Gairloch, and extending inland towards Loch Maree, there is another group of rocks that Graham Park has been studying for many years. They are deeply entrained within the Lewisian gneiss, and have been affected by the same, Laxfordian, metamorphism. But they are rocks that originally lay on *top* of the gneisses. Graham's careful mapping, published by the Geological Society of London, shows that they occupy two synforms within the Lewisian, one of which is now followed on the ground by a little stream running along the charmingly named Flowerdale. On the other side of the valley, volcanic rocks and sediments, now transformed by metamorphism, provide yet another window opening on to the ancient of days, for these rocks are almost as old as the gneisses they cover, though they embrace a different range of geology. They tell us about what was going on under the sea.

The Old Inn at Gairloch looked singularly welcoming in the rain, but we had to forgo the comfort of a pint of bitter by the fire to look at the rocks. Just behind the pub there are some heavy-looking, greenish-black rocks, which are densely foliated. These are amphibolites, which started out life as lavas with a basalt composition. They were erupted on the sea floor about 2 billion years ago and have been compared with oceanic plateau basalts. A little way up the road the rocks take on a kind of brownish cast—a little subtle for my palaeontologist's eye. Graham assured me that these were once sedimentary rocks lying atop the basalts: their chemistry is just what would be expected from sediments associated with an island arc. The alchemy of metamorphism has transformed them into different minerals, to be sure, but they have retained their identity at the most fundamental level of atoms and molecules. To follow the story further we had to walk along a muddy path up the river valley. On a good day it would be a delightful stroll; in the drizzle, the bracken-covered hills fading into distant mistiness retained a flavour of Celtic mystery that Sir Walter Scott would have relished. Even the crude stone walls were made from "fundamental" rocks. What better way to be led back to the beginnings of things?

After about half a mile we paused to examine some rocks in a slippery bank. The dark amphibolite was still with us, but interleaved with it was something different. To one raised on younger rocks, the black chunks of hard rock we levered out from these layers looked a little like chert, a form of silica, but in the hand they seemed much too heavy for this common material. Examined close up, fine banding was faintly displayed within the rock. These apparently humble pieces of stone proved to be one of the defining rock types of the Precambrian: the banded iron formation, commonly known as BIFs. Their characteristic heaviness is because of their high iron content—in some localities they could be even used as ore. BIFs are another sedimentary rock, but one that is unique to the early earth and that tells us much about its peculiarities. On the far side of the bank, we scraped out some pieces of lighter-coloured stone, which, if I had happened to have a bottle of dilute acid on my person, would have fizzed merrily on being doused. This was a marble, and would have been a limestone before its transformation by heat and pressure—the calcium carbonate of which it is made reacts with acid. No sculptor would have wished to expend his skill on these mean little pieces, but their significance is not aesthetic. Limestones are one of the commonest marine sedimentary rocks through the geological record, and evidently were still being formed 2 billion years ago. And just underneath the limestone there were other dark rocks with a familiar look that split into thin sheets. At first glance they resembled the marine shales they once were, but they, too, have been transformed—into chlorite schists. Among them are bands that chemical analysis shows have a high content of carbon, and the principal source of carbon in shales is organic, for carbon is the element most basic to biology. Is there a hint here of the presence of oceanic life?

Just on cue, the sun came out, as if to cast light on the mysteries of deep time. The yellowing leaves of the birch trees almost glowed. We stood discussing how to fit together rationally all the features that we had seen along Flowerdale. The rocks exposed in the flank of the valley seem originally to have been a piece of the sea floor: volcanic rocks at the base, with various kinds of sedimentary rocks accumulating on top. Limestones and shales are familiar enough today, so evidently the sedimentary processes happening 2 billion years ago were not different in some respects from those operating now. One striking fact is that the metamorphosed sedimentary rocks have patently been incorporated into the Lewisian "basement." Once tempered by deep cooking, they became one with the realm of old gneisses. So rocks derived from ancient oceans subsequently became *part* of the Archaean and early Proterozoic shields, which are the oldest and now most stable parts of the earth's crust. Cratons were created. Shields were sealed. Continental crust grew by a process of "plastering on"—by marginal accretion of bits of ocean crust and sediments. The places such "additions" occur lie along subduction zones. After all, younger zones of subduction still leave their legacy in great tracts of rock added to the edges of older continents, just as the fold belt of the Appalachian-Caledonian chain was soldered onto edge of the Canadian Shield. The igneous tonalites that comprise the oldest Lewisian rocks may be explained by a process of partial melting not so different in kind from that operating today at depth on active continental margins. It seems that plate explanations can, indeed, be applied to the Methuselahs of rock, the ancient of days. Thus the apparently endless and uniform gneiss begins to reveal its secrets. Those grey hills are not, after all, unreadable, although the script may be distorted. To understand processes that happened billions of years ago, the present is still—at least, up to a point—the key to the past.

Now we can create a vision of the Archaean world. Heat flow from the young earth was greater than it is now. What Charles Lapworth called the "great earth engine" cranked over faster. There may have been more numerous convection cells within the mantle. Recall how thick soup over a high flame breaks into bubbling fury, while a simmering broth swirls over in a few dignified turns. The nascent masses of continental crust had not yet congealed to their present size. Instead, smaller rafts of lighter rocks formed the nuclei of what would become more stable continental areas. The cycles of the earth—the generation and destruction of plates—probably happened *andante cantabile* rather than *largo*. The sea—and there was certainly an ocean—was whipped up by storms, reducing all land newly elevated by tectonics to sedimentary waste. Slabs of oceanic rock were covered with sediments derived from the rapid weathering of the proto-continents. Unprotected by any cloak of plants, the wind and rain worked fast upon the naked rocks. This was a world of tempests and flash floods, of jagged crags and dunes. Clays and grits, the mucky progeny of erosion, slumped down into deep water. Volcanic rocks were erupted over the sea floor—many of them associated with a multitude of "hot spots" that pocked the early earth like a plague of boils. Then sediments and volcanics alike were plastered onto proto-continents at the subduction zone*—or perhaps were sandwiched between two proto-continents that collided. It has been claimed that massive marine plateau basalts were added to the *base* of proto-continents, because they were reluctant to subduct, like sticky toffee that adheres to the gullet. Gradually, continental nuclei grew, accreted. Buoyed up by their less dense composition, the growing continents bobbed onwards, as would rafts of cork on a mill-pool, while heavy oceanic crust was created and then destroyed in the early cycles of plate tectonics. Smaller continents merged one with another. Thus cratons were born, which were to drift hither and yon for the rest of earth's history, the stablest partners in the dance of the plates. With further erosion, sedimentary rocks were laid down upon the cratons. A few portions of such sediments were to escape all the subsequent vicissitudes of continental collisions, to carry forward to our own era a fossil record of the earliest life. So the world acquired the first of its many faces.

Ancient Precambrian proto-continents often preserve tracts of volcanic and sedimentary deposits—now, of course, metamorphosed—along their edges. They may represent the line along which two proto-continents collided, like a Ural range of the earliest days—seams so ancient that they are often no more than suggestions sketched in pointed minerals, or linear bandages interposed between round masses of granulite. In these "greenstone belts," the rocks are green because of the abundance of the metamorphic mineral chlorite, as well as epidote and hornblende, more greens. The belts are often kilometres wide, as they are in the ancient shield regions of Canada, southern Africa, and western Australia. The rocks along Flowerdale exhibited features on the small scale that apply to greenstone belts on the large scale, except that they are often less metamorphosed and the geology is even more complicated. The belts are frequently rich in precious metals, distilled and refined from the early and turbulent earth: gold, silver, copper, nickel. South of Hudson Bay, a swathe of greenstone belts passes east to west across Manitoba, northern Ontario, and western Quebec. The Red Lake Greenstone Belt in Ontario is rich in gold, while the Abitibi Belt to the east not only boasts famous gold mines around the town of Tim-

* It has recently been suggested that subduction zones in the early earth dipped downwards at a lower angle than at present, producing the tonalite magmas, which have distinctive signals in their geochemistry (especially the element Neodymium).

A fossil stromatolite from Precambrian (early Proterozoic) rocks of eastern Siberia. The fine layers are typical of the rock type.

mins, but also has deposits of silver, copper, and tin elsewhere—gifts from the early Precambrian.

But it was a different world, this early one, for all that the same engines drove it. Most important, the atmosphere was initially almost without oxygen, while gaseous hydrogen sulphide and methane were abundant. Breathing this poisoned air would have killed us very quickly. At Solfatara, on the Bay of Naples, we had the merest hint of such a vile-smelling miasma, which would have been like a thickly suffocating blanket. Most of the oxygen in the atmosphere today is a by-product of life. Three billion years of photosynthesis, much of it achieved by very simple organisms like blue-green bacteria, added oxygen molecule by molecule to the air. The minute rod- or thread-like remains of the bacteria that changed the world have been found among the earliest fossils. They formed sticky mats, which left their record in finely layered, crimped, or cushion-like fossils called stromatolites, looking like piles of petrified flaky pastry. These survived to the present where they were formed on top of continental fragments that achieved early stability. A well-known example is the Fig Tree chert of southern Africa. Still earlier bacteria could thrive in the *absence* of oxygen—in fact oxygen is a poison to them. These most primitive forms of life survive, indeed prosper, in hot, sulphurous springs and in stinking mud. If we want to know how life started, it is these oddly tough organisms we should consult. Subsequently, life changed the air of the world—scrubbed out the early poisons—and in the process even modified the way rocks decayed, since oxygen is crucial to all sorts of chemical weathering.

Banded iron formations (BIFs), a few dark pieces of which we collected in the drizzle in Flowerdale, were typical of a period 2600 to 1800 million years ago. When these rocks are polished and examined through a microscope they are seen to be composed of alternating layers a few millimetres thick. Dark layers are iron-rich, often made of the mineral magnetite, Fe_3O_4, an iron oxide, alternating with lighter bands of common silica; stripes of a different colour. The fact that the iron is oxidized in this curious sediment indicates that there must have been oxygen around to do it. But the peculiar banding, and the occurrence of magnetite, demands a special explanation—and one of general applicability, too, because BIFs are found on virtually every Precambrian shield. Iron is very greedy to combine with oxygen. Iron artefacts quickly become horrid lumps of rust when buried in the ground. One explanation for the existence of BIFs that has found favour derives the oxygen from microscopic photosynthesizing bacteria (or algae) within the ocean. Iron is weathered from continental rocks, but there is not yet enough free atmospheric oxygen to gobble it all up, so it enters the sea in solution as positively charged ions. These combine with oxygen derived from the photosynthesizing scum, and at once heavy, insoluble magnetite is formed, which falls to the ocean floor as a gentle rain of minute black specks. Magnetite is the form of iron oxide that has both the most iron and the least oxygen—which implies that the latter was at a premium and had to be scavenged by the greedy iron ions. But biological productivity soon outstripped the available iron, and, so the theory goes, the burgeoning multitude of organisms became a victim of their own success, undergoing a "bloom" that became lethal. Similar plankton blooms still happen in the sea today. They might even have been poisoned by "excess" oxygen. During the quiescent phase that followed, only silica accumulated on the sea floor, until the whole boom-bust cycle started again. It is much more complex than I have described, for several different kinds of BIF have now been recognized, with hematite (Fe_2O_3), or siderite, iron carbonate ($FeCO_3$), as important components. Nonetheless, the theory does conjure up a picture of a simple ecosystem, a lurching, fluctuating one, without the checks and balances that many millions of years of organic evolution have since imparted to the marine realm. However, some plausible explanations of BIFs have been advanced that do not require the intercession of life. For example, it has been claimed that they can be explained by fluctuations in sea level, combined with a high supply of iron from hot fluids derived from the early crust and a low supply of atmospheric oxygen. The definitive explanation still eludes us. The earlier Precambrian was a strange place, with hints of what we know, but much more that is unfamiliar. The past, as L. P. Hartley said, is another country. They do things differently there.

However different they are from anything we know today, BIFs continued to be laid down for 800 or 900 million years—much more time than has elapsed since the beginning of the Cambrian and which embraces the entire proliferation of animal life—five times as long as the "Age of Dinosaurs." BIFs can be hundreds of metres thick and extend for a thousand kilometres. Our specimen from Flowerdale was the least impressive example. Just try to imagine how much iron is locked up inside them. The earth was remoulding itself throughout this long period, following the bidding of plate movements, and the atmosphere was changing at the same time. The whole planet was evolving. Life and earth are forever intertwined. The critic John Ruskin wrote in praise of rust, how it paints and dapples; but without oxygen there could *be* no rust. Iron meteorites, whose composition resembles that of the earth's core, arrive from space metallic black and pristine: it is only contact with earth's corrupting atmosphere that paints them rusty brown. Without the free oxygen generated by bacteria and plants there would have been no breathing, no animal life, no trilobites, no dinosaurs, no writer of this book, no reader. And no rust.

There are other rocks that help reveal the early earth. Komatiites are peculiar glassy lavas that are typically found in Archaean terranes. They cooled rapidly from magmas with compositions like that of the mantle. Inside these black rocks there are curious little bunches of olivine crystals that are said to display spinifex texture. The crystals radiate from a common centre into three dimensions, a little like the dark traces produced as mines explode. The name of the texture has a peculiar resonance for those of us who have worked in the outback of Australia. Spinifex is a ghastly, glaucous prickly grass that covers great tracts in the middle of nowhere. It, too, radiates from a common base. Each of its shoots, modified leaves, is tipped with a barb of silica that breaks off into your foot if you are foolish enough to stand upon it. If you light a match anywhere near it, spinifex bursts into flames and seeks to turn you into a crisp. But it does look exactly like the crystals within komatiites. Lavas with komatiite composition are proof of the higher heat flow in those primeval times; the peculiar olivine crystals grew downwards like stalactites from the cooled "roofs" of the lavas.

Archaean and early Proterozoic shields comprise the middle of continents—the Hebridean area that we examined is unusual in being close to the edge of an ocean. They are the old hearts, the obstinate cores, the unchanging and persistent bulwarks against which subsequent history has piled its afterthoughts. After all, in a world that is 4550 million years old, the last 1000 million years are less than Act Three in the drama of creation. Archaean and earlier Proterozoic rocks and greenstone belts make up the old Baltic Shield, underlying the territory where Lapps follow reindeer through lichen-covered and boggy wastes. They underlie the centre of Canada—the Canadian Shield, all stunted fir and endless lakes—and they extend southwards beneath the United States. The ground-down, acacia-sprinkled central veld of southern Africa, including the famous rocks of Barberton in the Transvaal; the dusty plains of much of the Indian peninsula; the rain-sodden northern parts of Brazil; many of the gold-rich regions of Australia, those plains carrying a few low knobs and ridges of fretted rock amidst the stony wastes, or rugged Pilbara in Western Australia—all these terrains are older than we can readily imagine. They display a similar pattern, though with many local variations: old igneous masses—often granulites—form nuclei, around the edges of which greenstone belts have been wrapped or plastered on. They show us the baked and exhumed remnants of splitting and clinching of former microcontinents from a time so long ago. When you are talking about 3 billion years, you could be forgiven for wondering what difference the odd 10 million years or so really makes. Yet it is likely that the earth in those distant days was as rich in events as it has always been, but that the detail has been fogged by metamorphism, or scrubbed away by erosion. Geologists interested in such remote eras have a task comparable to reconstructing a vanished civilization from a thumb and forefinger broken off a sculpture and a few shards of pottery; indeed, for many years geologically incredibly old areas seemed beyond the reach of ratiocination. Largely as a result of being thought about in terms of tectonics and geochemistry, and then being able to date them, such rocky Methuselahs are now capable of being understood, even though all our perceptions are inevitably blurred through the mists of time. As D. H. Lawrence wrote:

> *mists*
> *of mistiness complicated into knots and clots that barge*
> *about*
> *and bump on one another and explode into more mist,*
> *or don't*
> *mist of energy most scientific*

. . .

Plate tectonic processes evidently carried on from Precambrian into Cambrian times, and thence to the present day. By reversing time's arrow, the movements of continents can be tracked ever further back into the past. There is probably no harder exercise for the imagination within the pages of this book. We have already traced the wandering continents to the time before the assembly of Pangaea. We have seen how "palaeomag" can be combined with field geology to reveal another, still older world in which the continents were again scattered, as they are today, but with all the pieces in different places and arranged in a different way, as if the earth's surface were a collage reassembled by an amnesiac. To reach this stage, we have gone back perhaps 500 million years, only about one-eighth of the history of the rocks on the planet. There is so much further to go. We must travel backwards, much further backwards in time, following the bidding of the plates, if we are to attempt to find out how the "knots and clots that barge /about /and bump on one another" were laid out in the eons before the Palaeozoic. We have to let go of even the most rudimentary outlines of the continents we know, some of which accompanied us back as far as Cambrian days. We have to go to worlds without familiar signposts, lost worlds, uncertain worlds.

There is, though, a certain logic to it. If Pangaea split apart a couple of hundred million years ago, but was itself assembled from dispersed continents, then is it not likely that there might be a still older "Pangaea" when the continents were married on a previous occasion? There is time enough, and more. Perhaps there have even been a number of phases when supercontinents dominated the earth, so that the progress of the continents was like one of those eighteenth-century dances in which the participants move apart at one stage, only to repeatedly find one another again in the centre of the ballroom on an appropriate prompt from the music. There is now evidence for no less than four Precambrian supercontinents: at 2500 million years ago—that is, on the boundary between the Archaean and Proterozoic; then at about 1500 million years; again at approximately 1 billion years; and once more at about 80 million years before the beginning of the Cambrian period, approximately 625 million years ago, in the late Proterozoic. I say "approximately" advisedly, because it is very difficult to be precise. The assembly of a supercontinent is a time when most subduction has stopped, and hence most of the activity that might lead to a radiometric date has also become quiescent. It is a kind of "hole" in history.

Imagine being in the middle of the latest Archaean supercontinent, the first one on the face of the earth. Prior to this time, the continents had mostly been mere fragments, furiously jostling. But now, a crudely rugged landscape, unsoftened by any green vegetation, would extend to the horizon for thousands of kilometres in all directions. You would have to be alert to the possibility of being carried away by a flash flood emanating from the granulite hills thirty kilometres or so away. The ground around you is littered with angular boulders carried by the furious waters; feldspars glisten in the relentless sun. A hot spring bubbling and hissing at your feet is surrounded by brilliant orange and livid purple stains. They feel slick to the touch. The colours are painted by bacterial life that forms a slimy film; the lively hues are pigments that shield the tiny cells from harmful rays in the harsh sunlight. From here, you cannot see the tacky, green mats in the sea that are slowly transforming the earth's atmosphere; the ocean is a kinder place, a womb for the future. But here in the wilderness, you don't know which organisms are going to inherit the earth. The photosynthesis bug is no racing cert; it could equally well be a bacterium that gorges on sulphur and relishes boiling-hot conditions.

The early supercontinents juxtaposed parts of the present earth in ways that we would find most surprising. Just think about our small patch of Scotland. It was, of course, part of ancient North America (Laurentia). But in the Late Proterozoic supercontinent, Laurentia lay alongside, of all places, South America. The little patch with the Scourie Dykes was not so far from the South Pole. Evidence from matching geological ages and structures turns the world topsy-turvy. What, you wonder, would Professor Suess have made of it all, this world so stripped of fixity? Even those of us who have had a whole working life to get used to the idea of mobile continents have a struggle to remake the world so completely. But what we can accept is that continents cannot jump from one place to another. They have to move in a logical procession in order to make kinematic sense. So we know about the slow march of the polar wandering path for north-west Scotland, from near the South Pole to the northern hemisphere over a billion years, and it is a pattern that has to fit in with other evidence from separate continents and geology. The ancient supercontinents, once so speculative, have started to achieve the respectability that comes with a name. The continental mass at around a billion years ago has been called Rodinia, and now scientists are starting to bandy that name about with the familiarity that Pangaea attained forty years ago. There are other names that have yet to achieve widespread currency—"Kenorland" in the late Archaean is one—but who knows whether they will be part of scientific consciousness in fifty years? It is extraordinary how quickly names useful

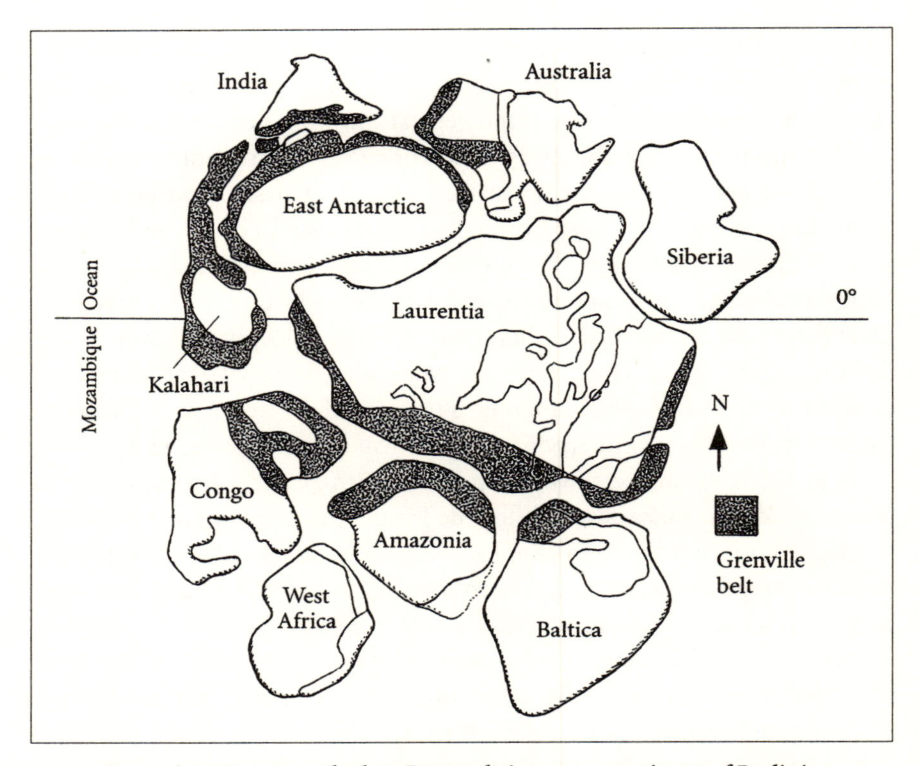

An ancient Pangaea: the late Precambrian supercontinent of Rodinia. Laurentia as its core is easily recognizable. The ancient mountain belt shaded marks the line of "stitching together" of the supercontinent from the two still earlier separated continents which collided. Toujours la même chose.

in scientific discourse become incorporated into the language. Soon, nearly everyone forgets who coined the name, or—more importantly—who first understood the concept that the name embodies. The first supercontinent to be named was Gondwana, but who now remembers that Eduard Suess was present at its christening? Yet without the story of Gondwana and Pangaea there would have been no vision of Rodinia, nor of any of the other continents united together deep into Precambrian times.

A mid-Proterozoic, 1500-million-year supercontinent is also becoming widely accepted as a fact, although at the time of writing it does not have a generally used name (if it were up to me, I would call it Suessia). It was remarkable for including masses of igneous intrusions, now mostly situated in North America, which do not seem to be related to orogeny in the usual way. Precambrian geologists, if they may be so termed, have had much amusement devising model systems to account for them. The igneous rocks include a widely used ornamental stone—the rapakivi granite, named for its type locality in Scandinavia. The polished stone can be recognized at once because it looks as if it is made out of sawn-through eggs floating in a dark background. Look more carefully and you will see that the centre of each "egg" is a very large, rounded-off crystal of pinkish feldspar which has been surrounded by a thin "shell" of a different mineral, usually a greenish-looking rim of another feldspar. It is quite easy to "read" this rock. The large feldspars must have crystallized out of the magma first, but then the surrounding magma changed. The early crystals reacted with the liquor to brew up the "shell" of surrounding (oligoclase) feldspars as the rest of the rock completed its cooling. A later event modified an earlier one: it is not so different in logic from the story of the Scourie Dykes. A rapakivi granite forms the counter of a bar at Paddington Station in London. I come through this station most days. Although Paddington is one of Isambard Kingdom Brunel's finest engineering achievements, the bar is a recent addition. If you have just missed your train, you can at least lean on a bar that is 1500 million years old and reflect that perhaps half an hour is not that serious a delay.

Your fellow commuters might be suprised to learn that the shiny surface of the bar was in fact a piece of a continent more ancient than Pangaea, but one that broke up in the same way as India did from Africa; and that when the pieces came together again, 500 million years later, it marked the formation of the *next* supercontinent, Rodinia, a billion years ago. The Grenville mountain chain was thrown up as a consequence of this rapprochement, which towered all along the eastern edge of the old Laurentian shield and made it possible eventually to build skyscrapers in New York on solid foundations.

Those parts of the rough mosaic of plates that were composed of continental crust moved, now together, now apart, in cycle after cycle. Gradually, these lighter continental slabs grew larger, as the squeezing and melting connected with subduction added material to their edges. The days, too, were getting longer as the Precambrian progressed, because the earth slowed on its spinning axis, just as oxygen trickled molecule by molecule into the atmosphere. All these changes, while not exactly pure Charles Lyell, are explicable in terms of modern processes, a transformation rather than a revolution. But there have also been events in the Precambrian that have been claimed as utterly unlike any that followed.

Of these, one that has achieved most notoriety is the "Snowball Earth" hypothesis, not least because it has such a memorable label. It joins the Big Bang, Black Hole, and Selfish Gene in the pantheon of memorable titles for scientific concepts. The idea is simple enough: that there was a time, or times, when the earth froze from end to end. We have already encountered the ice age of the Carboniferous, and there was another at the end of the Ordovician period; but no ice age after the Cambrian refrigerated the whole world. If this really happened, it would have had an obvious effect on the progress of life, though it is even more superfluous to note that life survived. The rocks produced during an ice age are rather distinctive, so that the evidence required of a global event is that such rocks should cover the whole world at the time. One of the chief proponents of the theory is Paul Hoffman at Harvard, one of those American academics who seem to have twice their fair share of energy. If expertise is defined as knowing more and more about less and less, I am at a loss to describe what it is to know more and more about more and more—but that is the Hoffman condition.

Along the bleak shores of the northern peninsula of Spitsbergen known as Ny Friesland there is little in the way of creature comfort but much to interest another brilliant obsessive—Brian Harland, from the Geology Department at Cambridge. Brian is the authority on the Arctic islands of Svalbard, of which Spitsbergen is the major island, and for many years he ran Cambridge University expeditions there. As befits a Quaker, Brian's expeditions were marvellously organized, but trimmed of all frills. I did my doctoral work in Ny Friesland on Ordovician fossils, so I know the area well. Thirty-five years ago we travelled by small boat towards the rocks that I was about to map, passing by a succession of late Proterozoic strata laid out like a spread deck of cards along the shore. We ascended, rock bed by rock bed, through geological time as we moved southwards, while screaming Arctic terns berated us from the air, urging us ever onwards. The rocks here were younger than the red sandstones of Loch Torridon and Canisp, and very different in character. They were carbonate sedimentary rocks—dolomites and limestones. We knew that they were laid down under the warmth of a tropical sun and that provided an indication of how much the latitude of Spitsbergen had changed over hundreds of millions of years. Rare fossils of marine algae had been discovered in the same rock formations. I knew, too, that the Cambrian and Ordovician strata that lay ahead were also limestones of the same general type, with the important difference that they contained myriads of fossil shells of extinct kinds of animals, including the trilobites that I was to study for several years. The sequence of rocks in Ny Friesland evidently spanned that (literally) vital time in the history of the earth between the later part of the Precambrian and the "explosion" of animal life that happened at the base of the Cambrian period. However, between the two "packages" of warm-water carbonate sedimentary rocks something very curious happened, something which Brian Harland had pointed out years before the "Snowball Earth" theory was proposed. For a brief interval the carbonates disappeared, and in their stead some utterly different rocks took the ground. Prominent among them was a reddish rock that might be best described as a kind of pudding, full of boulders and blocks of various kinds and sizes, all suspended in a pink mud. To a student of glacial phenomena this rock would seem quite familiar. When icebergs have "calved" from major glaciers to drift out to sea, they deposit just such a heterogeneous dog's dinner as they melt, dropping out the collection of stones they carried, along with the ground-up waste derived from the terrain they scoured. Now we look at an individual pebble we can see it even carries scratches acquired during its sojourn in the ice sheet. So here is the paradox: a glacial event extending so far into low latitudes as to drop its debris where, not long before, tropical algae and bacteria has flourished on a limy sea floor. There was no denying the facts: rocks do not lie.

Hoffman vigorously extended these pioneering researches to demonstrate that similar changes could be recognized globally in strata deposited at about 590 million years ago. "Snowball Earth" had been born. There was a vision of the whole planet engulfed in a shimmering ice sheet, from pole to pole. Since ice reflects light and heat, from space the earth would then have resembled nothing so much as a giant pearl. Furthermore, since the Cambrian "Big Bang" in the history of life occurred, geologically speaking, not so very long after the alleged great freeze had relented, it was tempting to link the two as cause and effect. The survivors, released from the grip of global refrigeration, evolved into new ecosystems, reset the evolutionary clock, were stimulated into unparalleled innovation, and so on. The shells that I collected from the hard limestones on the frigid shores of Spitsbergen were ultimately a consequence of an even chillier past. The explanation does, it must be said, have a compelling narrative drive. Of course, there had to be places for life to hang on through the icy crisis, and deep-sea hot vents and similar unfreezable but feasible refuges were cited. Algae that needed light might have hung out on the edges of hot springs. The "snowball" showed plenty of evidence of unfreezing rapidly: the speedy reappearance of tropical limestones in the Cambrian age rocks in Spitsbergen was one example. Hoffman's idea was that during the "snowball" phase, volcanic gases like carbon dioxide (for eruptions would not have been affected by the freezing of all that water) accumulated in the atmosphere, building up the greenhouse effect, until a critical level of global warming was reached. Then the ice sheets would have melted dramatically—even over a few decades—and seas would have flooded back over the continents, and marine life could get on with the job in the nutritious shallow shelf seas that followed.

As with many big ideas, the theory is not without its critics. They point to the fact that more than two-thirds of the earth's surface was covered with ocean, then as now, and since all 590-million-year-old ocean crust has long since been subducted to oblivion, there will never be proof that every part of that vast hinterland became deeply frozen. Others dispute the evidence of complete coverage by ice, and there are indeed places where the evidence is less convincing than in Spitsbergen. Then again, a few fossils of complex animals that antedate the big freeze have been recognized in a few localities, though these, too, are controversial. Some, but by no means all, molecular biologists claim that you need much more time in the Precambrian than 590–542 million years to account for the evolution of all the different designs of animals that appeared, apparently suddenly, close to the base of the Cambrian period. Currently, the debate is at that interesting stage where we don't know how the chips will fall: the next decade will probably decide. The intellectual brew has been enriched by Professor Hoffman's imaginative younger colleague Joe Kirshvink, from the California Institute of Technology, who has recognized up to four "Snowball Earths" between 900 and 590 million years ago. All these have deposits of glacial origin associated with them. Recently, Joe has upped the stakes still further by adding another, far older one—almost as old as the grey gneisses of Laxford and Scourie—at about 2 billion years. Glacial rocks that had miraculously survived from this distant era almost unscathed were discovered among the termite mounds and stunted trees in what is now the Kalahari Desert in southern Africa. What is amazing is that the "palaeomag" in the ancient boulder beds also seems to have survived the passage of billions of years. The readings indicated that the glacial beds were laid down near to the equator of the time: hence Joe's conviction that this was another global event, ice from pole to equator. Far from being a unique event, a frozen earth was allegedly a repeated risk during the Precambrian. I do find it remarkable that life weathered these frigid crises—if they really happened. We know that life existed before the earliest one, and traces of bacteria and algal mats appeared between the times when frigid conditions are supposed to have reigned. To adapt Oscar Wilde's famous line, to survive one snowball is fortunate, but to survive several looks like miraculousness.

We have moved back to earlier and earlier times, but we have yet to answer the question: where are the oldest rocks on earth? Finding the definitive answer is not straightforward. With regard to dates, a record is always liable to be broken by a whisker when a different zircon crystal from the right ancient locality is analysed. Then there is the question of the accuracy of measurement, which is serious if the "ruler" happens to be a radiometric date because different methods can give different results. The really important fact is that older and older dates are rarer and rarer. You would expect to have trouble finding ancient crust because, as we have seen, it is likely to have been remobilized by younger tectonic events as the old earth cranks over and over. Hence younger dates will be imprinted over older ones as the radioactive clocks are "reset." It has proved extremely difficult to discover rocks that are any older than about 3800–3900 million years. There are just a few places where the planet's plates have left small fragments of this antiquity relatively unscathed, scraps from early phases in the earth's history. Endurance may be a matter of luck, perhaps, of survival against the odds, like the grizzled old warrior who emerges from the battlefield with scarcely a scratch while his companions at arms are felled around him. This is the earliest time made tangible. We can put our hands upon the rock, stroke it, examine it with lenses or attack it with sophisticated analytical probes. It may look at first glance like yet more gneiss, but it is precious stuff.

A still more distant period lies prior to the oldest rocks. Dates from meteoritic material surviving from the early days of the solar system reveal that the earth formed at about 4550 million years ago. Hence it follows that there is a "missing" piece of history, an eon suited only for speculation and conjecture. Something like 600 million years is almost unrepresented by rocks. This period has come to be known as the Hadean, and perhaps appropriately so, for conditions during at least part of it were as furiously inhospitable as hell-fire itself. As the earth gained mass by accretion of planetismals, its increasing gravity attracted massive meteorite bombardments. The energy imparted by massive and relentless impacts eventually caused the melting of the planet. It became a sea of lava, but still the impacts continued. Most planetary scientists believe that the moon was the product of a massive impact. It spun off as a gob of debris. It continues to edge its way slowly further from the parent planet billions of years later, but still blatantly retains the legacy of the time in the meteorite impacts it received upon its surface so long ago. It is likely that both the moon and the earth continued to be bombarded by meteorites until about 3800 million years ago. The iron core of the earth formed while the planet was near molten, and at that point the elements were parcelled out according to their chemical fancies. This was a period when the ground itself was constructed, the dance floor set, before the slow dance of the plates could even begin. It remains a time of mystery, a morass of unanswered questions. However, hints towards some answers are preserved in the oldest rocks.

These great-grandfathers of all gneisses have attracted much attention from geologists, particularly those concerned with the early history of life. There are a handful of places from which rocks yielding ages up to a maximum of 3900 million years have been collected; and there are hints of still deeper time. These sites include the southern edge of the Churchill Province of the Canadian Shield in Montana; the most ancient part of the shield areas of Western Australia; and the Isua rocks of Greenland. The centre of Greenland is covered by a vast ice cap, but rocks poke out around the edges of the island, forming a rocky fringe, a kind of lithic tonsure around the bald centre. About 150 kilometres north of the capital Nuuk on the west coast, the oldest greenstone belt on earth at Isua takes up a narrow strip of barren outcrop about thirty kilometres long. It is a primeval sliver, underlying everything else. The belt was recognized more than thirty years ago, largely because it included the earliest banded iron formation, which gave a distinct signal to a gravity survey. The whole belt had been heated, of course—perhaps to some 600° C—but there were little pockets here and there that seemed to have escaped the punishment due to them after so long a sojourn on the planet. Even some of the "pillow" structures of the submarine basalts have been preserved. Like other greenstone belts, the Isua rocks also include metamorphosed sediments. They prove that there was already enough water on the planet when the rocks were originally laid down to erode still earlier formations; though old, they were not yet the beginning. A vision of still deeper time opens up, time beyond time. You can almost hear the sea breaking on some of the first shores, just as in summer it still tumbles against the margins of Greenland, and then you understand that of all things on earth the sea is most nearly immortal. Through all the reconfigurations of the face of the earth, the sea has endured. Beneath the frozen ice caps that may have once covered the earth, the sea still cradled life. It may be the sea that is the ancient of days.

Yet the sea itself originated, or was gathered together, during the Hadean. An ocean could not have survived during early, molten, violent days. Only when water could condense from hissing, steaming volcanoes—when the cooking-pot of the earth cooled sufficiently—only then could pools and basins survive. The impact on earth of many comets made of dirty ice may have contributed more water—quite how much is still debated. But one fact is certain: without water, there could have been no life. The origin of life on earth and the early history of the crust intersect on the barren, glaciated wastes of Greenland. Enough is now known about the most primitive living bacteria to suggest that all living organisms lying at the base of the tree of life relished high temperatures. A nearly boiling habitat was their land flowing with milk and honey. Recall that these hyperthermophiles do not need oxygen to grow—indeed, for many of them it is a poison. In short, they suited the early earth: hot and anoxic. But were they there? In 1996, a study was carried out upon minute graphite (metamorphosed carbon) specks within crystals of the mineral apatite preserved in some ancient Greenland rocks that have been compared with those at Isua. The negative values of the isotope of carbon (13C) obtained from this sample are typical of the carbon compounds produced by living organisms. Cooked though it was, the stamp of life was there—or so it was claimed. To look at one of the biggest questions about our planet, the scientist rootles around inside a single crystal with an ion microprobe. If life was there 3.8 million years ago, it would have experienced a day only five hours long, as a result of the earth spinning faster, and the moon would have hung large and close in the sky.*

As I write, still older dates from zircon crystals are starting to fill in the mysterious Hadean. The oldest of all is 4400 million years, from the middle of a crystal recovered from the Jack Hills of Western Australia. It is within a whisker of the earliest days. The ion microprobe has dug into this token from the aftermath of Hades. If the oxygen isotopes recovered from inside the tiny time capsule are to be believed, there was free water already present on earth even at this extraordinarily early time. The timetables for Hades might have to be rescheduled. There is still so much to learn. The intimate history of the early earth can only be recovered from hints and whispers

* The moon is moving away from the earth at 2.5 inches (6 cm) a year.

buried inside crystals. We are only just beginning to translate the story they have to tell.

The eternal sea breaks heavily over the Lewisian gneiss and the dykes of Scourie. The wind whistles through the gorse bushes, but even in October there are a few bright yellow pea flowers hidden among the spines. Looking across the bay, the greyness of the undulating ancient rocks continues onwards to the dense, tossing ocean. The horizon is obscured in mist, and the distant islets fade progressively. Scanning the distance is like trying to penetrate Precambrian time. The furthest objects are hard to see, and so much detail is lost. Generations of geologists have sat in much the same place and contemplated the same panorama, but what they have seen and understood, their very perception of land and sea, has changed with the knowledge and prejudices of the times. The fundamental rocks that were once so incomprehensible have been picked apart and placed in their proper place in the history of the earth. Supercontinents have come and gone, but the speckled gneisses endure. Now gulls shriek from a rocky promontory that once burned under a Proterozoic sun, before the Torridonian sediments buried the Lewisian again beneath a red blanket. These gneisses have seen it all, achieved a kind of repose. The assault by the sea today is no more than a postscript to a history that has seen these rocks glide around the world, roast in a deep furnace, witness silently the transformation of the atmosphere, and the rise of those presumptuous newcomers, animals. The rocks will still be here when the last organism has vanished from a parched earth.

11

Cover Story

The mule has its eyes set in such a position that it can see all four legs at the same time. This is just as well if the track it is traversing is only a metre or so wide and there is a sheer drop of 150 metres to one side. When it approaches a hairpin bend on the outside of a precipitous trail, the mule prefers to poke its head out into nothingness before jerking sharply round to follow the animal in front. I think it does it on purpose. "If you don't like it, close your eyes," says Ken the guide. "After all, that's what the mule does . . ." I find myself clinging to the pommel on the saddle with terrible determination.

The descent into the Grand Canyon in Arizona is one of the great geological journeys. It is more than a mile down (vertically speaking) from the rim of the canyon to the Colorado River: a journey through half the earth's history, rock formation by rock formation. As the mule treads delicately but relentlessly onwards, you are carried downwards upon its back through the geological column; you become a time traveller, borne step by lurching step back first through tens, and then through hundreds of millions of years.

You come across the Grand Canyon unexpectedly. In the approach to the south rim across the Coconino Plateau the road hardly undulates. You pass from a dry, flat plain into an endless open forest of piñon and ponderosa pines, mixed with feathery juniper trees. There is a sparse underscrub of spiny *Opuntia* cacti, and *Agave* bushes like green punk haircuts. You cannot see very far. It is all pleasant enough, and rather unremarkable. Then the forest suddenly comes to an end. Something has been slashed into the earth. The Grand Canyon opens up without any kind of scenic preamble—not even a gentle incline. It is at once familiar and alien: familiar, because you have already seen pictures of it many times—it is one of those universal icons, like the Mona Lisa or the Empire State Building. For an instant you wonder if this apparent chasm might be another, particularly clever, three-dimensional representation. Then, too, it is astonishing and strange, because no image conveys its sheer scale. The far side of the north rim is sixteen kilometres away or more; yet in the morning light you might imagine it much nearer—or maybe, in some other lights, much further away. Sometimes, indeed, the distant creeks and buttes look as if they were painted backdrops designed to present the illusion of distance, as in a theatre; wing after wing of cunningly crafted canvas: shades of cream, orange, rich-red, and umber stripes in broad swathes parallel to the horizon. All that scooped and fretted bare rock seems to have been carved, dug away, leaving massive turrets and castles and dark dungeons: architectural similes are irresistible. The appropriate comparison is with the Hindu temples of Ellora in the Deccan Traps, where the hand of man has created architecture by virtue of all the rock that has been removed, a kind of anti-architecture. In Arizona, the carver is the Colorado River, so deeply sunk in its dark ravine that it is usually invisible from the rim. The prime cause of all the erosion is hidden, hinted at only in a black-purple gash that dogs the bottom of the canyon. The dark Inner Canyon is often obscured by features developed higher on the slopes. When you do catch a glimpse of the river, it may be no more than a little silvery flash, and it seems absurd to suppose that what looks like a mere trickle could sculpt so vast an array of amphitheatres and promontories of rock.

In the winter, the crowds of tourists keep away. It is too cold for them at the top. You have the canyon almost to yourself. A trail follows the rim, dodging back and forth from the edge as it leads you from one viewpoint to the next. There is a light dusting of snow among the short pine trees. This is a dry climate, and the well spaced and rather stunted conifers take many years to reach maturity. It is quiet, for there are very few birds. Occasionally, a huge, black raven wings past on an inspection flight, faintly sinister. Just past dawn, the shadows are profound in the depths of the canyon, and as the sun rises higher, the shadows are lifted one by one. It feels more like an unveiling, a removal of a series of black opaque sheets from the sculptured forms below, than a progressive illumination. While the light is still at a low angle, the colours of the rock formations are at their most intense. The strata appear unwaveringly horizontal, like an infinity of stacked plywood worked with a giant fretsaw. There is a clear contrast between the cream-coloured layers and those that are richly russet, the colour of mature Italian sausage. From the rim, it is impossible to guess how thick these layers are, for depth and distance confuse your perceptions. You must go down through the great pile of rocks if you wish to reach the Colorado River, and then you will discover their true dimensions for yourself.

The Grand Canyon was not always perceived as one of the "seven natural wonders of the world," as the postcards put it. The first western visitor to reach it was the Spaniard Garcia Lopez de Cardenas in 1540. He was in search of the legendary Seven Cities of Cibola and, in the spirit of the times, a fortune in gold. He was unable to cross the great chasm, and, horrified by its impassableness, returned to Mexico with no treasure for his pains. Three hundred years later the great gash in the earth was still viewed with as little enthusiasm. In 1858, the surveyor Lieutenant Joseph Ives was to report to his superiors that the "region is altogether valueless. Ours has been the first, and will doubtless be the last, party of whites to visit this profitless locality." Exploration of the canyon had to wait for a one-armed Civil War veteran, John Wesley Powell, in 1869. What is now a ten-day "white-water adventure" along 446 kilometres (277 miles) of the Colorado River was a dangerous ordeal for the first explorers, blindly tumbling into what Powell termed "the great unknown." While shooting one of the rapids, one of their four boats was lost, along with much of their food. The party had to cope with meagre, damp rations, on top of the heat and danger. Three of the men never returned. Through it all, Powell continued to survey, and to collect rock samples and fossils. As was to be the case with the moon, it seems that the first way to establish the veracity of the experience was to bring home a piece of rock. Powell's account of the adventure, *Exploration of the Colorado River and Its Gorges,* established him as the pioneer's pioneer. The journals of one of his companions cast him in a slightly less heroic light, but no one disputed his bravery. Powell went on to found the U.S. Geological Survey. His memorial is on a promontory on the south rim commanding a panoramic view of the canyon, erected, it says, by order of Congress. A bronze bas-relief of the great man, bearded and splendid, somehow contrives to look like Charles Darwin.

The change in attitude to the grandeur of the Grand Canyon came about quickly. A transformation in appreciation of what made for beauty soon became pervasive: American painters like Thomas Moran began to respond to the aesthetic of the wilderness, creating emotionally charged landscapes of the untamed scene. Pristine acquired cachet. By the 1890s, a few enterprising souls were offering guidance and hospitality to the intrepid visitor to the south rim. Buckey O'Neill's cabin still survives, having been incorporated into the Bright Angel Hotel in 1898. It is a homely enough affair, lined with split logs. There is an appropriately laconic picture of him posted outside, lounging with his gun tucked in his holster, and a candle stuffed into a bottle behind him. He led Teddy Roosevelt down the canyon on mule-back in 1912. I doubt whether the President would have passed the weight limit of 200 pounds currently imposed on riders down the trail, but this visit doubtless had a part to play in the designation of the Grand Canyon as a national park in 1919. The Santa Fe Railway Company had reached the south rim as early as 1901, bringing those jaded with big-city life to seek a novel form of spiritual refreshment. A humming locomotive was still waiting in the station when I arrived just over a century later. All that was needed was the comfort of a good hotel to cushion the wilderness experience.

The Fred Harvey Company soon provided one. The El Tovar Hotel, built in 1904, is still rather stylish, in a knowingly rustic fashion. It settles down unobtrusively into the landscape, which is just as it should be. It helped to set a tone for North American national park architecture that you can identify from New Mexico to Newfoundland: it says, "harmonize with the natural environment." During the earlier part of the twentieth century you would have been attended to assiduously by Harvey Girls, neat and spruce and ever so respectable in their nice little uniforms. There were oysters on the menu, which might be served off monogrammed crockery. The splendour outside was still there, of course, but the middle-class visitors did not now have to suffer hardship to appreciate it. The essence of modern tourism had been defined.

All among the pine trees on the south rim a bedded rock the colour of rich cream crops out in blocky benches. This is a limestone, the Kaibab Formation. You do not have to search for a long time before you notice fossils weathering out on the upper surfaces of the slabs: the skeletons of corals and shells of several kinds. Evidently, this sedimentary rock was originally deposited under the sea, where such animals thrived: you may see the stem of a sea-lily, or crinoid, which would once have waved back and forth in the current. The waters must have been warm and shallow, too, because that is where calcium carbonate precipitates out of solution in sufficient quantities to form limestones. An inland sea evidently invaded this part of the early North American continent. Successive sea floors are laid out one on top of the other around the rim, as one rock bed succeeds another; here is a catalogue of the passage of geological time. Looking gingerly over the edge into the canyon, you can see similar, creamy limestones forming the steep cliff all around the top of the canyon: the former seaway must have extended for many kilometres. The fossils also tell us the geological age, for they are representatives of species long extinct: these limestones were deposited in the Permian period, some 260 million years ago. Dinosaurs had yet to achieve their hegemony when the warm sea engulfed this part of Laurentia.

Now that you can recognize the Kaibab, you know that it forms the uppermost layer for as far as you can see. Below it, other rock formations take their turn. So you look back further into the remote past as you look downwards. The Kaibab is only the beginning. The journey down to the Colorado River is a chance to see the ebb and flow of vanished seas over millions of years, to feel the slow pulse of the earth.

Journeying down by mule is remarkably safe these days, but still exhilarating. Going over the edge is something of a shock—like jumping into space. The trail winds steeply downwards for 100 metres from step one. Since winter reigns at the top, the narrow route is coated with ice. You cannot help but glance downwards and visualize a dramatic plummet into space. My mule is called Buttermilk. Soon, I realize that her step is sure and the quaking feeling in my innards subsides, though she lurches and stumbles and I keep my fingers tightly gripped on the saddle. My notebook remains unopened in my coat pocket.

The personality of the trail precisely reflects the geology. Hard and resistant formations make the steep cliffs and narrow trails, which are sometimes cut dramatically into the bare rock face. There is little vegetation clinging on to such formations, so you can trace them by eye as unscalable walls all around the canyon. Softer rocks produce gentler trails on gentler slopes, with less scary drops, but they also wash away more readily, so that the way is irregular and dotted with fallen boulders. The mules must pick their way very carefully.

The Kaibab Formation is one of the steeper ones. I tried to remember the names of the succession of rock formations that I would encounter on the way down. In this endeavour, I was helped by a slightly risqué mnemonic told me by a receptionist at the White Angel Lodge (*K*issing *T*akes *C*oncentration. *H*owever, *S*ex *R*equires *M*anoeuvring *B*etween

*T*empting *V*ariables: *K*aibab, *T*oroweap, *C*oconino, *H*ermit, *S*upai, *R*edwall, *M*uav, *B*right Angel, *T*apeats, and *V*ishnu, in that order). The list isn't quite complete, but perhaps that was to spare my blushes. That there was a trail at all was the result of a fault that has fractured the whole pile of sediments, a weakness along which erosion had preferentially worked its slow depredations. We are following the side of an enormous V-shaped gulley. I did not manage to scrutinize every detail of every formation as we descended: I was far too preoccupied with making sure that Buttermilk kept sniffing the trail in front. But I did notice that the Toroweap Formation marked a pleasant break in the downward zigzags for a couple of hundred vertical metres and that there were reassuringly solid pine trees alongside the trail: these rocks must be softer as well as somewhat older than the tough Kaibab Formation overlying them. I sit back in the saddle and relax.

This proves premature. For the trail next turns vertiginously along the sheerest cliff I could imagine: it hangs on by its lithological fingernails. This is the Coconino Formation, an almost pure sandstone, shining pale yellow. My geologist's eye takes in the sweep of the vertical cliff it makes all around the amphitheatre; surely, it must be at least 100 metres high. How improbable that anyone could ever have climbed such a barrier . . . it makes a rampart as far as could be seen. Then, the structures embedded in the sandstones catch my eye. There are billowing swathes of cross-bedding several metres high. The outer surfaces of the rock show etched faces scored with lines that sit at an angle to the horizontal lie of the strata above. I do not have to consult the geological guide to recognize this feature: we are traversing desert sands. Wind-blown dunes leave this distinctive cross-section. While the Coconino Formation was accumulating, the sea had abandoned this part of the world: another arid sea of sand dunes stretched from horizon to horizon; no fossils are here but the petrified tracks of scorpions, and the fingered traces of reptiles scampering across the ancient dunes seeking prey. They show as lines of little dimples on the sandstone surface, the kind of thing you might pass by without a second thought. This was a hard world, an ancient Sahara in the earlier part of the Permian period, when the earth was as dry as it has ever been. It seems curiously appropriate that clinging on through this part of the rock section was something of an ordeal.

Then, the rocks turn red. Immediately, the slope of the trail slackens off: gentler rocks, gentler gradients. The colour change marks our downward passage into the Hermit Formation, soft bright-red shales that are easily eroded back to make a sloping bench on the canyon's profile. This abrupt transformation from cream to red is one of the striking lines I traced out as I looked down from the rim, and when I flew in a Cessna from Las Vegas over the canyon I could follow it from the air for many kilometres. The mules now have to cope with an apparently endless switchback of slippery ruts: the problem here is that when it rains, the trail washes out. But the more dignified descent gives me an opportunity to look around at the richer vegetation that is able to grow on the lesser incline. There is no longer any snow. As we descend, the temperature rises so that we move into different climatic zones. A few of the conifers that lined the rim still grow here, but there are now also numerous small oak trees. The redness continues for a long, long time as we descend still further. At one point the track steepens again and there are tracts of sandstones making for a series of steep cliffs and exciting riding. We have passed downwards into the Supai group, which is itself subdivided into a number of thick, resistant sandstones alternating with softer rocks like mudstones. This variation produces a series of steps down the canyon side as the trail winds ever onwards, now plunging steeply and weaving from side to side, now sloping gently downwards for a longer, shaly stretch during which you can allow yourself to slouch down into the saddle like an old cowpoke. Before you know it, you have gone down another 300 metres. The red colour of these rocks, which lend the Grand Canyon so much of its drama, is just iron. The weathering of iron minerals in the presence of abundant oxygen assures that iron is in its ferric form, which paints russet, rouge, and rust. One of the most abundant elements in the crust is also the most versatile at decorating scenery. Such red rocks often indicate deposition under terrestrial conditions: the influence of wind was still important. Had we been able to scramble off our mules and tap at the rocks with geological hammers (strictly forbidden), we might have found the fossils of horsetails or the tracks of amphibians. There was life in this place between 270 and 320 million years ago, while the varied Supai rocks were accumulating; the strata take us all the way back to a coastal plain of the Carboniferous period.* The ancient landscape has been compared with the Gulf Coast today. There were occasional little groves of primitive plants, streams with their own levees, and, elsewhere, more sand dunes. All these have left a legacy in one rock bed or another. Thus was one more vanished landscape built into the walls of the Grand Canyon, another blanket of sediment covering up the still earlier histories that lie beneath it. Another cover story.

* U.S. readers will know this as Mississippian and Pennsylvanian.

Any such musings are abruptly brought to a halt by a sudden lurch downwards. It is the Redwall Formation. The name is entirely appropriate for what it does: it makes a red wall—an implacable vertical barrier 150 metres high. John Wesley Powell christened it, and feared it. Nothing grows on the bare rock face. The trail winds over the cliff face thrillingly downwards, with the random inflections of a bolt of lightning. The trajectory demands total concentration. As I round one switchback I notice from a broken surface that the rock constructing this geological Jericho is not red at all, it is pale grey, almost white. It is, in fact, a massive limestone, like the Kaibab where we began. The red colour is a superficial stain derived from the Supai immediately above, a kind of iron colour-wash that renders the mighty rampart pink. In the early morning light it seems to glow. There are caves excavated within the upper part of the Redwall limestone that show up black and inaccessible high on the wall. Like almost all limestones, this one was marine, laid down under a warm sea. It contains numerous fossils that betray its origin: shells of brachiopods and nautiloids, skeletons of corals and bryozoans. They also tell us that the limestone was deposited in early Carboniferous (Mississippian) times, when a shallow sea flooded over a large part of the early North American continent. This segment of history makes my legs stiff, and poor Buttermilk comes out all in a lather.

Our guide Ken, however, does not allow us to dismount just yet. He whistles for Buttermilk to catch up and she breaks into a trot. I scarcely notice the Temple Butte Formation, a thin marine formation of Devonian age, but I do notice getting down to the Cambrian rocks below it, for the trail eases up more than a little and the strata are finer-bedded and softer. This is almost like coming home for me, because the Cambrian was the time when trilobites were most abundant, and I have been studying trilobites for most of my life: 520 million years ago is my kind of time. Trilobites only lived in the sea, so when the Bright Angel shale was accumulating the Grand Canyon area was again immersed beneath the waves. It is a time so remote that no creatures had yet ventured on land, and neither the corals nor the nautiloids we saw in the rocks at the beginning of our journey had yet evolved. We are getting back towards the beginnings of animal life. I feel a pang, because there is something missing. There are no rocks of Ordovician or Silurian age in the canyon, and I have always been an Ordovician man. Nobody has discovered a rock record in this locality representing nearly 100 million years of geological time. The sea withdrew to somewhere else. Between the Devonian Temple Butte Formation and the Cambrian Muav limestone below there is an unconformity. We shall never know what happened in the Grand Canyon during this great period. Without a rock record, there is no way of reading the book of time. It is lost to us more definitively than the secrets of the Aztecs, or the rituals of the Easter Islanders. We pass downwards through the Cambrian, passing the Muav Formation, and onto the outcrop of the Bright Angel shale. This is the softest of all the rocks in the canyon; as a result, it underlies the large bench interrupting the cascading series of cliffs. This is called the Tonto Platform, which can be seen very clearly from the top of the Grand Canyon. It stretches out far below as a kind of dark plain, its sombre colours contrasting noticeably with the warm hues of the steep Redwall limestone above it. It provides a landing upon the relentless stratigraphical staircase, and in the old days it also permitted the principal route to run along the length of the canyon. A trail still winds along it. From the top, the Tonto Platform also obscures what lies beyond it and below it. As my mule ambles on more comfortably, I have a chance to see that the endemic vegetation is black scrub, a kind of prickly bush, devoid of leaves in the winter. At this depth in the canyon, all the trees have disappeared: we are effectively in a semi-desert. The layers of clothes we donned on the frigid rim should be discarded soon.

Here, at last, is a chance to restore our legs to their proper shape. We arrive at a little glen known as Indian Gardens—a few trees, and reeds, and a couple of low buildings—where a spring emerges from the ground. Suddenly, the scale of everything is rather domestic. There is a hitching post ahead. Ken kindly helps us off our steeds, and we stagger around for a few minutes, bow-legged. The Pueblo Indians who once lived here would have grown corn, beans, and a couple of varieties of squash. Water gushes out because the Bright Angel Fault throws rocks of different permeability together. The water for all the buildings on the south rim is pumped from this resource. It seems curious that in a mighty landscape eroded by water it should be necessary to pump upwards for 1000 metres to get a decent drink. Evidently, what little precipitation falls upon the Coconino Plateau drains quickly down into the underlying porous rocks, until emerging again deep down at the spring line.

With more reluctance than elegance, I remount Buttermilk. The little stream that emerges at Indian Gardens has carved a charming little canyon of its own a few metres high, with the mule track running along one side. The clunk-clunk of hooves on stone is accompanied by the splashing of the brook. This trail cuts through the outcrop of the Tapeats Formation. I ask Ken whether it is safe to drink the clear water, but he warns me that even here there are unpleasant micro-organisms and it is as well not to taste. The trail is gentle now, almost a rustic footpath, and it is pleasantly sunny and warm. The sandstones through which we are passing are yellow and slabby, lining both sides of the little gorge as if we were passing through a sunken passage leading to some antique temple; for once, I have time to look at the rocks properly. Like those making up the Coconino sandstone, they show cross-bedding, but of a steeper and more delicate kind. These are the kind of sands that were swept along by strong marine currents close to the Cambrian sea-shore. For a moment the sun is shining under an ancient sky upon a seascape of the primeval earth. Trilobites scuttle over the sandy sea floor. There is no bird in the air, nor any tree upon the distant landscape. Now I notice a braided track made by the archaic animals as they plough their way through the substrate. There is the feeling of getting back to the beginning of things.

Which, indeed, we do with some dispatch. Once again, the trail begins to plunge downwards. There is the kind of sinking feeling inside brought on by one of those fairground attractions that does something dramatic just when you think the ride is over. We have entered the Inner Gorge, and all the rules have changed. How accustomed we have become to horizontal sedimentary rocks—that great staircase of strata that we descended level by level. The pink rock now by the trackside is quite different; for a start, it shows no signs of stratification. The little stream cascades over a hard mass into a tumbling waterfall, where the rock has been polished by years of erosion. Why, it is obviously a granite! Before there is a chance to take in the implications of this discovery, the trail starts upon its most contorted descent so far; turn after turn descends implacably. Ken tells us that this is called the "Devil's Corkscrew," and I reflect briefly that the Devil is always attached to geology's more sinister works: there are any number of Devil's Punchbowls, Devil's Canyons, Devil's Towers, and Devil's Staircases around the world.

The Inner Canyon is a darker world. The rocks have changed tone: some of the strata are almost black. Gone are the horizontal rulings that scored out the upper part of the canyon: the layered strata, the tiered amphitheatre. Now the rocks are all on end and twisted: they have been squeezed into convolutions and tempered into contortions. The deeper, inner gorge is narrow and steep-sided. It makes a narrow V at the bottom of what we now see to be the massively wider cradle of the upper part of the canyon. It is so deep down that in the winter the sun slinks in there furtively for a short while, like an archaeologist shining a torch inside a burial chamber. When we get a chance to dismount and hobble around at the bottom of the steep descent, I pick up some pieces of the dark rock that makes up much of this inner sanctum. It is a flaky, rich-green stone, streaked with black, and shiny on its flat, broken surface, unlike anything we have seen up above. This is the Vishnu schist. The shine is produced by one of the mica minerals, probably chlorite. The Inner Canyon is made of metamorphic rocks, which have been baked and turned vertically in some ancient paroxysm of the earth. It is another Newfoundland, another alp—but far older. We have taken a further excursion to the ancient of days. In the cliffs, I can see that the schists are intertwined and interwoven with pink veins, which sometimes produce blobby masses, and in other places are crimped like tresses of hair in a turbulent current. It is all so complex that it seems difficult to make any sense of it. The pink rock lying around in lumps is mostly feldspar. It is evidently an igneous rock, injected in magmatic veins into the metamorphic rocks that surround it. The granite that we crossed earlier must also have intruded into the metamorphic mass during an ancient mountain-building episode. I recall the similar rocks seen in the mobile belt of Newfoundland. Further up the Grand Canyon the granite makes a terrifyingly narrow gorge.

Looking upwards from the bottom of the canyon, what we have seen makes a little more sense. We can see the corkscrew trail winding its way up towards the younger rocks. The Tapeats Formation lies on top of the dark, twisted rocks of the Vishnu schist as if it were a laid blanket. It is the oldest of the many horizontal, undisturbed formations that we crossed on the way down from the Kaibab limestone. The profound events that heated the Vishnu rocks and intruded the granites into them must have been over and done with long before that first Cambrian sandstone was deposited. Everything thereafter was part of the cover story. An ancient tectonic belt had already been worn down, the mountains had been laid low and the rough places had been made plain, before the Cambrian sea crept over its ground-down remnant. Beneath the Cambrian sandstones of the Tapeats Formation there lies what has been described as the Great Unconformity, a great hole in the rock record. In fact, the radiometric "clock" tells us that the Vishnu metamorphic rocks were squeezed in the earth's vice some 1700 million years ago, in the early Proterozoic, an age not far different from that of the ancient rocks of north-western Scotland. Elsewhere in the Grand Canyon a series of later Proterozoic rocks intervenes between the Vishnu schist and the Tapeats sandstone, rather as the Torridonian rocks filled in the gap between the Lewisian gneiss and Cambrian rocks in Scotland. They serve to "fill in" some of the missing time but they were tilted and worn down before the Cambrian was deposited, testifying to yet another phase of earth movements. Era after era is laid out before us in this greatest of canyons. When we remember that a thousand years of erosion has been calculated to wear back the walls by no more than a metre, and that a huge fallen stone before us has probably not moved for centuries, we begin to get a feeling for the vastness of geological time; for the ages required to lay down and then uplift the cover rocks; for the still more ancient cycles that were dead and gone even before the first trilobite scurried on its way to extinction over the first Cambrian sandstone; and for the slow procession of the plates that underpinned it all.

Then, suddenly, our trail turns out upon the Colorado River: it is powerfully rushing and swirling with eddies, with a few rollicking masses of waves. Can this be the same river that made such a thin streak when viewed from the rim? It must be the same river that Powell described, fed by water after the spring melt that "tumbles down the mountain-sides in millions of cascades. Ten million cascade brooks unite to form ten thousand torrent creeks; ten thousand torrent creeks unite to form a hundred rivers beset with cataracts; a hundred roaring rivers unite to form the Colorado, which rolls, a mad, turbid stream into the Gulf of California." Today, the "red river" lives up to its name, its water thickly pink-buff from the load of sediment it carries. So, this is the motor of erosion, the only begetter of the Inner Canyon. The Vishnu cliffs plunge giddily several hundred feet towards the water. At this point the river seems almost stately in spite of its strong and relentless flow. But we know its course is punctuated with dangerous rapids where rockfalls have impeded its smooth running, where boulders can smash wooden boats into matchsticks, and where whirlpools can suck you to your death. The admonitory names of the rapids speak for themselves: I particularly like Sockdolager Rapid—in the mid nineteenth century, "sockdolager" was slang for a knockout punch. It was one of the last words Abraham Lincoln heard, in the play he was attending at the time of his assassination.

The Colorado River has not so much carved out the canyon as carried it away. Its waters have flooded or gently flowed according to its different moods, moving weathered rock from higher ground to lower, carrying particles, rolling cobbles. The whole Colorado Plateau has been uplifted around the river, allowing it no peace from grind and toil. The river has merely tried to maintain its position while the world has risen around it. It is now trapped within its own canyon like a donkey strapped to a wheel. If you watch a little runnel draining over a sandy beach you might see a micro-canyon form in the sand; soon, it will recruit small side-streams, and these in turn smaller tributaries; in an hour or two the beach will roughly replicate the design of 6 million years of uplift. The rise of the Colorado Plateau was in its turn a further consequence of the Pacific Plate impinging on the North American Plate. There are few places in the world where the tectonic circumstances conspire to make a Grand Canyon.

There is no rest for a while. The trail turns to the right and runs parallel to the river. We pass a footbridge. Mules will not cross this one because they can see the river beneath, and this spooks them badly. Instead, a new track takes us along to a more robust bridge at the South Kaibab trail. Much of this track has been blasted out with dynamite, and it hangs on by its fingernails in the middle of a cliff of Vishnu schist, with the Colorado River swirling far below. The trail might be scary, but by now there is no anxiety that the mule will fail to carry you safely through. A few cactus plants cling to the rock walls, apparently subsisting on nothing. If I were here after rain, brilliant scarlet flowers would burst out of their spiny columns. A few skinny mesquite bushes poke out of the gentler slopes, scraggy proof of the near-desert climate at the bottom of the canyon. We have come so far down that it is almost warm, even in January; in the summer it must be like a furnace. We pass a disused mine-shaft in the cliff. Further along the canyon another abandoned mine was formerly the major source for uranium in the United States. Its rusting pulleys still stand on the south rim.

Before the first bridge was constructed across the Colorado, visitors to the Phantom Ranch were hauled across the river in a small cable car, completed in 1907. Mules went the same way, reluctantly, one at a time, and it is easy to imagine the discomfiture of these animals dangling above the torrent. Even today, they scarcely approach the crossing with enthusiasm, and mules and riders alike reach the north cliff with gratitude. There is something of a bank on the north side of the river, where the Bright Angel Creek joins the Colorado. The creek has found out the geological fault, exploiting the natural weakness in the strata to produce its own side canyon, and thereby furnishing a natural trail from the north rim. For once, a little soil can accumulate in the valley bottom. We pass the ruins of an Indian

In the early days, passage across the Colorado River to the Phantom Ranch was an exciting adventure on Rust's cableway, built in 1907.

One of the huts designed by Mary Colter at the Phantom Ranch makes good use of geological materials to hand.

hamlet, no more than a cluster of dwellings crudely built from the boulders that litter the narrow shore. There is evidence that people lived here for more than 1000 years, gardening and hunting in due season. Is there no corner of the earth where a seed will sprout that has not been found out by our opportunistic species?

We approach the Phantom Ranch, tucked into the valley of the Bright Angel Creek, where we will finish our downward journey. The mules amble easily now; our legs are promised relief. We dismount at a small corral, and a young woman greets us with a glass of lemonade. The ranch is an oasis in the umber-coloured, dry landscape of the inner canyon. Cottonwood trees planted here in the early years of the twentieth century now make a grove that rustles in the breeze; in the summer, the shade they provide must be a blessing. Cabins are dotted unobtrusively among the trees. The one alloted to us is a single-storey, rectangular building, with walls partly constructed from the same kinds of boulders that were used by the Pueblo Indians; the rest is green-painted wood. Wood also lines the interior, and frames the windows. There are bunk beds, and a kind of wicker-work chair and table: all very simple and functionally effective. The ranch was designed in the 1920s by Mary Colter, who worked for the Santa Fe Railway—a rare woman in John Wayne territory. She evolved her own style based on natural materials and designs, and inspired by the region's traditions, part Indian, with a touch of Spanish. She was also responsible for some of the memorable buildings on the south rim. The Phantom Ranch works very well. Using morsels of the geology in the buildings helps them settle gently into the landscape, and nurtures the ambient calm.

In the morning, I examine some of the boulders and pebbles down at the streamside, the kind of materials that Mary Colter recruited. There is a selection of cobbles from the formations through which we had passed the previous day, many of them well-rounded. Any rough corners have been knocked off in their passage down the side canyon during stormy crises, when a wall of water might sweep everything before it, rolling and tumbling stones downhill. Buff-coloured sandstone pebbles are probably the commonest ones (Coconino perhaps), and I select one that fits comfortably in my grasp, reminding me of Seamus Heaney's "Sandstone Keepsake":

> *. . . sedimentary*
> *and so reliably dense and bricky*
> *I often clasp it and throw it from hand to hand*

It might also serve as a reminder of the perpetual state of change of the world: for this ancient pebble is made of grains of sand derived from still earlier cycles of the earth, and now it is halfway towards being worn down again into its constituents—sandgrains that will, eventually, find their way to the sea. Once there, the grains will be incorporated into another sandstone, which will in turn be elevated to form another cliff, another butte. And thus, turn upon turn, until the world ends.

The mule journey back to the rim is along a different trail, and the tape of geological time is wound in again, upwards through history. There is a little more to see of the late Proterozoic formations in the top of the inner gorge. Above the hidden, steep-sided Inner "Canyon within a canyon," the familiar rock formations jog past in reverse order: the steep slopes repeat one after the other in due course. Buttermilk needs more frequent rests on the steep return climb, and a lather of sweat soon congeals on her flanks. On the South Kaibab Trail, the ascent of the Redwall limestone takes a narrow turn out beside a precipice called Skeleton Point. Eighty years ago, ladies had to be revived with smelling-salts at this prospect. (Ken remarks that if you fall from here, you would have time to roll a cigarette and repent all your sins on the way down.)

From this height, you have a clear view of the sculpted rock masses that flank the canyon. We are just about to pass around O'Neill's Butte, named for Buckey O'Neill, which is a small platform held aloft on a stunningly steep natural pedestal, a remnant left behind when all the strata surrounding it were eroded away. O'Neill is unusual in having a butte (pronounced "beaut") named in his honour. Most of the erosional features that you can see from the rim of the canyon are dubbed "temples," "shrines," "thrones," and the like. Then they have something borrowed from classical, Scandinavian, or eastern culture further to identify them. The Cheops Pyramid looks towards the Isis Temple and Shiva Temple. Wotan's Throne, adjacent to Freya Castle, contemplates Krishna Shrine and Sheba Temple. Solomon, Venus, and Zoroaster are all somewhere in the canyon. The architectural monumentality of the mesas and buttes is indicated by their association with gods, or at least with other famous monuments from the Old World—an ecumenical lucky dip. This lavish mythography was the responsibility of the geologist Clarence Dutton, whom we have already met in Hawai'i. Dutton's *Tertiary History of the Grand Canyon District,* published in 1882, is regarded as a classic account. Dutton did not stint on drama either, as when reporting the prospect from Point Sublime on the north rim: "The earth suddenly sinks at our feet to illimitable depths," he wrote. "In an instant, in the twinkling of an eye, the awful scene is before us." Indeed it is. His antiquarian eclecticism inspired the choice of names for the amazing features he remarked in the "awful scene," as if by piling together the great names of the Old World he might validate a claim for the Grand Canyon as the wonder of wonders in the New.

I soon realized where I had seen the same grandiloquent multiculturalism before: Las Vegas, in the adjacent state of Nevada. In downtown Vegas, Luxor (thrill to the ancient Egyptian experience!) rubs shoulders with Excalibur (jousting knightly!), while a stroll away Caesar's Palace (the glory that was Rome!) competes with Paris (Mais Oui—Paree!) or New York! New York! Each gigantic hotel has a kind of symbolic approximation to the source—columns and porticos for ancient Rome, lots of shields for Arthurian romance. Otherwise, the casinos seem much the same wherever you are. It is all most absurd and endearingly exuberant, and entirely inappropriate to the Mojave Desert in which Las Vegas lies, soaking up water. Yet five times as many people visit Las Vegas every year as visit the Grand Canyon.

The peculiar geological circumstances surrounding the Grand Canyon afford us a chance to see deep down into history. However, there are numerous parts of the world where a cut down into the earth—a slice through the upper part of the crust—would reveal a similar succession of cover rocks laid over deep basement that dated from when the world was young. Although these buried strata are concealed beneath younger rocks overlying them they are accessible to boreholes probing for oil, or seeking sources of geothermal energy. A diamond-tipped drill makes our journey for us. The rock cores retrieved from such boreholes are stored as vast lines of stratified poles, laid out on shelves in the archives of surveys and oil companies. These cores may seem rather dull to the eyes of the casual observer, yet if the strata they represent had been exposed to the forces of weather and time they, too, might have made another Grand Canyon. If we could wave a tectonic magic wand and gently elevate southern England, the River Thames would excavate a canyon of its own, another magnificent thing—and, deep enough, there would be the equivalent of the Vishnu schist. If we could do the same in northern France, the Seine would carve through a sequence of hard and soft layers back to a deep and ancient metamorphic foundation. The same goes for Texas, or the Pirana Basin, or the Arabian Peninsula, or western Africa, or much of Siberia. In China, the Yangtze Gorges are perhaps the oriental equivalent to the Grand Canyon, displaying mile after mile of nearly horizontal strata, though never creating a canyon quite on the scale of the Grand. Sadly, much of it will soon be flooded when a new dam is completed. Geologists are now collecting what they can before the rocks become inaccessible.

Stratigraphers love places where the rocks are laid out horizontally, as in the upper part of the Grand Canyon. As Eduard Suess observed of Dutton's exegesis of the canyon's rocks: "So Nature writes her own chronology, and we may well envy the observers who are called upon to read this history in the original." There is no mistaking geological order, as you might among the scrambled and confusing folded rocks in a mountain belt. The strata are laid out for all to see. They are usually referred to as showing "layer-cake" stratigraphy—lasagne stratification might be another way of putting it. Yet there may be "holes" in the time actually represented by strata—periods when no rocks at all accumulated, or if they once had, when they were subsequently stripped away by erosion. Caution about dating rocks in any one place is therefore essential. Reading down the rock succession is like turning back the pages in the diary of the earth. The geologist observes when seas flooded continents, or when they retreated, leaving the world to migrating sands and howling winds. He knows from the features of the rocks and fossils when fresh-water lakes and rivers made an environment hospitable for horsetails, thriving alongside cockroaches, fish, and millipedes. He discovers when the climate was tropical, and when it was frigid. He knows from studying sediments when violent storms once raged, or whether the sea floor was so calm as to leave the carapace of a moulted trilobite undisturbed, the feathery arms of a sea-lily unbroken.

All these cycles of change relate to plate tectonics. Once the continents had formed, parts of them stabilized and were ready to receive a cover of sediments. Whether or not sediments were preserved there depended on the delicate balance between land and sea level. It is now quite certain that there were times when the global sea level was relatively high, flooding the continental interiors, and others when the oceans tended to drain off the continents back into the ocean basins. Times of particularly active ocean-floor spreading seem to coincide with times when the oceanic waters invaded the continents. If global sea level were to rise now—as it may when the polar ice sheets melt away—it is easy to imagine waters drowning the plains of Australia, creeping over the Nullarbor Plain and leaving Ayers Rock's standing out as no more than an island above a glittering inland sea, the Blue Mountains an archipelago. Parts of India and most of Bangladesh, the Netherlands, the Mississippi Basin . . . all drowned. In the Grand Canyon, the Tapeats sandstone represents a series of shore sands early in the Cambrian when the sea flooded over an ancient, planed-down, Precambrian landscape. I have stood on similar Cambrian sandstones on the Arctic island of Spitsbergen, with the cold wind whistling through my parka. I have described them already under rain in northern Scotland. I have seen them in Sweden, in Newfoundland, and in Australia. Nobody could doubt that the early Cambrian was a period when the earth was inundated. Contrariwise, that other sandstone, the terrifyingly beetling Coconino, provides testimony to a time of aridity, when prevailing winds drove dunes in shoals across a parched landscape, when hardy animals scuttled across the wastes as fast as they could. Similar sandstones are common around the town of Penrith in the English Lake District. The Permian was the time when the continents assembled together in Pangaea. Deserts spread far and wide. The shape of the world made the climate, and seas were all but banished from the interior of the united supercontinent. What we see in one canyon is but a sample of a hundred places where similar aeolian rocks were accumulating under a fierce sun. The history of plate movements provided the plot line for the intimate narrative of the rocks.

So much for the story of two sandstones. The same arguments could be rehearsed for all the rock formations encountered in our descent of the Grand Canyon. Any episode recorded on the stratal staircase of the canyon has to take its place in the narrative of the whole earth. As the continents are carried around the world, so the kinds of sediments that can accumulate upon them are constrained; after all, climate changes profoundly as the plates move from one end of the earth to the other over a billion years or so. All the pulses and retractions of the sea are under their deep instruction. Everything connects.

Everything we have seen falls under the control of forces still deeper, geology more profound, which no canyon will ever reach even if the workings of erosion carry on till the crack of doom. To see further inside the earth we need experiment and intuition. There is an end to intimate exploration, to what can be felt from the back of a mule or tapped with a hammer. That is what we must now investigate.

I say goodbye to Buttermilk with a certain regret. The flight back from the Grand Canyon takes us over other gorges along the Colorado River, where the distinctive markers of the rock formations with which we are now familiar can be traced along the cliffs: cream to red painted up in the evening light. Streams are still seeking out weaknesses in a hundred little valleys, still eroding away what deep time has constructed, essaying to lay low what geological forces once made high. Thousands of years count for nothing, for this is a process with a pace not to be measured against any human chronometer. Now there is Lake Mead below, a giant blue amoeba fingering out into the desert. The Hoover Dam that created it seems little more than a temporary impertinence from this height, although it is doubtless an engineering marvel. I am reminded again of playing on the beach, of temporarily backing a trickle in the sand with a little clay, in order to pond up a transient pool for the amusement of my children. Dusk is falling. The little aeroplane heads towards Las Vegas. Now there are temples and castles, the whole world dressed up in lights. A laser beam shoots vertically upwards from the apex of the pyramid of the Luxor Hotel, routinely dissipating its energy upon the firmament.

12

Deep Things

It would be wonderful to take a trip to the interior of the earth, to see for ourselves the engine-room of plate tectonics. In practice, it is easier to go to Mars or Venus than venture down to the Mohorovičíc Discontinuity, let alone cruise to the mantle. Earthquake waves can reach there, and it is as well that they do—for they send messages back from places we will never see or touch. We have already encountered a few situations where rocks from the depths have been brought to the surface. This may be by wholesale obduction of slices of oceanic crust, preserved in the "lunar" ophiolites of Oman and Newfoundland. Or else the millstone of geological time, grinding away over many, many millions of years, has exhumed deep layers inside ancient mountain belts, such as those surviving in Archaean shields. Rarely, small boulders from the depths—xenoliths—have been brought to the surface, entrained in vigorous mantle plumes, and torn from their natural home kilometres beneath our feet. *De profundis,* these rocks speak directly of places we can never visit, where no subterranean bathyscaphe can ever cruise; they give us glimpses of the deep underworld.

The endless voyage of the earth's plates is controlled from beneath, by processes operating in profound regions that lie beyond our direct apprehension. In this respect, modern geology comes to resemble chemistry or physics more than its image as a vigorous, field-based, "hands-on" science might suggest. The properties of matter are in thrall to what is happening at the sub-atomic level, which sometimes seems to resemble a series of Chinese boxes: for every particle discovered, another is mooted. So we progress from what is observable and solid to what is elusive and quantum mechan-

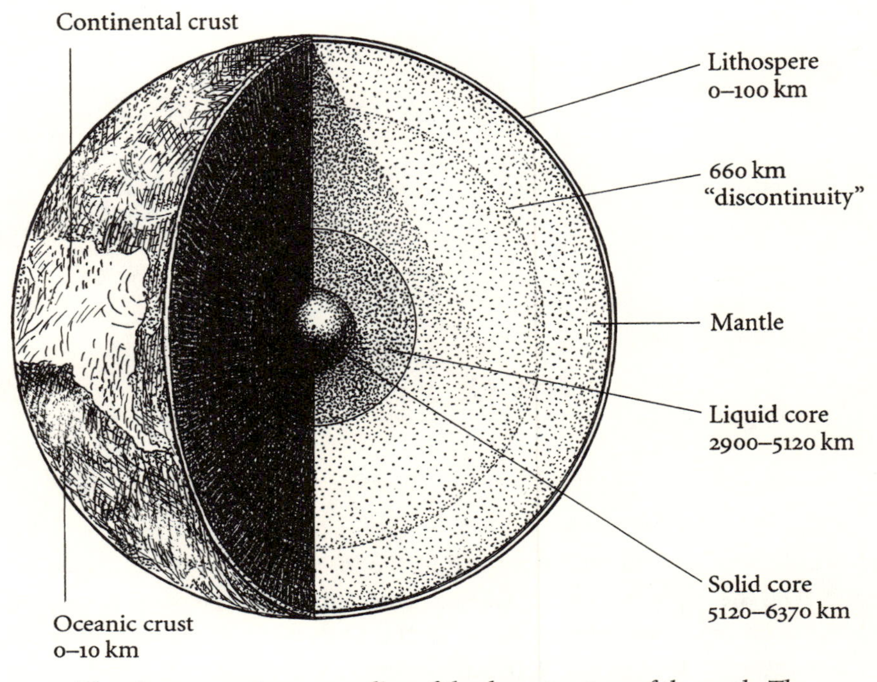

The planetary onion: an outline of the deep structure of the earth. The 660-kilometre discontinuity within the mantle is indicated.

ical—but also more fundamental. This book has been a portrait of the surface of the earth, where the all-too-solid flesh of the land has proved to be the physical expression of a deeper geological physiology. I have emphasized processes observable in the earth's crust, but they in turn lie atop a deeper motor; to try to understand it, we are forced to move further and further from things directly observable; from the surface, deeper and still deeper into the interior of the earth. As we do so, certainties fade, and mathematical models of what might lie below become progressively important. What are these equations, but a way of writing out dreams in feasible numbers? Experiments in the laboratory take the place of journeys across mountains, or descents into deep mine-shafts. Such tests are not so different in kind from those in nuclear physics conducted in particle accelerators purpose-built to explore the contents of the latest Chinese box. In both cases, the experimenters are trying to reach a place where no one has been before. The difference is that atomic accelerators cost more than the export budget of a small country; in comparison, geology comes cheap.

Although we cannot visit the depths, what we *can* do is observe one of the experiments for ourselves.

In the West of England, the Earth Science Department of Bristol University is to be found inside a very large and slightly pompous edifice called the Wills Memorial Building, which towers over its neighbours like a cathedral. W. D. & H. O. Wills built their fortune on tobacco—and like others who made large amounts of money in ways fraught with moral ambiguity, they spent a lot of it founding one of the best universities in the world. In the basement of Earth Sciences there is much experimental equipment that tries to reproduce what happens deep inside the mantle of the earth, hundreds of kilometres below the surface. This is beyond the roots of the deepest mountain range, and it is a strange world where nothing is as it is under sunshine and rain at the surface. Pressure and temperature increase enormously and progressively deep into the earth. All the properties of materials change utterly under these conditions. The very atoms arrange themselves in different ways, in the manner described previously concerning the structure of diamond. The effects are pervasive; everything is squeezed together in a common regime governed by the "equation of state." If we want to know how the earth is made it is important to know what happens to minerals in this strange world—but how to simulate the conditions?

The guru of squeeze at Bristol is Professor Bernard Wood. He is interested in reproducing the deep earth in the laboratory—I should say in a very, very small part of the laboratory, because the amounts of material that can be used as subjects are tiny indeed—a few milligrams at most. I had to put on my extra-strong glasses to see the mounted specimens encased in their ceramic beds. There are different intensities of temperature and pressure as you go deeper through the lithosphere and into the mantle; hence different equipment is appropriate to examining various depths. One problem is that the sample has to be heated, even as it is simultaneously squeezed. Another is that you do not want the sample to explode. All the apparatus looks robust, with thick steel frames, solidly screwed to the floor.

For less extreme conditions a piston cylinder apparatus is suitable. It copes with pressures of up to 40 kilobars and temperatures to about 1500–2000° C, the kind of conditions appropriate to the upper part of the mantle and base of the lithosphere. The equipment is quite simple in principle: it simply squeezes a small sample, encased in a cylinder, with a piston using a hydraulic pump. The sample is surrounded by a graphite "sleeve," which is turned into a furnace by passing current through it. A thermocouple attached to the sample measures the temperature reached in the test. Thus the experimenter can control both the pressure and temperature (P–T) of the sample under investigation. Then, when it has been tortured sufficiently, the sample is quenched very rapidly. This serves to "freeze" the mineralogy in the form it took up under high P–T.

So now there is a tiny sample of a mineral preserved under particular deep-earth conditions. The next thing to do is analyse its structure and chemistry. These days, the classical approaches of petrology using thin sections have been superseded by automated methods: years of experience have been replaced by the twiddling of knobs and the flickering monitor in a darkened room. Modern instruments can measure the elemental components of a mineral more or less directly, and they can do so using a minute sample. The principal chemical elements in a sample are measured by an "electron probe." A sensitive detector measures the different energies of the X-rays generated from a specimen under examination when it is bombarded by electrons. These energies are characteristic of the various chemical elements that go to make up a given mineral. They are displayed on screen as a spectrum of X-ray lines. Sitting down in comfort in the darkened laboratory, you can watch the various elements under analysis sort themselves out into different "piles," the heights of which are more or less in proportion to the quantities of the elements present in each sample. Various correction factors have to be incorporated to allow for the absorption and fluorescence within the machine itself, but the analysis is essentially automated.

Although it is indispensable for identifying mineral species, an electron probe is not sensitive enough to measure very minute quantities of the rare trace elements that are more and more important in understanding deep processes. There are other machines that can do this. The current state-of-the-art technology is the Laser Ablation Mass Spectrometer. This can work on exceedingly small quantities of sample with extraordinary accuracy: an interesting speck *within* a small crystal is all grist to its high-tech mill. The laser punches out a tiny sample, which is then analysed by an advanced kind of mass spectrometer that "sorts" individual atoms by weight. This method is also suited to identifying and counting isotopes of the same elements, which differ in atomic weight, rather than in their chemistry. The accuracy of these instruments makes the mind reel, since there are calibrations approaching picograms—that is, one million–

millionth of a gram. We are talking about identifying and weighing one grain of sand in a desert. These machines have transformed the scale of study of earth's chemistry and the thermodynamics of rocks at depth.

To take temperature and pressure experiments to a higher level—and deeper into the earth—a multi-anvil apparatus is brought into play. Anvils, of course, were those notoriously heavy pieces of steel upon which blacksmiths beat out horseshoes. The multi-anvil device in the Bristol laboratory is a massive and serious piece of "kit" weighing 18 tonnes. Despite its size, the sample used is less than 0.1 of a millilitre, a tiny scrap on which to concentrate the fury of the massive press. The sample is embedded into an eight-sided ceramic sample-holder, which is then fitted inside a battery of eight cubes—each the size of a child's building brick—made of tungsten carbide, one of the toughest substances known. Each cube has one corner cut off to accommodate one face of the octagonal sample-holder, so that the sample is held in the middle of the eight cubes as they are pressed together to form one larger cube. This in turn is placed inside six massive steel "squeezers," which press upon the six faces of the large cube. The sample heater has to be immune to the temperatures of the experiment, which rules out any materials familiar to us in the home. Lanthanum chromate ($LaCrO_3$) has just the right properties, when encased within a zirconia (ZrO_2) sleeve. Tungsten carbide's atomic structure has something of the three-dimensional fortitude of diamond; a trace of added cobalt improves the toughness. The equipment takes three or four hours to reach pressures of about 250 kilobars—a long, slow squeeze. It takes even longer—fifteen hours—to decompress. It is worthwhile reciting all these technical details for once, if only to point up that there have been places in this book when I have leapt from idea to conclusion without properly doffing my cap to those with the patience and skill to design equipment. Advances in understanding of how the earth works have tracked the means of measurement almost step for step.

The multi-anvil system is very useful for experiments relating to the mid or deep mantle, below 660 kilometres. Current experiments in Japanese laboratories are trying to push the specification beyond 400 kilobars to 1000 kilobars, approximating to conditions at the core/mantle boundary. With single minerals as the "squeezee," it is possible to determine their atomic structure at variably high pressures and temperatures. This models what happens to real minerals at great depths beneath the earth's crust. By heating and pressurizing "mixtures" appropriate for mantle composition it is possible to show under what P–T conditions melting begins, and what kinds of products are produced. An obvious limitation is the minute quantities of material that can be used at any one time. But it does seem likely that conditions "down there" are much more uniform than they are nearer the surface. The microcosm of the sample might reflect the macrocosm of the world at depth far more faithfully than it would in the crust. Once you predict properties you can look for their expression in nature—how they will transmit S-waves from earthquakes, for example. So the fidelity of your model can be double-checked.

It is possible to push up the pressure even further, by using diamond anvils to press up to 2000 kilobars. Diamond anvils have been used in research programmes based in the University of Mainz, Germany, by Professor Böhler and his group, and in the Geophysical Laboratory, Washington, D.C. Diamonds do not crack under pressure, but the sample that can be squeezed by them is extremely small—only a microgram. This is rammed between two diamond anvils and contained within a gasket. The sample is heated by a high-powered laser beam and the temperature attained measured by a ruby fluorescence spectrometer. The whole process can be viewed through transparent diamond windows. Hence the progress of the tiny sample through its transmogrifications under pressure can be monitored directly using a suitable microscope, or by placing the machine within an X-ray diffractometer. The X-rays reveal how the stacking and arrangement of atoms change under the conditions found deep within the earth. Everything—matter itself—alters under pressure. Increase in temperature takes a metallic element towards its melting point, but simultaneous increase in pressure serves to raise the melting point still higher. At high temperatures, natural alloys can also alter the melting properties: a small percentage of another element changes the temperature at which melting occurs. It is exactly this ambiguous and inaccessible world that the high pressure/high temperature experiments in laboratories around the world are designed to investigate. The latest machines employ shock waves rather than "squeezers" to blast a specimen to extinction, literally shooting at it with a projectile. During the last nanoseconds of its existence, the properties of the sample are captured at P–T conditions close to those pertaining at the centre of the earth. High technology takes us to the innermost regions of the underworld.

It is surprising to discover that experiments on melting rocks were carried out from the earliest days of scientific geology. The Scottish scientist Sir James Hall (1761–1832), (not to be confused with the North American James Hall we met in Chapter 5) was a friend of James Hutton, whose *Theory of the*

Earth (1788) is considered the bedrock upon which all subsequent work, including Lyell's, has been constructed. Hutton had adduced good evidence for the formation of igneous rocks from hot magma, and thereby eventually disproved the earlier ideas of the so-called Neptunists, like Professor Abraham Gottlob Werner, who favoured precipitation of such rocks from solution. In the spirit of the muscular scepticism that ruled in Scotland at that time, Hutton wrote: "The volcano was not created to scare superstitious minds and plunge them into fits of piety and devotion. It should be considered as the vent of a furnace." James Hall set out to test Hutton's ideas by experiment. Hall was dissuaded by his senior Hutton from experimenting during the latter's lifetime; possibly, Hutton was nervous that the experiments might not come out quite as he would have wished, or perhaps he thought that the differences of scale between nature and laboratory were too important. Hutton warned of "those who judge of the great operations of the mineral kingdom from having kindled a fire, and looked into the bottom of a little crucible." What would he have thought about using a fraction of a milligram of material? After Hutton's death Hall did, indeed, melt rocks that had been collected from active volcanoes and showed that when they cooled they yielded products identical to ancient volcanic rocks. Even more remarkably, he used sealed gun barrels as pressure-vessels, which could then be heated to give the first simulation experiments of deep-earth conditions. (In the Grant Institute of Geology, in the University of Edinburgh, Hall's broken barrels are still kept in a display case at the top of the main staircase.) Hall showed that marble was produced by heating limestones under pressure, a persuasive reproduction of the processes of metamorphism.

So how to describe a journey through the deep earth? Perhaps the best way is to move upwards from the middle of things, from the "still centre of the turning world," and travel towards the surface—from the unfamiliar towards the more familiar. We will take the opposite journey to Jules Verne's entirely fabulous *Journey to the Centre of the Earth* (published in French in 1864), equipped with a modicum of facts and a modest map based on reasonable speculations. Facts first. The centre of the earth is some 6370 kilometres from mean sea level. The earth's core extends to 2900 kilometres and takes up somewhat less than 20 percent of the total volume of the planet. From the study of seismic waves reflected back after major earthquakes it has long been recognized that the outer part of the core does not transmit shear (S-) waves, and is therefore liquid. Only solids will transmit such waves.

The inner part of the core begins at 5120 kilometres below the surface and is in a "solid" condition, but at the enormous pressures at those depths it has a special kind of squeezed solidity. The velocity of pressure (P) waves through the inner core suggests a density appropriate to iron, but with a proportion of some lighter element. This is where high-pressure experiments come into play, because it is possible to investigate the solubility of various elements in iron under core conditions. It is equally possible to make theoretical predictions of the properties of these alloys using quantum mechanical calculations, and thus attempt, in the spirit of Sir James Hall, to match prediction to observation. High-pressure experiments have suggested that a small percentage of sulphur might be the appropriate addition to the inner core. Silica—so important in the outer shell of the earth—is not significant in the core. The outer part of the core, too, is mostly iron, but the reduced pressure, compared with the inner core, and an appropriate temperature (greater than 3000° C), enables it to exist in liquid form. We all sit atop a melted metal sphere. It, too, is lighter than it would be if it were pure iron, and there has been much speculation about what element serves to "dilute" the liquid metal. Again, ambient conditions at great depth are so different from those we are used to at the surface that we have to forget our everyday ideas of common physical properties conditioned by kettles boiling at 100° C and sugar dissolving in water. Instead, we have to conceive of substances "dissolving" in liquid iron, to produce strange solutions. High-pressure experiments can play around with sulphur, or carbon, or oxygen to find out their solubility in iron under core conditions. This weird world is not altogether beyond our grasp. Most experiments have suggested that oxygen must be present—probably in the form of iron oxide dissolved in molten iron. It is also apparent that some elements are "attracted" to the iron in the core because of their solubility in its molten phase. These include tungsten, platinum, gold, and lead: together, they are known as siderophile elements. The result is that these same elements are relatively depleted in the mantle above the core. They were stolen away in the early days of the earth's formation, and sequestered in the very depths of Pluto's kingdom. A further consequence is that such valuable metals require special geological circumstances if they are to occur in commercial quantities near to the surface of the earth. They have to be doubly refined to find their way into veins and skarns. This is one entirely practical reason why we should care about what goes on so far down beneath our human ken.

Another reason is that the outer core is the source of the earth's magnetic field, the result of its behaving as a geodynamo. Without this magnetization there would have been no "palaeomag," and tracing the history of plate movements would have been much more difficult: no polar wandering if no pole. So the earth's magnetic field is essential to the tale told in this book. If the magnetic field were turned off, compasses would immediately be useless, and some species of migrating birds would wing around in circles, hopelessly disorientated. The magnetic field is an invisible map drawn over the face of the earth: many organisms have sought it out to read their route. It is worth remembering that the magnetic field is also very weak—more than a hundred times weaker than the field between the poles of a toy horseshoe magnet. The development of accurate instruments to measure natural magnetism is another example of technique and theory advancing cheek by jowl. Such scientific progress can be a slow business—after all, the first suggestion that the world *did* behave like a magnet was made by William Gilbert before 1600; it took nearly four centuries for science to investigate Gilbert's insight in detail.

Much still remains mysterious about the geodynamo. It is clear that it has been around for more than 3500 million years because remnants of ancient magnetism have been measured in rocks of this enormous antiquity. Nor does it seem to have altered in strength over that time. So it seems likely that magnetism is an innate property of the liquid outer core. The metallic core is, of course, a very good conductor of electricity, and also a fluid capable of movement; a magnetic field presumably must be generated by the interaction of these two properties. Nonetheless, its mathematical modelling has proved very difficult. Since the dynamo has to be driven by energy, much depends on the nature of the energy source—which must also have been rather constant for a long time. Thermal convection in the outer core is one possibility: a kind of deep, simmering turnover of the molten layer providing a motor of magnetism. Another possibility is harder to understand from our perspective as surface animals—that the inner core grows by liquid iron "freezing" onto it at its outer boundary (can you imagine freezing at thousands of degrees?). When this happens, it leaves behind a "light fraction" in the outer core that then rises, leading to another, but compositionally driven, form of convection. Nor is the magnetic field as simple as that generated by an ordinary bar magnet; in detail, it is very complex indeed, and varies over historical time. A geophysicist at Leeds University, David Gubbins, has mapped variations in the magnetic flux, not at the surface, but at the core—mantle boundary, using mathematical manipulation of historical records of field observations dating back several centuries. It is surprising how the patterns of magnetic flux have changed over this geologically short timescale. Although some "bundles" of flux are stable and intense (e.g., under Arctic Canada and the Persian Gulf), others drift westwards. Gubbins relates these to columns of liquid within the inner core. Fluid spiralling down through the column creates a dynamo process that concentrates flux at the sites of the columns. It seems that there is a swirling activity "down there" that is almost indecently fast in comparison with the stately movement of plates at the surface. It is astonishing to be able to peer into such inaccessible regions in intimate detail. However, everything we know is based upon inference, and we have seen already that changes in perception have happened more than once in the story of our understanding of the earth. New thoughts may well change how the moving innards of the earth are modelled.

Then there are the magnetic reversals—those times when north and south poles "flip over." This was first suggested in 1906 for some volcanic rocks in the Massif Central in France, a region of classically conical extinct volcanoes known as *puys.* Today, nobody seriously questions the reality of such reversals, since magnetizations have been precisely dated, using evidence provided by the signature of characteristic fossils and confirmed by radiometric dates all over the world. Recall that recognition of compatibly magnetized "stripes" to either side of the mid-ocean ridges was one of the crucial discoveries that hoisted the flag of plate tectonics over the bodies of its rivals. It proved that new ocean crust—appropriately magnetized at birth—was moving away from the creative centre of the ridges. Hence reversals are another property of the deep earth that constrain the narrative of this history. It is clear that they are by no means as regular as clockwork. There may be a million years or more of one "north," followed by a short-lived reversal—north to south—of perhaps a few tens of thousands. Certain time periods are characterized by more reversed than normal fields—and vice versa. The Cretaceous was a long period of normal magnetism, though nobody knows whether lumbering dinosaurs used it to negotiate paths through their reptilian world. The periodic switch is like that shown by one of those cheap battery torches that blink on and off, apparently with a maddening will of its own. The short-lived "reversal events" within a longer period of opposite magnetization are distinctive time markers in geological history—so distinctive, indeed, that they have been given names. As we have seen, the Jaramillo Event is a short period of normal polarity about 900,000 years ago within a long period of reversed magnetism. It can be used as a precise geological tie-line between events on the sea floor and lava extrusions on the continents, a signal of great prehistoric utility, as might be the issue of a particular coin during the historical era. The "switchover" from one polarity to another is completed within 4000 years—a mere blip in geological time. Detailed studies on rocks that preserve a record of the "flip" show a decrease in the intensity of the field for a thousand years or so before the switch, and then short-lived, irregular swings of the magnetic vector, before the opposite polarity is established, weakly at first. It is a subtle thing; no animal species felt it as it might the jolt of an earthquake. In fact, switching poles is a comparatively easy thing for a dynamo to do, and the "flip" may be controlled by relatively small changes of the fluid motions in the core. It is intriguing that our notions of direction are at the whim of a swirl of liquid iron thousands of metres beneath our feet.

Upwards: to the mantle. If the earth is likened to one of those rounded avocado pears, the mantle is the edible flesh above the core: the crust might be the skin of the fruit. The motor of the earth churns over in the mantle; it is where mountains are born and plates die. It is the deep unconscious of our planet, the hidden body whose bidding the continents obey. The expanding oceans ride upon it. The mantle is the well-spring of tectonics. The remodelling of the face of the world that happens when plates move is ultimately a consequence of the power residing in the mantle.

The possible causes for the shape of the earth have been reflected upon for millennia. Creation myths often invoke the distillation of order from chaos—the ancient Greek word for primeval confusion, according to Hesiod's *Theogony.* The separation of the heavens from the earth and of land from the sea in the biblical account could almost be an historical scenario, if you interpret it generously, and without regard to the time involved. The Kono people of Guinea believe that *Sa* (Death) created endless mud, from which God refined the solid earth, in a kind of sedimentary genesis. The Egyptians described the world as originating from the endless and formless "sea" known as *Nun;* out of *Nun* an egg was formed from which light emerged. In turn, the sun-god *Atum* arose from *Nun* to form the dry land. On Easter Island, a bird-god laid the egg from which the world hatched. Cosmic eggs figure in many creation myths in cultures stretching from China to South America. In general proportions, the egg—with yolk, white, and a thin outer skin—is a reasonable model for the structure of the earth (especially if you allow it to be a reptilian sphere); for the relative dimensions of core, mantle, and crust, respectively, are not so different from the layers of an egg. To extend the comparison further than it ought, you could say that that shell of the global egg has been repeatedly broken and annealed by tectonics. As for the underworld, in ancient China it was, according to one mythology, the realm of the emperor of dragons: not exactly a tectonic model, but one at least associated with the vigorous heat of a fire-breather.

The mantle extends from the outer part of the core to the base of the earth's crust, which is at about eleven kilometres below the ocean floor and, on average, about three times as thick beneath the continents. The mantle thus comprises the bulk of the earth. If our planet were observed by means of a clever instrument from a distant galaxy, an alien astronomer would probably report to his superior a planet composed of the elements silicon, iron, magnesium, aluminium, and oxygen: the elements of core and mantle. He might, if his instruments were sensitive enough, notice traces of carbon from the living gloss on the surface, and detect the components of the atmosphere. Such is the lightest brush of life on the exterior. The uppermost part of the mantle is incorporated, with the crust, into the lithosphere—literally, the "sphere of rock." The lithosphere is the part of the earth that comprises the rigid, tectonic plates—hence the most important part for the natural history of the visage of the planet, and the part of which almost all this book has treated. The lithosphere is *not* just the crust, although such is a common misconception. We should add to the catalogue of useful names the layer underlying the lithosphere, which is known as the asthenosphere—and which, of course, lies entirely within the mantle. *Astheno* means "weak," or "feeble," in Greek, so this is a weak sphere underlying a rocky one. It is hardly surprising that the junction between the two is where the surface of the earth goes a-wandering. This is the layer of weakness that allows the physiognomy of the planet's surface to change with the slow pulse of ocean-floor spreading. You will see by now that crude analogies with avocado pears or eggs have already been surpassed: in detail; there are just too many layers in the mantle. There was an egg I bought in St. Petersburg in which several eggs were cleverly tucked one inside another, which is a more accurate analogy. The mantle layers are nested within one another like the skins of an onion. Some layers are much thicker than others. To understand the interior of the earth it has to be exposed layer by layer.

. . .

You may wonder why the mantle is described as a single unit at all. The answer is that it has a generally comparable composition: iron magnesium silicates are the dominant materials. At least the upper 160 kilometres of the mantle have the composition of the rock type known as peridotite. We have already met this rock in Hawai'i as rare lumps in the lavas, brought up from deep in the earth by the ascending mantle plume. It is a dully lustrous green-black rock, coarse in texture, and heavy—its density is obvious in the hand. The crystals of the minerals olivine (peridot) and pyroxene that make it up can usually be discerned with the naked eye. Because *ma*gnesium and iron (*Fe*) are so dominant, rocks of this kind are known as ultramafic. There is evidence that the deeper layers of the mantle may be somewhat different in composition. One line of research looked at the heat lost at the surface of the earth: to return to the egg analogy, imagine a freshly boiled example cooling in the palm of your hand. Those who work in deep mines experience this heat flux at uncomfortably close quarters. Oliver Goldsmith described the phenomenon in *A History of the Earth* in 1774: "Upon our descent into mines of considerable depth . . . we begin by degrees to come into warmer air, which sensibly grows hotter as we go deeper, till, at last, the labourers can scarce bear any covering as they continue working." At the present time, the heat flux at the surface of the earth has been estimated at 44×10^{12} W, derived from within the body of the earth. Unlike the egg, this is heat that replenishes itself. Most of it is derived from the decay of radioactive elements—a kind of internal bonfire that smoulders away in the manner of a fermenting compost heap. The decay of uranium isotopes is the most familiar—they are, of course, one of the sources of the radioactive "clock"—but potassium and thorium are implicated, too. You have to imagine heat generated by countless, scattered atomic "sparks" in the subtle bonfire of the earth's interior. However, the content of the radioactive elements in the upper mantle is not sufficient to account for the known heat flux. Recall that the lavas at the mid-ocean ridges are generated by partial melting of higher mantle rocks, so they provide a standard for the precise measurement of isotopes in that part of the mantle. There are not enough of them—only about one-eighth of the total required for the observed heat flow. It is likely, therefore, that the deficiency is made up by a radioactively "hot" lower mantle enriched, at least at some time in the past, in uranium and its sizzling fellows. So the lower part of the mantle may be subtly different compositionally from the higher.

I can now describe the several layers within the mantle. Because it is so thick, the mantle spans a range of pressures and temperatures that increase with depth. These constraints force its constituent minerals to change their atomic structure, goaded beyond endurance as they lie progressively deeper. The mineral olivine undergoes more than one transformation in its journey towards the centre of the earth. Deeper, and its atoms are forced closer together, jostling for accommodation like crowded passengers on a rush-hour subway train performing subtle readjustments as more and more of their fellows join them. At about 410 kilometres depth, olivine changes into wadsleyite, and at about 520 kilometres wadsleyite transforms into ringwoodite. The corresponding temperature is approximately 1600° C. These minerals both have the same composition as common olivine, but different structures: like those subway passengers, pressure has forced them into a new accommodation with the conditions in the underworld.

This is where the experiments with which this chapter began come into the argument. Olivine can, indeed, be squeezed in the apparatus until its atomic pips squeak. When the pressure is piled on, and if it is also sufficiently hot, olivine can be persuaded to change its personality. In the technical jargon, it undergoes a phase change. I have seen artificially made ringwoodite, and it has a blueish hue: only Pluto would be able to admire its colour deep within the earth. Ringwoodite has a distinctive atomic configuration known as the "spinel structure." Spinel itself is an attractive gemstone (chemically, magnesium aluminium oxide) that has been employed as fake ruby—even finding a place in crown jewels. Ringwoodite shares its atomic design, despite its different chemical composition. Wadsleyite is a different variant on the spinel structure, one which has "room" for some water within its atomic lattice. The mineralogist J. R. Smyth calculated that there could be more water locked up within the wadsleyite in the mantle than in all the world's oceans: a sea beyond seeing. Neither ringwoodite nor wadsleyite is stable at the surface of the earth, in the realm where olivine is normal; but, like diamond, once made, these minerals endure for a long time. Wadsleyite *has* been discovered on earth—or, I should say, off earth. It occurs in the Peace River meteorite, which fell in Canada during the 1860s, possibly a piece of a shattered planet that preserved this deep mineral for posterity. So in the laboratory you can look under an eyeglass at a minute piece of earth's interior, forcibly brought to birth in an alien place between carbide jaws of man's invention.

Nor is this the end of the story. Deeper still in the earth, at a depth of about 660 kilometres, ringwoodite comes to pieces. Its atomic framework cannot hang together under the increasing pressure and it splits into com-

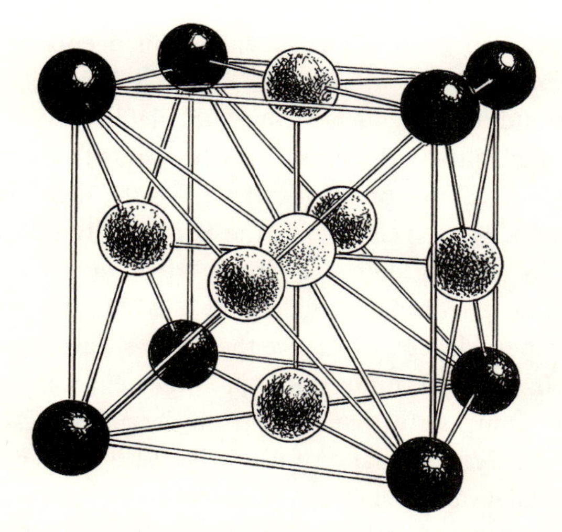

Diagram of atomic structure of perovskite type which is dominant below the 660-kilometre-deep boundary layer.

ponent parts. "Things fall apart; the centre cannot hold," as W. B. Yeats wrote. This results in two mineral phases coexisting together: one with perovskite structure, the other a magnesium/iron oxide (magnesiowustite).* If you were on a game-show and were asked the question, "What is the most abundant material on earth?" the answer would be: "the perovskite phase of the lower mantle." The applause should be resounding. A simple calculation tells you why your answer is true. The layer of earth's "onion" containing perovskite stretches from 660 kilometres to the boundary layer at the outer edge of the earth's core, 2900 kilometres below the desk on which this is being written. It is a vast slice of the earth. Since the perovskite phase comprises something like 70 percent (perhaps more) of this layer, it is an easy calculation to show that this unseen, unheralded, and, some would say, obscure mineral phase is the bed upon which the whole of our multifarious biosphere ultimately lies.

* For those who like details, the equation for the transformation is not complicated:
$(Mg, Fe)_2SiO_4 = (Mg, Fe)SiO_3 + (Mg, Fe)O$
Ringwoodite Perovskite Magnesiowustite

Magnesium and iron can replace one another in olivine molecules in a continuous way. This is signified by the brackets (). Ringwoodite was named as recently as 1969. At the molecular level, the stacked octahedra of perovskite are quite different from the "pseudocubic" structure of spinel. A. E. Ringwood and A. D. Wadsley were notable mineralogists.

What, it reasonably might be asked, has this deeper reality got to do with us? After all, it is a long, long way down and far removed from the sunshine and clouds under which we live our brief lives. We cannot see into the earth with our light-bound eyes, and who wants to travel to this realm of crushed molecules and phantom phases? But other probes *do* reach there. Earthquake waves respond to the profoundly concealed mineral phase changes by altering their velocity in "steps" as they pass from one layer to another deep within the earth. They "see" that the earth is mantled by successive mineral layers, which are themselves subdivisions within the larger mantle. In a wonderful example of congruent science, it transpires that the seismic changes are *predictable* from the properties of those deep-earth minerals made in the laboratory: it is just a matter of making the appropriate calculations. Hence one field of investigation independently confirms and constrains another. At the end of it all, the journey from the core towards the crust can be recited zone by zone like a travelogue: proceed through the perovskite layer upwards to the ringwoodite layer, and thence to the wadsleyite layer, and onwards and upwards into the olivine layer. Through these layers pass the convection cells that drive the plates that in turn sculpt the face of the world. In this way the depths intercede in our superficial lives: there are unseen and unbidden forces, as indifferent to the fate of the sentient organisms living above them as the distant stars.

The glass in a medieval church plays tricks on you. The world seen through it is a distorted one: swirling, refracted, blurred. Its original transparency has been compromised by some transforming agent. That agent is flow. It operates so slowly that its progress would be hard to see in a century. Over hundreds of years the surface of the pane distorts from the true, crinkles and falls into lobes under the force of gravity, until we see but through a glass, weirdly. Yet we know, too, that glass is a brittle solid. It seems that glass can be simultaneously both solid and flow like a viscous liquid. It is possible to see the same phenomenon over a shorter time scale with a block of asphalt. Follow the smell to where a road is being resurfaced and you may see a dark block of crude asphalt lying in the roadway. It has a fractured edge, so it must have broken like a solid. Yet you will probably notice a side bellying outwards, as if it sagged under its own weight, viscidly flowing. Again, glaciers creep slowly down mountainsides, yet an ice-block taken from the freezer often cracks and splits when dropped in a glass of whisky at room temperature. Glass, tar, ice: these are all "solids" that also show long-term fluid behaviour. They creep, they flow. They help us visualize what happens within the mantle, for this is how the inner earth moves.

The Swedish scientist Anders Celsius was a professor in the earlier half of the eighteenth century at the University of Uppsala, the most ancient establishment of learning in Sweden. It is a place of austere, four-square and solid academic buildings, with a fine cathedral. Celsius is immortalized for having given us the scale that interposed 100 degrees between freezing and boiling water. He also noticed that around the Gulf of Bothnia on the Baltic Sea there were strand lines—former sea-shores—that were now marooned above the present sea level. These observations were published as early as 1744. As Eduard Suess described it 150 years later: "At Tornea, [Celsius] had been shown to his astonishment that the harbour constructed in 1620 was already useless." The implication must be either that the land was rising, or that the sea was retreating. From a boat on the gulf you can observe the raised beaches as if they were drawn against the land with a ruler. Celsius himself had concluded that the seas might be shrinking on a global scale, draining away to leave their old shores high and dry. Charles Lyell observed in 1834 that the elevation of the old shores in the north of the Gulf of Bothnia was greater than that in the south, and carefully argued that the former strand lines should be attributed rather to a rise in the land's surface. Curiously, Eduard Suess did not follow Lyell but instead went to some lengths to diminish the rather strong evidence for elevation of the land that had accumulated by the end of the nineteenth century. This may have been because Suess preferred to think of land movements as periodic convulsions, as in his beloved Alps, rather than the steady tinkerings of Lyell's classic explanation. He thought that the Gulf of Bothnia was draining out by "*an emptying out of the water, not a rise of the land*" (his italics, as usual when he signalled what he regarded as a crucial point). All the more curiously, at the time Suess wrote, it had been accepted universally in scientific circles that a huge ice sheet had covered Scandinavia in the last great ice age, disappearing less than 10,000 years ago. It seemed a reasonable inference that the massive burden imposed by "loading" of the ice had depressed the land, which had rebounded in a regular fashion after the ice had melted, thereby carrying upwards former shorelines beyond the reach of the sea. Even with the evidence to hand, Suess was determined to reject the flexibility of the earth's response that this implied, and recruited the ice to his own purposes. Referring to strand lines in Norway that were similar to those in the Baltic he wrote: "The great majority of terraces in the fjords of western Norway, must be regarded as monuments of the retreating ice, and not as evidence of oscillation of the sea level, and still less of oscillation of the solid land." You could not wish for a clearer statement of his priorities.

It was not often that Eduard Suess was so in error. The rise of the glaciated country of Finland and Scandinavia is now an established fact: half a century after Suess, Arthur Holmes reported that in the northern part of the Gulf of Bothnia an uplift of about 270 metres (900 feet) had occurred and that it was continuing at a rate of about a thirty centimetre (foot) every twenty-eight years. The degree of uplift is in proportion to the thickness of the ice overburden that is removed: the more ice there was, the more uplift there will be. The uplift has even been mapped in contours. Modern laser techniques can measure the rebound with considerable accuracy. The elastic recovery of the land is produced by compensation in the mantle: it flows in to "fill the space." Imagine the world as a balloon filled with honey: the skin of the balloon might be an appropriate surrogate for the earth's lithosphere. Prod the balloon with your finger, and a dent remains. The honey slowly flows in to compensate the depression, which rebounds at a fixed rate in an elastic response. It becomes a comparatively easy calculation to determine the viscosity of the honey from the rate at which this happens. So it is with the mantle. It can be understood as a liquid which is 10^{23} times more viscous than water—trillions of millions times as sticky, but a liquid nonetheless: think, rather, of that asphalt. The exquisite sensitivity of today's instruments shows just how responsive the lithosphere can be to loading: even the weight of a building can produce a detectable downward deflection. Is it possible that even a heavy footfall produces a twitch in the mantle? We live on an elastic, quivering earth, buoyed up on the responsive layer beneath the lithosphere.

The mechanism of convection currents within the mantle will now be understandable. It is flow on a dignified scale. The mantle includes a few huge convection cells that transfer heat from the core-mantle boundary to the base of the lithosphere. A slowly simmering cauldron of porridge, roiling and rolling, is both useful as an image of convection but also misleading as a metaphor for the nature of the flow. In the boiling porridge, the oatflakes are merely carried by the convecting water between them; in the mantle, it is the rock itself that flows. The hot, rising limbs of the convection cell part beneath the mid-ocean ridges, providing the energy source for the lavas that create the surface of the world anew. The lavas are developed from partial melting of deeper peridotites. We have seen that occasionally a lump of peridotite is entrained in an erupting lava: it always contains olivine, and never wadsleyite, showing that it must have originated from the outer mantle layer. The work of the rock-crushing theoretician and the geologist finally meet on the rocks. What mantle convection has created, it also destroys at the subduction zones at the edges of oceans, where the downward-plunging limbs accompany slabs of comparatively cold oceanic lithosphere back towards the foundations of the earth. Described thus, the great convective cycle seems devoid of turmoil: yet it is the destroyer of cities and the generator of mountain ranges. When viewed on the scale of the whole world, however, an alpine system is the merest surface puckering, whilst an earthquake that might destroy a civilization is no more than a momentary shudder. The intimate earth history I have described in this book is like the adventure of a flea on the back of an elephant.

It is difficult to conceive of structures so vast as the convection cells that turn the underworld. One school of earth scientists is persuaded that there are *two* sets of such cells within the mantle, one above the other, turning in tune like linked cogs in a clockwork mechanism. They envisage that the boundary between upper and lower cells is the profound, and profoundly important, 660-kilometre discontinuity, below which the perovskite structure predominates. Recent reviewers have preferred the model that convection operates through the whole mantle, which would make it the greatest single phenomenon on our planet, turning and turning in tune with the slow beat of geological time. The properties of shear waves (S-) show that the lower mantle is not differentiated elastically from the upper mantle, so that convection cells are able to pass through the layers of mineralogical phase changes. They change personality, if you like, but they do not change their purpose. Recall also that there are mantle plumes, like the one underlying the Hawaiian chain, that pass through much of the mantle to form "hot spots" below the lithosphere. Furthermore, the slabs of oceanic crust that are carried down by the *descending* limb of the convection cell can be detected as heterogeneities in the mantle: they plunge down into its lower part. The downward plunge of the slabs is measured by an earthly equivalent of the CAT (computerized axial tomography) scans that are now routinely used as a more precise alternative to conventional X-rays for "seeing inside" the human body. Seismic waves provide the data needed to nose into the deep innards of the earth. The method is known as seismic tomography: the earth itself is the patient. Where a subducted slab, which is relatively cold, is pulled downwards into the mantle, it subtly alters the "speed" of the seismic waves—up to plus 2 percent. The effects can be mapped with remarkable precision, as, for example, beneath the Tonga-Kermadec Trench in the Pacific. The fast velocities produced by the descending slab can be mapped down to depths of perhaps 1600 kilometres—deeper by far than the epicentres of earthquakes caused by the movement of the slab itself. Under Japan, the 660 layer is depressed by about thirty kilometres where the cold slab plunges downwards. It seems that even when the convected slab is beginning to merge back into the mantle from which it was ultimately born its history is retained as a kind of "ghost" that the earthquake waves alone can see. The same methods also reveal that the mantle is less homogeneous than was once believed: it includes "blobs" of different composition on the scale of a few kilometres, which may be the last remnants of subducted lithosphere slabs.

What moves the lithosphere plates above the mantle convection cells? Are they pushed up at the mid-ocean ridges, or are they dragged by the subducting oceanic crust, which is denser than the mantle through which it sinks? The two theories are known as "ridge push" versus "slab pull" respectively. Either way, the effect is the same. The mid-ocean ridges are elevated above the sea floor, where the heat flux is greatest on the upwelling limbs of the convection cells. Do the plates slide away from the ridges towards the trenches? Or do the subducting slabs of "cold" crust—which, as we have seen, go down deep—tug at the slabs somewhat as one might pull a tablecloth from one side? Intuition would perhaps suggest the former, if only because there seems to be something inherently vigorous about the creation of new crust at the mid-ocean ridges, as if the earth were heaving itself up by its own bootstraps. In detail, however, the volcanic eruptions at the ridges are found to be rather gentle, whatever the impression given by lines of fiery chasms along fault zones. The new crust almost insinuates itself into the spaces generated by adjacent plates moving apart. Furthermore, when different plates are compared, it seems that the velocities* at which they travel are related to the length of trench-edge that they possess: the longer the subducting margin, the faster the plate moves. This supports the notion that slab pull is more important than ridge push. Plunging plates turn the engine of the earth.

* This absolute motion can be deduced from the rate of movement of plates above "hot spots" like Hawai'i, as described in Chapter 2.

The turnover of the oceanic crust also means that much of the earth above the core is recycled. The current rate of subduction down trenches is about 3 square kilometres per year—the area of a large village, perhaps. This implies that at least 18 cubic kilometres of oceanic crust and something like 140 cubic kilometres of the peridotite layer beneath it are sucked downwards into the maw of Pluto's realm every year. Then, too, there is a quantity of sediment that is scraped off and directed Hades-wards along with the oceanic plate. The mantle begins to look more like a thick and varied soup rather than a bowl of porridge. One set of published calculations indicates that almost the entire mantle has been through the recycling process over the course of geological time. The different subducted sources also retain their chemical element signatures, which can be precisely measured to parts per billion with modern mass spectrometers. Isotopes of rarer elements such as uranium and thorium are particularly useful in identifying sources of subsequently erupted lavas. Some oceanic island "basalts" stand revealed as having originated from recycled ancient ocean crust.

The crust of the continents tends to stand apart from much of the activity at subduction zones. Eduard Suess long ago commented that the continents were of "trifling density (2.7 grammes per cc) compared with the density of the Earth as a whole (5.6 grammes per cc)." The continents are composed predominantly of what both Suess and Arthur Holmes would have termed "sial," light rocks with alumina, an important component compared with the dense, basic rocks of ocean floor and mantle: sial included all granites and gneisses, and many sedimentary rocks. A long catalogue can now be added of minor elements also sequestered into the continental crust. The light, thick continents rise above the troublesome subduction zones. They have not been continually recirculated like the ocean basins: they rise above things, which is why they include the ancient survivors, the Methuselahs of rocks. But wherever subduction zones plunge downwards against a continental area, the rocks of continents and oceans interact and fuse. Lavas generated by frictional heating include the gummy magmas that make the catastrophic volcanoes of the "ring of fire" through Japan, Indonesia and adjacent islands, and the Philippines. They dog the downward-dipping subduction zones around the Pacific, like boils around an infected wound. The trace elements recovered from their volcanic rocks reveal the proportions of the mixture between magmas derived from ocean and continent, respectively, almost in the manner of a genetic marker revealing human ancestry. The volcanoes burst to the surface landwards of the ocean trench, rising as they do from the depths of the subduction zone. Suess had recognized this fact with characteristic clarity. "The volcanoes which accompany the folded festoons *never* stand in the foredeep," he wrote, "but belong entirely to the folded cordillera." He attributed this to subsidence of crust in the trenches. Would that he had known that the area beyond the trench was where a forced marriage between the kingdoms of Pluto and Zeus had been enacted deep in a subduction zone, and where hybrid magma is the result of their union. These offspring generate splenetic eruptions, spawning pyroclastic surges that wreak destruction indiscriminately on man and beast. The hollow mummies of Pompeii were the tragic consequence of such a magmatic marriage. The same wrathful issue engenders volcanoes along the length of the western rim of the Americas, where the oceanic plates of the Pacific impinge against the unyielding continents to the east. From Popocatépetl to Mount St. Helens, destructive volcanoes are the consequence of ocean meeting continent and the furious fulminations that follow.

The boundary between the lithosphere and the asthenosphere beneath it is the top of the Low Velocity Zone. This is a region where both kinds of earthquake waves are slowed down. It lies within the upper mantle, some 40–160 kilometres below the surface. Experimental work suggests that the change in seismic behaviour happens because some small part of the mantle melts—the result of the release of pressure at these particular depths, at the appropriate temperature. The partial melting of mantle rocks is sufficient to reduce the coherence of this layer in the earth: it is where "slippage" occurs between rigid lithosphere and weak asthenosphere. We have arrived at the base of the shell of earth's egg.

Upwards, ever upwards. The name "crust" has a nice, homely feel to it, carrying images of pies fresh from the oven. Compared with the alien interior from which we have travelled, it *is* almost like coming home. The rocks of the crust are those familiar to us; they comprise the stage upon which my intimate history has unfolded. We have seen that everything turns upon a deeper geological reality, powered by leisurely rotations and passing through strange mineral phases that cannot exist on the surface. This deep world, accessible only through experiment, or probing earthquake waves, is overlain by one that can be understood by geological mapping and geological tapping. This is our home, our rich and complex habitat, laid out upon a geology as various as human culture, and as challenging to understand; where the history of more than 3 billion years of life is recorded in ancient sedimentary rocks; where life and the earth have evolved together in an intimate collaboration that is a marvel in the galaxy. Welcome back to the surface, almost.

The base of the earth's crust is drawn at the highest of the seismic lines that I shall mention: the Moho. This is a somewhat anonymous abbreviation for Mohorovičíc Discontinuity, a name that justly commemorates its discoverer, the pioneer Yugoslav geophysicist A. Mohorovičíc. In 1909, he demonstrated that earthquake (P) waves "jumped" in velocity from about 7.2 to 8.1 kilometres per second along a discrete surface. Evidence for this boundary was not known to Eduard Suess, but familiar to Arthur Holmes. In the upper part of the crust, rocks fracture brittlely: this is the realm of earthquakes, whose shudders have done so much to help decode the rest of the earth. The junction between crust and mantle is where the dense peridotite layer of the upper mantle circumscribes the earth. This heavy rock speeds earthquake waves. The crust is thinnest under the oceans, especially towards the mid-ocean ridges. We have already met sections through such crust in the obducted ophiolites of Oman and Newfoundland, eight kilometres or so thick, each with their own distinctive layering. Beneath the lighter continents, crustal thickness is thirty to forty kilometres in the quiescent, ancient, stable parts of the earth, where, all passion spent, the surface of the earth lies calmly on its deeper foundations. Contrariwise, the continental crust is thickened—often doubled again—along mountain belts, especially where continent has collided with continent in ranges like the Alps or the Himalayas. In some places one continent has even underplated its neighbour—pushed bodily underneath—in the ultimate heave. The consequence of earthly engagement is elevation of icy peaks above deep tectonic roots. We have seen how granites are sweated out deep in mountain belts; and how continental rocks are changed by heat and pressure. And then the forces of air and rain and ice seek at once to wear down the elevated peaks, just as the Book of Isaiah describes it: "every mountain and hill shall be made low: and the crooked shall be made straight, and the rough places plain."

The face of the earth has its character scoured upon it by the elements, but they can only work on what has been set upon the surface by forces operating in the hidden depths.

The journey from the centre of the earth may seem a more theoretical enterprise than the rest of this book, yet deep forces that we can infer from surface observations and experiments move the plates that make up the world. To the extent that human history has always been in thrall to geography and climate—both of which depend on geology—our very character could be described as geological. We may all ultimately be the children of convection. Mountain ranges—gross thickenings of continental crustal rocks—have divided one race from another, and have helped generate the diversity that is one of the chief glories of humankind; they have even provided refuges for peoples extinguished elsewhere. The great peninsula of India south of the Himalayas displays common cultural roots that dig deeper than modern religious divisions; the same area was the home to some of the earliest civilizations. Greater China to the north of the same range comprises one of the largest areas of cultural idiosyncrasy, abode of the Han people, whose sculptures of lions and dragons show cultural persistence over more than four millennia. On a smaller scale, the Pyrenees both guarantee the character of Iberia and maintain a refuge for Basques. The Alps seal in Italy on its northern side—a barrier famously pierced by Hannibal across the Little St. Bernard Pass in 218 B.C. My Scottish friends attribute their Celtic hardiness to that bare country being underlain by gneisses and schists of the ancient Caledonian range. My Norwegian friends show how the northward continuation of the same range has generated a variety of distinctive dialects in isolated communities along their coast. Even today, modern culture retains an inheritance of character and language from these ancient limits: it may even survive the onslaught of global capitalism, for it has already survived centuries of trading. Low sea levels associated with the last ice age allowed humans to penetrate into North America, perhaps 13,000 years ago, and into Australia before that. Some of those people dispersed into jungles lying along the passive margin of South America, into Amazonia, where the South American continent gently sagged alongside the widening Atlantic Ocean, and a huge river system parted tribe from tribe. In western Africa, the Congo Basin provided a mirror image on the corresponding eastern, passive margin. Cultural diversity was bred by separation at all levels. On the other hand, it is an uncomfortable fact that some of the areas of greatest human conflict, such as central Europe, are those where there are no such natural barriers, where one group after another has pushed for, and achieved, temporary dominance. Hungarians, Romanians, Russians, Poles, Slavs, Austro-Germans, Danes, and so on, have fought viciously over gentle European hills and plains. The sad litany of a hundred wars and skirmishes attests to the grim underbelly of diversity: intolerance.

All the intimate details of landscape and culture that I have described in this book are rooted in geology. Many of the older gods were related directly to landscape and their propitiation was often a way of rationalizing natural phenomena, whether invoking Madame Pele's uncertain temper in Hawai'i, or Vulcan's smoky workshops on Mount Etna. Time was tamed in myth, for humans have always wondered about their beginnings, inventing scenarios spanning primal eggs and sculpted mud. Geology provided a rational timescale, but one so incomprehensibly expanded that our own species might seem no more than an afterthought, a conscious postscript, rather than lying close to the essence of things. There are those who still regret the loss of gods who might intercede on our behalf. The true measure of the earth could be that slow overturning of the mantle that calibrates the march of the tectonic plates. James Hutton's famous statement in 1788 about the earth, that he found "no vestige of a beginning, no prospect of an end," is now at least constrained by a figure for the former: 4.5 billion years. The record of rocks on earth is consistent with plate activity over some 4 billion years of that span. During that time the planet has changed from barren to prolific. The photosynthetic activity of blue-green cyanobacteria and plants transformed the atmosphere to an oxygenated one that animals could breathe. Life moved from sea to the land, and thence into the air. All the while, oxygen altered the way the rocks themselves weathered, just as vegetation altered the rules of erosion thanks to their obstinate webs of roots that cling to and process soil. Earth and life became progressively interlinked: the "greenhouse" effect that worries us today is just the latest illustration of this obligatory wedding. And during all this slow transformation the motor of the earth reshuffled the continents and oceans, now spreading the continents apart, as at present, now locking them together, a leisurely procession to which life had no choice but to respond.

There are two possible reactions to the tectonic history of the world. Either you might despair at the apparent randomness of it all, at our insignificance in the face of gigantic forces that are as indifferent to our species as is a torrent to the fate of a mayfly that rides upon it; or you might wonder at the extraordinary richness of history, and feel privileged to be able to understand some part of it. Were it not for a thousand connections made through the web of time, the outcome might have been different, and there may have been no observer to marvel and understand. We are all blessed with minds that can find beauty in explanation, yet revel in the richness of our irreducibly complex world, geology and all.

. . .

With his own nod towards Indian mythology, Eduard Suess completed volume 2 of *Das Antlitz der Erde* thus:

> As Rama looks out upon the Ocean, its limits mingling and uniting with heaven on the horizon, and as he ponders whether a path might not be built into the Immeasurable, so we look over the Ocean of time, but nowhere do we see signs of a shore.

13

World View

. . . let him not say that he knows better than his master, for he only holds a candle in sunshine.

William Blake, *The Marriage of Heaven and Hell*

The marvellous thing about the face of the earth is that it is such a mess. It is an impossibly complex jigsaw puzzle of different rocks. Like Gilbert and Sullivan's wandering minstrel, it is "a thing of shreds and patches." More than 3.5 billion years of history have stitched it together. It has been modelled and remodelled, split asunder and rejoined in tune with the waxing and waning of oceans. Here, continents have been inundated by shallow seas which have then drained away, leaving a legacy of sandstones or limestones, shales or gravels. The Painted Desert was painted long ago. Elsewhere, erosion has exhumed baked rocks that once lay deep within the crust. The multifarious world wears geological motley. Mediated by climate, the underlying rocks influence the form of the landscape, the vegetation, the crops that can be grown—even the particular stones and bricks for making cities before steel-and-glass became universal. We have all grown out of the geological landscape, and perhaps unconsciously we still relate to it. Human beings seem to be programmed to love their home territory. Russians pine for the open steppes; Aborigines for the endless Australian interior; shepherds for rolling country dotted with sheep;

Shropshire lads for blue remembered hills. Most of us appreciate the world in this intimate way, rocks and all. Our patch is our home. Few shepherds, perhaps, pay much heed to the strata over which their charges scramble, although the rocky underlay is ultimately in control of their livelihood: other rocks, other livings. We respond more affectionately to landscape at the local scale, just as we register tragedies most acutely within our own family; broader issues of society do not tug at our sleeve in quite the same way, although we are reluctant to admit it. Whether we acknowledge it or not, geology is important in the most visceral way.

I have tried to explore the influence of geology on the ground—its intimate ramifications, its pervasiveness. I have let the particular place speak for the general case. I have no doubt that there are a hundred other locations as suitable to make the point as any I have chosen, and any selection will inevitably miss out important things. But then, there was never any intention to write out the whole world in the fashion of Professor Suess. Always in this story—far beneath us and largely unacknowledged—there is the bidding of the plates. The tectonic unconscious may provide a rationale of the earth, but perhaps it seems too remote to interest the shepherd lad with one eye on a lame ewe and the other on the girl from the next village. Nonetheless, an understanding of the deep tectonic driver has unified the world. Our patch is actually linked to every other patch. Perhaps realizing that we are all small creatures riding pick-a-back atop our own tectonic plates located within the irregular chequerboard of the earth might enforce a proper sense of humility upon our arrogant little species. Somehow, I doubt it.

Knowledge advances and old ideas are shed, sometimes with reluctance. Enough has been said already to convey a picture of what has happened to geological thought over the century since Suess' masterly global survey. There is no question that we are now close to a unified theory of the earth. It seems unlikely that many of the recent fundamental advances in understanding of the deep structure of the earth—the lithosphere, the mantle, the movement of plates—will be undermined in the future. Equally, it would be astonishing if there were not new discoveries that will render my account an antique within a few decades. We have seen enough of cherished ideas in the previous hundred years foundering against a new thought or a critical observation to make further enlightenment a sure bet.

However we try to retain a proper modesty in deference to the past, it is difficult to record the growth in comprehension of the earth in the light of plate tectonics without occasionally sounding smug. "Aren't we the clever ones?" we seem to imply. Nothing could be more unjust to those who went before. The pioneer geologists who provided intellectual ammunition for Suess and his successors were nothing short of heroes: they trekked across remote areas with little more than a hammer and a notebook. There was nobody to airlift them to safety; they had few maps to follow. The information they provided was seminal. Suess gave them the recognition they deserved. He praised W. T. Blanford, pioneer in the mapping of much of India and the Himalayas, and "Przewalsky's adventurous travels [which] have opened up a large part of Tibet"; and then there was Bell, who mapped the ancient shield areas of Canada (1877) at a time when they were unexplored, mysterious, and dangerous. These were courageous men of rare intelligence. Przewalsky is familiar to some because the wild horse that he discovered still bears his name—*Equus przewalskyii.* But many of these pioneers gave their lives to explore places that can now be reached in a helicopter in an afternoon, and their names are forgotten except to a few aficionados. They made their contributions, which are now woven as obscure threads into the tapestry of science. Most of us will have lesser memorials.

J. Tuzo Wilson is the exception. His name is built into the very fabric of the earth. He has been mentioned several times in this book, as a kind of seer for geological processes. He it was who recognized Iapetus. The slow progress of the plates, with initial continental splitting and formation of a new ocean, eventually followed by its closing again with the generation of an orogenic belt, is known as the Wilson, or Wilsonian, cycle. It is just about the biggest generalization that can be made about the world. As we have seen, the whole process may take more than 200 million years to complete. Of the many historical cycles expressed upon the earth—whether of chemical elements, sea level, or temperature—the Wilson is both the most stately and the most transforming. We have visited just a few of the mountain belts generated by the end of these cycles: the youthful Alps, the old Appalachian–Caledonian chain, the ancient scars of the Precambrian. The face of the world can be seen as the consequence of a succession of Wilsonian events—dominated, naturally enough, by the present cycle—but with the memory of previous cycles preserved as streaks and blotches like old wounds. The traces of ancient cycles far back in the Archaean are sometimes difficult to detect, but they are still there if you search hard enough. And the further back in time, the more the rocks betray the processes that occurred deep in the lithosphere, for these most ancient rocks have been scrubbed clean by erosion of all their overburden, it might be for thousands of millions of years. The motley of the world could well be explicable after all: there is time enough to patch it all together.

After earth was formed some 4550 million years ago, a pattern of plate movements can account for much of what is preserved on its surface—only the first few hundred million years have left little record. As for the end, it is predictable that the present Wilson cycle will continue, so that the Pacific Ocean will be subducted away in a hundred million years or so, producing a new Pangaea, another supercontinent when Asia and America finally conjoin. Beyond that? It is impossible to say exactly where it will happen, but it is certain that this future supercontinent will itself fragment again, and once more the continental blocks will be on the move: our world will be forever in transit, forever reinventing itself, until the internal motor of the earth finally turns no more.

Eduard Suess began his great work by inviting his readers to part the clouds and take a view of the earth from the vantage point of a visitor from another planet. It was a gesture of omniscience on his part. You are reminded of the Temptation of Christ, when the Devil laid out all the kingdoms of the world from the prospect of a high mountain. Geology will master the world—Suess supposed—and bring it to heel. This lofty panorama might have been a bold thing to attempt at the end of the nineteenth century, but we are now thoroughly familiar with the satellite viewpoint. That NASA photograph of earth from space might prove to be the definitive image of the twentieth century, the moment when the finite bounds of our planet entered common consciousness, so fragile it appeared, so blue and dappled—more soap bubble than mountain and mineral. As Suess began, so I shall end. After exploring intimate geological details in just a few places on our cobbled-together planet, it is time to change the focus briefly to look at the whole. We have seen how even a single crystal links to profound circulations of the earth; pull away upwards from that crystal until the granite holding it is just a tumid blemish on the greater landscape; and higher again until the granite is lost within the wrinkles of a mountain belt; from this altitude you can see the oceans and continents together. We will turn our narrative upside-down: from the details of a stone wall, or one lava flow named for a half-forgotten god, we will move to the outlines of oceans and ranges, pausing only fleetingly. Like Puck, we will "put a girdle round about the earth / In forty minutes."

Start in my office in the Natural History Museum in South Kensington in the West End of London, on a site where there have been geologists working for more than a century. Imagine flying higher and higher, until

ABOVE: *Lowest of the low: Death Valley, California.*

RIGHT: *"The loneliest town on the loneliest road in North America." Eureka, Nevada: a gold town.*

BELOW: *Fault-controlled scenery: a road across the Basin and Range, Nevada.*

RIGHT: *One side of the Great Rift Valley in Kenya, the ocean that never was.*

BELOW: *Flamingos on Lake Natron, in the African Rift Valley.*

Northwest Scotland. Ancient rock relationships: dark Scourie Dyke cutting through even older pale gneiss on the north shore of Loch Torridon.

The view across the alluvial flats at the south-eastern end of Loch Maree. The Kinlochewe Thrust forms the prominent line on the right, slicing through Cambrian quartzites, which in turn overlie Torridonian rocks. The Kinlochewe Thrust is part and parcel of the earth movements that produced the Moine Thrust, but is not so historically celebrated.

Liathach, northwest Scotland: pale-coloured Cambrian sandstones on the hilltops overlie pinkish Torridonian rocks.

The ancient of days: typical countryside in the territory of the Lewisian gneiss, with traditional dwelling built of local materials, island of South Harris.

Difficult rocks: the Lewisian gneiss south of Loch Tollie. The rocks hereabouts are in the amphibolite metamorphic facies, indicating that they have been affected by an event younger than Scourian, in the Laxfordian. Later events have erased earlier ones.

A Precambrian banded iron formation rich in the iron oxide hematite, showing the fine layers typical of the rock type. South Africa.

ABOVE: *The Grand Canyon. Sedimentary strata are horizontal and undisturbed. The Colorado River lies hidden far below within the Inner Canyon.*

LEFT: *Mules are the favoured means of transport for reaching the bottom of the Grand Canyon. They are rigorously selected for amiability and surefootedness.*

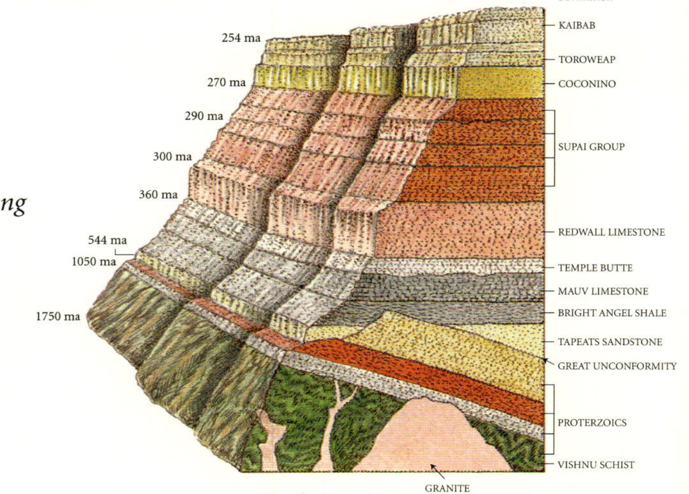

A profile through the rock formations encountered during the descent into the Grand Canyon along the Bright Angel Trail, with an indication of how they affect the topography. For a useful if indecent mnemonic, see pages 325–26.

The Inner Canyon is a different world from the higher horizontal strata: Precambrian Vishnu schists make precipitous cliffs plunging into the Colorado River, here living up to its name.

Winter at Phantom Ranch at the bottom of the Grand Canyon. The Bright Angel Creek has cut out a valley providing a trail to the north rim. Notice the rounded boulders to the right, proof of slow erosion.

The Gulf of Bothnia has "rebounded" since the last ice age at a measurable rate, allowing for calculations on the viscosity of the asthenosphere. On the High Coast World Heritage Site, the cultivated fields often lie upon raised beaches as in the middle of the picture.

ABOVE: *A sulphur miner on Kawah Ijen volcano, Java—the toughest trade on earth?*

ABOVE RIGHT: *Blue Mountains of eastern Australia: an escarpment without equal—and valleys beyond choked with eucalyptus.*

RIGHT: *Antipodean Alps, where plates collide: Mount Cook in New Zealand.*

BELOW: *One of the driest places in the world: the Atacama Desert, Chile. Its aridity is a product of the Andes, and the Andes are a consequence of the meeting of plates on the Pacific margin.*

we can see that all the fine hotels and monuments and endless suburbs of London lie in a bowl of strata of Tertiary age. The River Thames is now no more than a silvery line following the centre of the bowl. Beneath these strata—mostly soft sands and clays—there are older rocks again; the white Cretaceous limestone known as the Chalk reaches the surface north and south of London on open downs, where sheep were once universal. From his parsonage built upon this rock the incomparable naturalist Gilbert White recorded his intimate history of just one village, Selborne. White's book has sold nearly as many copies as the plays of Shakespeare, and part of its charm is its dedication to the particular place, the particular geology. South of the London Basin, the Chalk frames the Weald, which was the major source of iron in medieval times and now is thick with groves of sweet chestnut burying ancient hammer ponds. From high up, most of what you see is forested. Climb higher still and we can see that the Chalk, again, forms the white cliffs of Dover—to many English people perhaps the most sentimentally significant piece of geology there is. From this height we can see that the Dover cliffs are of a piece with facing cliffs in France on the other side of the English Channel, which is nothing more than a geological afterthought, breached by the eroding sea just a few thousand years ago. Geology knows no national boundaries and from here we can even make out the Chalk extending far across France, to underlie the endless plains in the north, where the grain that goes into making 100 million baguettes is grown in fields that have neither hedgerows nor apparently any end. And could we but follow the Chalk around the world we would find similar white limestones stretching from the Canadian Shield "all the way through to Texas and Mexico," as Suess said, to the Black Sea and well beyond in the Middle East. Chalk rock records one of the great transgressions of the sea onto the continents, one which happened close to 100 million years ago, and which painted great slabs of the world white for eternity with the sediment it left behind. Chalk once extended northwards over much of the British Isles, though erosion has removed nearly all of it; for below us now we see areas underlain by older rocks—Jurassic, Triassic, Permian—north of the affluent south of Britain; and then the blatant spine of the Pennines bisects the north of England along its length. In the south-west, the bare reaches of Dartmoor show where granite comes to the surface, wild still in an island so long tamed and settled. The Cornubian Peninsula is the legacy of an old Wilson cycle, the Variscan orogeny, that cut through Europe more than 200 million years before the Alpine convulsions. Much of the mineral wealth of Europe, like the old mine in Joachim's Valley, is the legacy of this earlier episode, which was as momentous in its

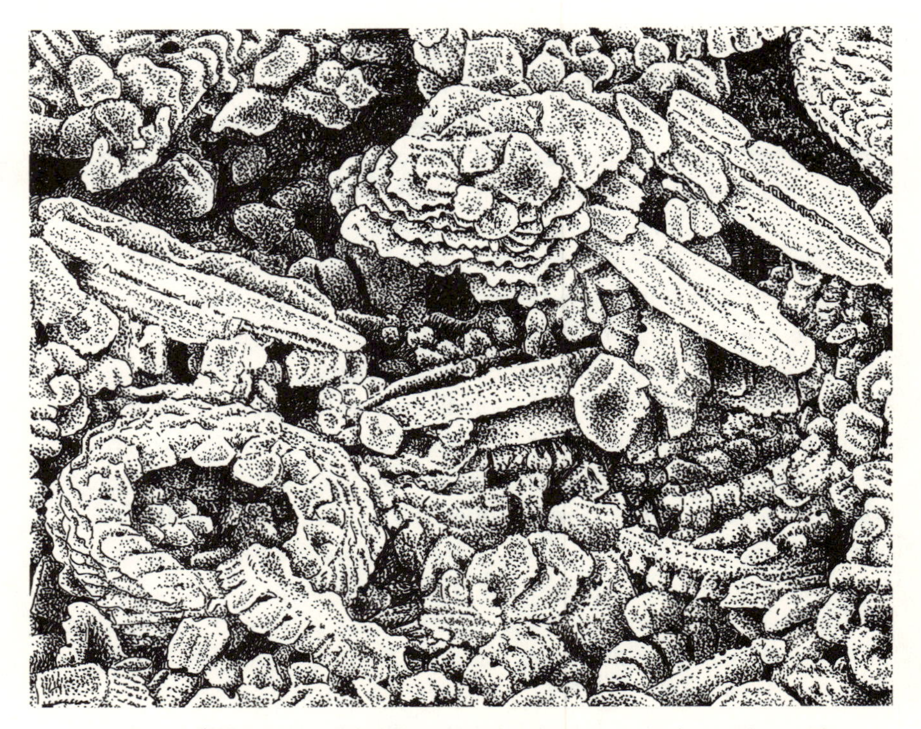

Cretaceous Chalk at very high magnification is seen to be made up of myriads of coccoliths a few thousandths of a millimetre across. Chalk seas spread widely over the world 95 million years ago.

day as that which threw up the Matterhorn. Carboniferous coal basins that developed on either side of the Pennines powered the Industrial Revolution and fed the steam engines of empire. Now virtually all the shafts are flooded and disused, and where you see an old piece of pit tackle it is rusted and useless, like a skeleton of an organism we can barely recognize; human cycles are unconscionably short, after all, compared with the revolutions of the earth.

Now we see the Cambrian Mountains, comprising the rocks that gave their names to the divisions of the Lower Palaeozoic era: Cambrian, Ordovician, Silurian.* These mountains are wilder by far than in any part

* These subdivisions were worked out by a succession of pioneering nineteenth-century geologists, but it was many years before they were universally adopted. The derivation of Cambrian is obvious; Silures and Ordovices were tribes that inhabited the Welsh Borderland in Roman times.

of England, and provided a formidable redoubt for the Welsh King Llewelyn to keep the Anglo-Saxons at bay. These rocks, too, were folded in part of the Caledonian chain, in the Wilson cycle before the Variscan: for cycle is piled upon cycle. From our high vantage point we can see how the rocks making high ground continue into the Lake District and further north again into Scotland. Can the leisurely bite into the British coastline that makes the Solway Firth really mark where the great ocean Iapetus disappeared and two ancient continents conjoined? The nations of Scotland and England have been much at odds in the past, but would any Borderer have dreamt that the line of national schism was also set in stone at a time long before he could imagine? But there is more: on the coast of Antrim, in Northern Ireland, crops out black basalt, set in columns. From this height, the basalt's perfect polygons look like the lenses on a fly's eye. This is the Giant's Causeway, a sight once deemed so amazing that a special railway line was built to carry Victorian sightseers to admire its geometrical perfection, its countless hexagonal "steps." Now we can no longer fantasize that it is an ogre's staircase, for we know it is a volcanic eruption connected with the latest Wilson cycle to leave its mark on the British Isles: the opening of the present Atlantic Ocean. This patterned basalt is the afterbirth of the creation of a new ocean, not so far from the line of old Iapetus. What giant could inspire more awe? The same eruptive events led to Fingal's Cave, on the island of Staffa, which inspired Felix Mendelssohn to some of his most delicate sound-painting, and William Wordsworth to three sonnets. Would they have written something fiercer had they pictured the cracking and parting of the two sides of the Atlantic Ocean, the beginnings of the world we know?

On Charles Lyell's own home islands, then, where much of the compass of geology as a science was defined, cycle after cycle belonging to the greater earth has left its mark; seas have advanced and retreated many times; volcanoes have had their day of fury and sunk into quiescence. If this small corner of the world boasts such complexity, we need not be surprised at the richness of the whole—an impossible concatenation of details, you might think. But as we move still further from the surface of the earth, the vision simplifies; distance loses many of the details but leaves the broader outlines clearer. See, there is the line of the Caledonian Mountains continuing across the North Sea far beyond Scotland all along the western seaboard of Scandinavia. You can almost hear Professor Törnebohm's attempts to stuff the ancient nappes onto the still more ancient foreland of the Baltic Shield. See how the rivers drain from the heights that mark the boundary between

Norway and Sweden, running eastwards through the birch and pine forests into the Gulf of Bothnia; and beyond the gulf the lake-speckled, undulating wastes of Finland, where the exhumed remains of Wilson cycles far more ancient and arcane are interwoven between gneiss and granite and now support bogs of sphagnum, cloudberries, and reindeer moss.

We are moving westwards, over the North Atlantic Ocean. The sea conceals its geological foundations. We have to remind ourselves that the continental shelves continue beyond the shorelines that we can pick out far below us. The true edges of the continents lie at the edges of the shelves, and there have been times when more of the shelves have been exposed, just as there have been other times when higher seas have drowned so much that is now dry land. Shores are just a temporary line on a shifting earth. Beyond the continental shelves: the deeps. For a moment, perhaps, looking at the endless unbroken blueness, we can empathize with Eduard Suess and his attempts to read global signals from concealed depths. We might for an instant imagine, even, that the oceans were the site of great subsided masses of the earth; that pieces in a global pattern might be missing because they had sunk to the abyss. From where we are, anything might be possible. Then, the sight of Iceland reminds us that the oceans really *are* different—that they are the realm of basalt and hold the genesis of the skin of the world. Why, we might even just make out the Rift Valley cutting southwards across the island, and those little cones that look like mounds left by solitary wasps—they must be the traces of volcanoes. As we pass by southern Iceland a larger cone is Maelifell on the edge of the Myrdalsjökull glacier, already transformed from primal black to green by the moss *Grimmia*. Now might be the moment to visualize again the slow creation of new crust at the mid-ocean ridges and recall that it is the surface expression of deeper things. Try to imagine a mantle plume rising from deep within the earth and spreading out beneath Iceland; try to imagine all those subtle changes in chemistry, in the way atoms are bonded together, including varied permutations in the silica frames, and understand that even a whole island—a whole ocean—can be written out in changes in molecular structure.

Westwards again, towards the coast of Laurentia. From this height it does not seem so far away from western Europe. Beyond us lies the frozen edge of Greenland. Here the oldest rocks on earth have stayed on the earthly carousel for turn after turn, escaping the recycling that has been the fate of almost every other scrap of rock on our second-hand world. An ice cap obscures much of the huge island, and now it may not be so difficult to imagine much of the northern hemisphere encased in ice as thick or thicker just a few thousands years ago, the latest mighty scrape of erosion stripping away time to the depths of an ancient earth. Newfoundland now lies below us; we can see how Viking navigators might readily have found their way from Iceland to "the Rock," with all its geology laid out upon its naked shores, like leaves from a crumpled book that must be restored before its text can be deciphered. We can see the grain of the land, the trend of almost every bay and promontory from the north-east to the south-west. Every pilot whale should be able to navigate it. We follow the trend southwards, past the Gulf of St. Lawrence and into the deeply-forested Appalachians. There are breaks in relief—like shrugs in a green carpet—following ancient faults that might have been the San Andreases of 400 million years ago. Now is the time to admire Marcel Bertrand's insight in stitching together the Appalachians and the Caledonian Mountains into a single system; the Atlantic Ocean then no more than an unexplained rent between the two halves. We can imagine the Caledonian to Appalachian Belt rising and wasting away into James Hall's geosynclines. And there is molasse, too, in the Catskill Mountains, where peaks are formed from the rubble that torrents dragged and tumbled down from the elevated mountains at about the same time as higher plants were colonizing the land from the sea for the first time. Life evolved and changed and the face of the earth greened with it, but for mountains it was more a case of "what goes around, comes around." And there is an earlier Wilson cycle, too, a billion years old, entrapped alongside the Appalachians: the Grenville, which rises to the surface in Central Park, New York, to remind us that the human and urban is no more than foam on the sea of the past.

Across northern Canada, close to the boundary with the Arctic tundra, the ground is sprinkled with a million lakes. They seem to be spattered over the surface as if they had been flicked from a paintbrush. The last ice age may have left behind these water-filled scrapings and scourings, but the land was reduced to a plain long, long before that. The Canadian Shield around the Great Slave Lake was ancient even when the Appalachians were young. From this height there are swirls and cracks to be made out upon its surface—like doodles almost, or the scribbled sketches of Paul Klee. These are the worn-out corpses of still older cycles, taking us back into the Archaean, thousands of millions of years ago, to the time when bacterial life alone spread over the face of the earth. "Unimaginably old" the textbooks once described this period, as if the imagination were subject to the laws of time. All those scribblings might be the foliation of a rock metamorphosed deep in the earth, but they also write an invitation for us to try to understand. What lies below us is almost as impenetrable a territory now as it was when Bell made known its mysteries to the world. In Canada, to go back to the beginnings of things you have to start beyond the end of the road.

The Canadian Shield forms the ancient, triangular heart of North America, to which everything else was added round about the edges—like the work of a slapdash plasterer. There are ranges on the east, ranges on the west, and even an ancient mountain belt running across the north through the Canadian Arctic islands and North Greenland. Each was the product of one or more major tectonic events, and each added something on to the edge of Laurentia. Then former seas washed over on to the shield, leaving their sediments behind. Southwards of where we cross the continent, we might be able to make out the northern part of the Great Plains, where the ancient seas draped blanket after blanket of sedimentary rocks through more than 600 million years. From our great height, the view is much as it was after the Pleistocene: we are too far away to see whether there are buffalo on the plains, or, more likely, over-sized modern farm machinery perambulating up and down. West of the plains, we have seen how the Colorado Plateau has been uplifted so that erosion over millions of years has cut down again through the sedimentary blankets to the ancient core beneath.

The Rocky Mountains contort the face of the earth from Alaska to Mexico, and west of them other ranges, the Cascades and Coast Ranges, parallel the coastline in ruck after ruck. From our northerly vantage point we can see the Alaska Range reach out towards the tip of Asia. It is not difficult to imagine, when Pleistocene ice locked up much of the world's water and sea levels were lowered, a land bridge connecting the Americas and Asia across which horses, bison, and humans could invade eastwards. The relatively recent withdrawal of the sea made much of America what it is today. It is more testing to imagine the mountainous land beneath us as a jigsaw puzzle of pieces shuffled and tethered together over many millions of years. It looks too solid and too unified. Nonetheless, a map published by the U.S. Geological Survey in 1992 shows no less than 200 terranes making up the Cordilleran Orogen. Just to list them with pertinent details would take up the rest of this book. The mountain chain has been put together in phase after phase, a litany of orogenies through time: in order, the Sonoma, Nevadan, Franciscan, Laramide, the whole stretching through 200 million years. What they share is accretion from the west as island arcs and other pieces of crust were plastered on the edge of the huge continent, and inward-dipping subduction zones developed at its edge. Naturally, the oldest events would be in the interior of the continent, and progressively younger ones outwards, range by range. Nor was it all push on push: from time to time stretching happened, which let down chunks of crust. Granites welled up, some of them huge. From where we are we can see snowy dissected peaks of the Columbia granodiorite, which Eduard Suess said "is of unsurpassed magnificence," on average eighty kilometres broad, and in length "almost fourteen degrees of latitude." It is hard to think of a ready analogy for all this muddled-up tectonic complexity.* An approximation might be a mill-race, where a steady stream carries pieces of wood to bank up one by one against a grille when it plunges into the leat; from time to time a variation in current or eddy might pull the aggregation briefly apart or shuffle the pieces against their neighbours, but the overall onward pattern is inexorable. Today, the confrontation between the Pacific Plate and its abutting continent continues, and what we have seen of the northward shift of the crust west of the San Andreas Fault is one consequence. Another is the periodic violent eruption of volcanoes; we can still see the scar left when Mount St. Helens blew out one of its sides in 1980, flattening millions of trees and a few humans in an extraordinary pyroclastic surge. The perfectly circular Crater Lake to the south in Oregon was the site of an explosion 8000 years ago that some geologists reckon to have been the equal of fifty Mount St. Helens. My first schoolmaster was wont to instruct us, following the Book of Common Prayer, to "read, mark, learn, and inwardly digest"; but, as we soon discover, any fact about the earth involving amazingly large quantities is easy to read, learn, and mark, but impossible to digest. To underscore the point, before we leave the north-east of the North American continent we get a glimpse of the Snake River basalts: plateau basalts as black as my grandfather's bowler hat, 5000 metres thick and erupted "only" 3 million years ago—they are a kind of postscript to the story of Laurentia. Yet they may be fifty times older than *Homo sapiens,* and a thousand times older than science, whilst the oldest rocks on Laurentia are more than 3 billion years old.

Out into the Pacific Ocean, and soon we are far enough away from the North American coast to look back and appreciate how the mountains run parallel to the shores of the continent we have left behind. The ocean below

* Recent seismic research on deep structure indicates that the whole Cordilleran system may be under the control of a failure in the lower part of the crust—a so-called decollement.

Crater Lake in Oregon, a legacy of the volcanically active western coast of the United States and Canada. Mount St. Helens was recent vigorous proof of the consequences of plate moving against plate.

is calm and brilliantly blue, and apparently goes on forever. It is hard to believe that the ocean basin is actually shrinking. Yet to the north we can just make out the line of the Aleutian Islands continuing westwards from Alaska. What we *cannot* see from above is the deep ocean trench that exactly dogs the line of the island arc on its oceanward side, nor that the trench links in turn with the Kuril Trench, and thence to the Japan Trench: the long, thin line of the zone of subduction. Mountains in various stages of their evolution are consequent upon the activity of the northward-dipping subduction zones: they are to be found in Honshu, Hokkaido, and the Kuril Islands. Mount Fuji, which Lafcadio Hearn expatiated upon and Hokusai delineated delicately, is merely the most emblematic example of the associated volcanic activity.

Suess had already distinguished such a Pacific continental margin from one of Atlantic type. The borders of the Atlantic Ocean are as distinctive in their own way. What he called "table-lands" of flat-lying sedimentary rocks commonly approached the coast—a good example would be the Karroo Series of South Africa—whereas in the Pacific type "no table-land reaches the shores." Mountain belts are truncated at the edge of the continents in the Atlantic type, as we saw in the Varsican chain running westwards across Europe. As usual, Suess had spotted important facts that subsequently became explicable in the light of plate tectonics: Atlantic coasts were "passive margins" where severed continents were drifting apart: an ancient mountain belt might be "cut," or a once-contiguous "table-land" split in the process. Pacific coasts were active sites of subduction, where the parallel fringing mountain belts reflected the continuing processes of tectonic activity, as explicit in their way as sputtering volcanoes and frequent earthquakes, the twisted roads of Kobe, or the fiery wreckage of San Francisco. It is a paradox that Pacific coasts should be anything but pacific.

As we continue westwards we are able to make out one of the northern islands of the Hawaiian chain—one of the Midway Islands perhaps—and it comes as something of a shock to find the cerulean perfection of the great ocean disturbed at last by little white circles of foam where a sea mount breaks the surface. We have to imagine a line of hidden, submarine mountains continuing northwards below us, each imprinted in turn by the same "hot spot" that is forcing out gobs and lobes of lava on the Big Island even as this is being written: the blow-torch beneath the moving plate. We can imagine Polynesian sailors in insubstantial boats feeling their way from island to island—singing songs about Pele, whose volcanic power might be demonstrated only too dramatically upon a sudden whim. Perhaps we should follow their example, and head south-westwards now across Micronesia towards the Equator, an area where many more sea mounts come to the surface. Almost every one was colonized. There below us are the Gilbert Islands, indelibly portrayed by Arthur Grimble in *A Pattern of Islands.* This is a place where the dead walk, and where shamanistic curses still count. When we reach the Mariana Islands—a classic island arc forming a beaded necklace of islands—we have already crossed the Mariana Trench, which includes the deepest hole on the planet at 36,198 feet (10,915 metres) below sea level. We recall the *Challenger*'s half-bewildered plumbing of such depths, and the realization by ocean scientists that the surface of the world goes downwards further than it soars upwards. Westwards again lies one of the most complicated and tortuous bits of the planet, where several subduction zones have succeeded one another across the Philippines to the South China Seas. Somehow, you might guess its complexity from its appearance. There is the island of the Celebes beneath us, a place that looks all leggy and out of place. It is properly called Sulawesi today, but for some reason Celebes is the name that sticks in my mind, with its curiously classical sound, as if it were the name of some monster that ravaged Thebes instead of 197,000 square kilometres (76,260 square miles) of tectonics in the Far East. There are 10,000 islands in the Sulu and Celebes Seas, many without names. Pirates still lurk there. Furthermore, there are several tectonic plates: it's a place of mystery. As we continue westwards, we encounter a simple island arc—Indonesia—stretching through Sumatra, Java, and Bali and onwards towards the Banda Sea. The island of Komodo is part of the chain and is famous among zoologists as the home of the "dragon"—the world's largest lizard, a creature that bites its prey and has saliva so foul it then hangs around to wait for its victim to die of infection. The conspicuous arc is fronted by the Java Trench, and active subduction along the junction is proved by the eruptions that regularly trouble the superstitious inhabitants of Indonesia. Oceanic crust meets continental crust directly along this line, and some of the resulting magma mixtures produce very violent explosions. On Mount Ijen, Sumatra, emaciated workmen carry on their backs great blocks of native sulphur scooped from the bowels of the volcano: hard labour in places that would reduce most of us to helpless tears.

The Java Trench is at the edge of the huge Indian–Antarctic Plate, whilst the Mariana Trench to the north-east lies at the edge of the equally gargantuan Pacific Plate. The mangled and complex area of South-east Asia in between—including Sulawesi's arachnoid silhouette—is the result of shuffling and adjustments between these two behemoths and the continent beyond, titanic jousting that has engendered short-lived basins, evanescent seas, and many successive tectonic events. The tremendous movements have continued for 30 million years and are still going on. Even the continental "foreland" of Thailand, Vietnam, and China comprises separate, fault-bounded blocks that have shuffled about like dominoes on a tin tray. The curiously shaped Malay Peninsula obviously follows the grain of the region. Its shape is rather like the thorax and abdomen of an ichneumon wasp, so narrow at one point that you can see the sea on both sides of the road that runs along its length through grove after grove of oil-palms before it bulges out into Malaysia. In a small boat I once dodged between limestone islands rising sheer from the Malaysian tropical seas in search of trilobite fossils, dining on crabs plucked fresh from the coral reefs and mangoes fresh from the tree. The shallow waters of South-east Asia are one of life's cornucopias, for its tectonic complexity has encouraged a thousand ecological niches under the beneficent gaze of the sun. The more varied the habitat, the more species it can support. Marine life, too, can "read" a geological map, from the Grand Banks of Newfoundland to the complexities of Borneo and the South China Sea; the thousands of tiny fishing boats that dredge out tropical riches from this sea are ultimately in thrall to the adjustments of plates moving more slowly than limpets grow. From our high viewpoint we can see that the hills of the Malay Peninsula extend northwards into Burma, and thence curve westwards into the mountains that wrap around to the north of the Himalayas. They are just a part of a system that extends through the world as far as the Alps, draping over the

Alfred Russel Wallace (1823–1913), pioneer of biogeography.

irregular collection of continental fragments to the south like a dust sheet over the clutter in an attic.

But we will not pass that way yet. Instead, we will drift southwards and eastwards over New Guinea. A precipitous mountain range divides it lengthways, and we can just make out steep valleys cutting inwards. We begin to understand how this one island supports so many separate languages among the aboriginal peoples that live there. It is a Stone Age version of the Alps: geology and philology acting in synergy. In the short distance from the Banda Sea we have passed over one of the great biological barriers on earth, and one that is completely invisible. This is Wallace's Line, its name celebrating Alfred Russel Wallace, who first recognized it. Wallace is mostly known today for his co-presentation with Charles Darwin of the theory of evolution by natural selection at the Linnean Society of London. He was also the founder of the discipline of biogeography, stirred as he was by wonder at the variety of life he found on his travels through the Malay Archipelago and beyond. He saw that almost all nature changed abruptly across his "line": westwards lay the Asian faunas and floras; southwards, almost everything spoke of Australia. It was as if an impalpable wall had been erected between the two. The root of it all—of course—is plate tectonics. Wallace's brilliant insight had identified the boundary between the Australian Plate and what lay to the north in the Philippines complex. Tectonically speaking, New Guinea *was* Australia.

The Australian continent had formerly been further away from Asia, and during its long isolation since the break-up of the land of the Gonds evolved its own special flora and fauna—notably dozens of marsupial species and a varied endemic flora, including eucalyptus trees and bottle-brushes like *Banksia.* It is one of the world's biological treasure-houses. Northward movement of the Indo–Australian Plate at last brought the hermetic continent into proximity with the rest of the world, and across the narrow gulf that now marks the line of contact a few daring species jumped—including humans. Most species, however, stayed where they were. Wallace himself was acutely aware of the aesthetic value of the rich fauna he studied. Describing the considerable difficulties in the way of collecting birds of paradise, he wrote, in *The Malay Archipelago* (1869), "It seems as if nature had taken precautions that her choicest treasures should not be made too common, and thus undervalued." Several of the "choicest" species of Paradisidae are now on the edge of extinction. Wallace would have been appalled.

Australia is so vastly brown. From a great height it seems to be a limitless, ochraceous to burnt-sienna expanse of almost nothing: stony semi-desert for the most part, dotted with eucalyptus and spinifex clumps that are unrecognizable from this altitude. The centre of the continent is nearly flat, or at least its gentle undulations hardly register from far above. It is possible to persuade yourself that you might be looking down on ancient Gondwana itself. In a sense, you are. Occasional ridges of rock break the monotony. Uluru, or Ayer's Rock, is the archetype of a piece of geology left behind when all around it has been removed by erosion. On the ground, it has the curious property of seeming both small and massive: something about the scale of the great interior expanse of Australia distorts the feeling for distance. Walking towards this small lump—which takes longer than you expected—you are taken by surprise to discover that it is actually enormous. Much of the western two-thirds of Australia is underlain by ancient Precambrian rocks, Archaean and Proterozoic in age, which emerge in battered hills in the interior. We have no time now to look at them in more detail. Australia has been around for a very long time, and it has been worn down over and over again by hundreds of millions of years of erosion. Its interior is dog-tired, out for the count, all passion spent. Over the last 600 or 700 million years, the sea has made inroads onto it, and sedimentary rocks overlie the ancient basement in many places. I was in the Toko Range in search of fossils of animals that lived during one of these episodes, 500 million years ago: you could readily imagine the ancient sea lapping over the already worn-down topography and leaving a legacy of sandstones—

sandy shallows turned to stone. Other fossil occurrences are world-famous. In the Flinders Range are some of the most important fossils known anywhere, that record soft-bodied life forms in the latest Precambrian; and in Western Australia are found fossil bacteria that are among the oldest traces of life on the planet. As we continue drifting eastwards, we see the exception to the exhausted stretch of the interior: towards the eastern coastline the Great Dividing Range includes several areas of high relief, richly forested. It divides rivers that run fast to the coast from those that drain westwards and sometimes peter out in salt pans in the desert. This is another ancient mountain range, which was folded at the end of the Palaeozoic and then deeply eroded; its present-day uplift is much younger. Even from high above it, the line of the Great Escarpment running through the Blue Mountains makes a gigantic step in the surface. I once sat in an Art Deco hotel looking outwards from the top of this mighty vertical cliff; beyond the balcony lay ridge upon ridge of stately eucalyptus far below: even today, secret valleys are still being discovered in the hinterland. Who knows if fearsome Bunyips—the Bigfoot of the Antipodes—might lurk there? (Rather more prosaically, Australian scientists believe that a giant marsupial called *Diprotodon* is a possible origin of the Bunyip legend: it certainly overlapped in time with the colonization of Australia by those who made it across Wallace's divide.)

The fold belt at the Great Dividing Range makes sense when you fit Pangaea back together. Antarctica slots back adjacent to Australia, as neat as you like. In this altitude the marginal mountains of Australia run on into a similar belt adjacent to the Ross Sea in western Antarctica. It is rather like the Appalachian–Caledonian story over again: what time has ripped asunder, imagination can join together. Hence the eastern Australian fold belt was produced at a subduction zone that developed at the margin of the ancient supercontinent. The legacy of Pangaea lives on in forests of ironbark and blue gums. Further along the north-eastern edge of Australia, the Queensland Plateau is a continuation of the continent out to sea, a platform that has provided a foundation allowing the Great Barrier Reef to flourish. From above, we observe a broken white line drawn by crashing waves marking its course along the Queensland coast: these ramparts of coral and algae support one of the most diverse habitats on earth. Today there are distressing stories of coral die-back and bleaching on reefs. The death of these structures would be a unique tragedy.

Across the Tasman Sea to New Zealand, the furthest point on the globe from my desk in London: the two islands look as if the highest parts of the European Alps had somehow been excised and dropped bodily into the wide expanse of the Pacific Ocean, so steep are the mountains stretching along its unpolluted shores. Eduard Suess and Elie de Beaumont would have felt at home here—the latter, as previously mentioned, has a mountain named for him. New Zealand is a rival to Hawai'i for loneliness: it has been a place apart for so long. In the absence of voracious mammals, until humans arrived, flightless birds like moas, kiwis, and the kakapo proliferated there. Millions of sheep are now the commonest creatures, and from high over the North Island you see them scattered like a dusting of white pepper over the hillsides. The dark green line of rain forest fringing the west coast of the South Island is where tree ferns sprout in profligate profusion. It is a special place, no less than Hawai'i, and its origins go back to Gondwana and beyond. A sliver of Gondwana broke off from the disintegrating supercontinent to form the core of New Zealand. It carried with it an extraordinary primitive reptile, the tuatara, which survives only on New Zealand, a memory of Triassic times preserved against the odds. Subsequently, younger crust accreted to the growing island. At the present day, New Zealand sits astride two plates: the Australian Plate to the west, and 10,000 kilometres of the Pacific Plate to the east. The conflict between the movements of these plates generates mountains, earthquakes, volcanoes, and geysers alike: for the Australian Plate is headed north and the Pacific Plate westwards, at about forty millimetres a year. The result is a twisting of the crust, a rending of the fabric of the earth. We can see the surface effects of accommodation to this movement in the Alpine Fault in the South Island, which is a kind of antipodean San Andreas. Even in the 1940s the far-seeing geologist Harold Wellman had deduced that hundreds of kilometres of displacement must have accrued along the Alpine Fault over many millions of years, for the rocks to either side of it are so utterly different. With modern GPS systems, comparable with those installed around the Bay of Naples, deformation can be continuously monitored to an accuracy of one or two millimetres over forty kilometres. They reveal that some parts of New Zealand are moving faster than others—Christchurch nearly forty millimetres a year, Auckland hardly at all. It is an irregularly fidgety bit of the crust.

So we continue eastwards across the southern hemisphere from New Zealand: beyond the Chatham Islands nothing breaks the waters of the Pacific Ocean. It is apparently the emptiest part of the world, an intensely blue continuum. We cross one of the truly arbitrary lines that humans have drawn on the face of the earth—the International Date Line. Tucked out of the way here, it will not cause trouble to those who have difficulties with Sunday suddenly changing to Monday. If we were able to see through 5000 metres (15,000 feet) of sea water we would observe that the sea floor is anything but featureless, for here is the ocean ridge that generates the Pacific and Nazca plates: the East Pacific Rise, running northwards to the Equator and beyond, but side-stepping at each transform fault, like the onward progress of a determined drunk. It leaves the surface of the sea untroubled, except for Easter Island, which is the merest pinhead from our vantage point. The people who called themselves Rapa Nui were the advance guard of all the Polynesian colonizers and may have reached this island about 1500 years ago. They celebrated its basalt geology in the most blatant way, hewing out from it huge and inscrutable heads, *Moai,* which now dot the island in heavy-featured profusion. We cannot read the meaning of their pursed and determined lips, nor their recessed and sightless eyes. But we know that the islanders ate themselves to extinction, over-population and isolation having done its worst: the original Great Ecological Disaster.

From where we are, we cannot quite make out the volcanic Galapagos archipelago further to the north. These islands were Charles Darwin's natural evolutionary laboratory, whose finches and tortoises figured prominently in *The Origin of Species.* They have consequently become a place of pilgrimage for modern biologists. Even here, the creative isolation afforded by the Pacific Ocean has been violated by our own species, and by the animals we brought with us. What geology has made over millions of years, humans, it appears, can undo in a century or so.

The coasts of Chile and Peru appear as relief (in both senses of the word) after the limitless seas. The Andes rise sheer from the coast—well, not quite sheer, because as we approach the shores of the southern continent we see that there is a narrow line of desert that borders the sea. Aridity rules in this rain-shadow of the great mountain range to the east. Below us, unseen, lies the ocean trench that follows the South American coast so closely. Here, the continent deflects the easterly ocean current of the southern Pacific so that it turns northwards along the Chilean coast as the Humboldt Current. This brings cool Antarctic waters to warmer climes, and supplies rich nutrients to a profusion of marine life. Anchovies abound—unless it is an El Niño year, when they suffer with the rest of the world. So on this part of our journey the most perniciously dry Atacama Desert and the most ebullient ocean life are found cheek by jowl. We can see rivers sprinting a few tens of kilometres to the sea from the high Andes; but those draining to the east from the far side of the range will have to wander 6000 kilometres to reach the Atlantic Ocean. From far above, the Andes do not seem that complicated. They comprise two mighty ranges running in parallel: at the point we are crossing these are called the Cordillera Occidental and Cordillera Real (or Oriental), west and east respectively, separated by a broad depression—but this still at a high altitude. A great belt of Cretaceous and Tertiary granites and other intrusive rocks runs along the Cordillera Occidental: they make some of the most inaccessible peaks, on which alpinists risk their lives. As we have seen in many other mountain belts, older rocks are caught up in the folded rocks of the present range. We have seen enough by now to recognize the median basin as a fault-bounded graben, as Suess would have called it. Conical volcanic cones line both sides, recalling Iceland or the Great African Rift Valley, however different their tectonic circumstances. The German explorer Friedrich Heinrich Alexander, Baron von Humboldt had already recognized this "avenue of volcanoes" by the beginning of the nineteenth century. We pass over the Altiplano—literally, "high and flat"—where the valley has been filled in with a succession of ashes and molasse: it is a dusty, brown, tough, cold place, to which llamas and alpacas are peculiarly adjusted. The indigenous people who live there probably have the blood of the Incas running in their veins. The last redoubts of this dead civilization were perched improbably high in the fierce terrain of the Cordillera, at Machu Picchu. Today's inhabitants still grow potatoes on Inca terraces. Lake Titicaca below us now is the highest great body of water on earth. No doubt, if we could swoop downwards and collect a fragment of rock from the side of a volcano, we would find it to be composed of andesite (what else?), a volcanic rock resulting from magma "cooked up" by the interaction of subducting slab and continental margin. For this whole coast is the southerly continuation of the active margin of the eastern Pacific Ocean—trench, mountains, volcanoes and all—and perhaps rather simpler than the Californian story to the north, for there is no San Andreas Fault here. The Andes are squashed and thickened against the edge of the stable, ancient South American shield to the east. The volcanic "avenue" is interrupted only where the relatively thick Juan Fernandez and Nazca ocean ridges approach the subduction zone: they prove rather more to bite off than the tectonic mill can chew.

The eastern part of the Cordillera includes old terranes that accreted to a primitive South America long before the present Andean range—indeed before Pangaea had been assembled. I once went to visit one of these strange slabs of country near the town of San Juan, in western Argentina, where the country's wine is made. The theory has it that a chunk of Argentina now in the Precordillera had originally broken off Laurentia in the Ordovician and drifted across a vanished ocean to meld with South America. To reach the critical rocks, our party had to go in a bus along a terrifying zigzag road carved out of a mountainside, so narrow that it had "tidal flow"—you could go one way (upwards) for part of the day, and the other way (downwards) for the rest of it. From time to time I glanced downwards to see—hundreds of metres below among the cactus—the back ends of coaches past that had plummeted off the edge.

The thickened, light crust of the Andes is rising—and fast. Eduard Suess described it thus: "Throughout Peru and Chili . . . signs of negative movement, and particularly terraces, present so striking an appearance that they have attracted the attention of observers for many years past." Darwin had already connected such uplifted beaches with volcanic causes in a paper published in the *Transactions of the Geological Society of London* in 1840. Suess himself typically downplayed it—recall that he habitually attacked uplift as the cause of mountains. He went so far as to describe the habits of sea-birds lifting edible sea-shells to great heights to form "kitchen middens" and tried to cast doubt on some of Darwin's allegedly elevated beds as no more than old dining-tables. But even he could not explain away the uplifted guano deposits in the Morro de Mejillones near Antofagasta, Chile, with shore-lines at around 500 metres and at various heights below, arranged like a series of steps.

We are lingering too long: we must press on. Our journey takes us onwards beyond the Andes, towards the Amazon Basin. The ancient core of the South American continent beneath us is floored by Precambrian granites and gneisses of similar ages to those in Africa, draped variously by younger sediments. We glimpse mountains to the north where the old basement rocks come to the surface. Those must be the sandstone cliffs of Mount Roiraima on the border of Venezuela and Brazil, a weird, isolated, cloud-covered plateau 3000 metres high with a unique biology, including endemic frogs and carnivorous plants, that is said to have inspired Sir Arthur Conan Doyle's novel *The Lost World.* The Amazon Basin itself is a sea of green forest below us that seems almost as vast as the blue Pacific. It has been the site of an inland sea at various times in the last 100 million years, but now every waterway drains ultimately into the mighty river: the basin holds nearly two-thirds of the world's fresh water. The individual rivers and streams are beyond counting, adding up to a total of perhaps half a million miles of running water. From above, we see the "white-water" rivers draining from the Andes to the west—actually more coffee-coloured from suspended silt—meeting the Rio Negro from the north, which appears to us black as a stream of tar because of its dissolved tannin, produced by an abundance of partly decomposed organic matter. The differently coloured waters are reluctant to blend, running side by side for seven kilometres before combining into the true Amazon—*O Rio Mare,* the river sea, as the early explorers called it. The rain forest fills in all the space that is not river: it is a temple of biodiversity, whose congregation is composed of insects, plants, and birds that belong to so many species that they have yet to be numbered, or, in some cases, named.

The Amazon is no less than 200 miles wide at its mouth. More water passes through it in a day than does in a whole year through the River Thames, where our journey started. It is, to use a devalued word, awesome. Sediment brought in suspension by the brown Amazon River is then deposited over its wide delta, or carried far out to sea: the ultimate fate of the eastern Andes is to become mud at the bottom of the Atlantic Ocean. In 1500, a Spanish sea captain called Vincente Pinzon was sailing over open sea far from shore when he observed the sea changing colour from blue to a kind of bronze. The reddish water was quite fresh. The lighter, sediment-charged waters of the Amazon evidently "floated" above the salt sea for some distance offshore. We can see it for ourselves: as we pass over the edge of the continent we can see the Amazonian efflux like a vast, brownish wraith gliding into the Atlantic beyond: a ghost of mountains reprocessed, one wisp displayed in the cycle of the earth.

As we continue eastwards, we cannot avoid the reflection that everything we have seen in South America is a consequence of plate structure. Were it not for the passive margin in the east, at the edge of the opening Atlantic, the Amazon could not debouch so freely into the Guiana Basin: we have to imagine an apron of sediment stretching out far, far beyond the last mangrove towards the deep sea. The crust sags to accommodate the new load deposited upon it. Conversely, the western active margin guarantees the drainage pattern of the continent, so that the great rivers will continue to run eastwards until the Andes are brought low. But when will this be? For the inexorable operation of subduction along the Peru–Chile Trench equally guarantees that mountains will be reborn until the end of the current Wilson cycle. The world continually changes, but it changes according to rules. And one day—who knows?—the sedimentary pile

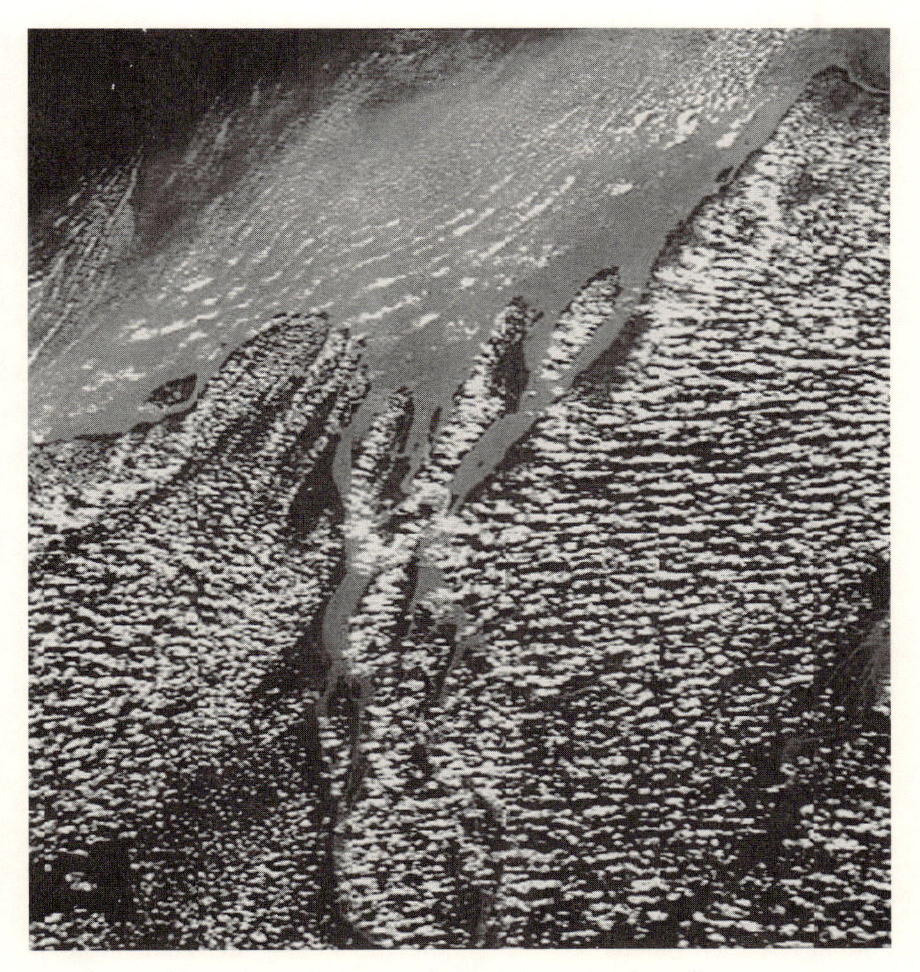

The greatest river on earth: the Amazon carries sediment (and fresh water) far out into the Atlantic Ocean.

dumped at the Atlantic edge will be revived as a mountain chain in some future cycle. No human being will witness it, that much is certain, for the whirligig of geological time moves too slowly for our ambition, our brief hegemony.

Passing urgently eastwards across the Atlantic Ocean there is little to remind us of the presence far beneath us of the Mid-Atlantic Ridge—no Iceland decked with recent cones—unless it is the solitary island of Ascension rising thousands of metres from the rugged sea floor but appearing to us no more than a pimple breaking the surface of the waters. We have passed over the unseen rift that profoundly marks the change in spreading direction of the oceanic crust. The African coast lies beyond. You could argue that the Congo Basin is the eastern equivalent of the Amazon Basin, a distorted mirror image—populated by different species, surely, but ecologically similar and likewise spilling off a passive margin. Its dense, green endlessness looks familiar: somehow, we know that some African equivalent of the giant anaconda will lurk there beneath an inexhaustible canopy of trees.

Africa is the continent that the others left behind. India, South America, Antarctica, Australia, all drifted away from it when Pangaea split apart. Africa has passive continental margins, except in the north, where it engages with Europe—as we saw among the sweet chestnut trees in Italian Switzerland. Fully to understand the African continent, its scattered progeny must be restored to it, for it lay at the heart of Gondwana and its personality was shaped long, long ago. Beyond the Congo rain forest, rainfall is strongly seasonal and a vast savanna stretches out before us. It looks uniformly straw-coloured from above—an ocean of dry grass makes the background, while scattered dark dots are umbrella-shaped acacia trees spaced just so far apart as the annual precipitation will allow. This is grazing land, hunting land, herding land, and the cradle of humankind. Mounds, sometimes towers, of rock emerge from the general plain. Can that low hump, layered like a poorly split onion, be all that remains of a granite? Can that low ridge covered in scrub be a Methuselah among gneisses? The resemblance of the scene below to that of Australia is no coincidence, for this, too, is land with an ancient core that was already old when Pangaea was young. The sea invaded over what is now Africa many times in its history, or ancient lakes and streams left a patchy cosmetic, a veneer of sediments, but which may be plastered on thickly in places. We have met the Karroo in South Africa, which Suess used to build his picture of Gondwana; I have studied sedimentary rocks 200 million years older again, in the shadow of Table Mountain, bearing trilobites and other extinct animals with jointed legs; and my colleagues have brought back much younger dinosaurs from the desert regions of the north-west. These fossil finds are all from rocks draped over an older geology, the Precambrian basement that passes into the core of South America and Australia: the heart of darkness of the ancient continent, the primeval layer, providing a vision of how things were even before the first fish, the first trilobite.

Combined with new fieldwork, the ability to date rocks in Africa was like suddenly being provided with a microscope after subsisting for years with a magnifying glass. So now we have a picture of the core of Africa (and the whole of Gondwana) as a series of truly ancient "islands," or cratons, of Archaean age—greater than 2500 million years old—surrounded by, and interwoven with, younger Proterozoic fold belts. The latter mostly ignore the present boundary of Africa, which, after all, is just an arbitrary shape imposed by a later Wilson cycle, another slicing of the map of the world. Hence, with Gondwana restored, an ancient fold belt might well pass into Madagascar, or perhaps into Arabia and India, or Brazil. The youngest Precambrian events are so pervasive that they have been called Pan-African, 900–550 million years old: one such, the Mozambique Belt, runs along the eastern margin of Africa. When you put Gondwana back in place, a single shear zone runs from Africa through Madagascar to southernmost India. The Mozambique fold belt is interpreted today as a result of the *original* collision between two older continents comprising eastern and western Gondwana—a record of the coming-together of the ancient supercontinent that has figured so prominently in this book. Nothing is so massive on earth that it has not been created by stitching together smaller pieces of lithosphere.

Archaean memories are perhaps fresher in southern Africa than anywhere else in the world. There are cratons of greater than 3000 million years old, such as the Zimbabwe Craton and the Kaapval Craton around Johannesburg. Sedimentary and volcanic basins (such as those that are part of the Pongola Supergroup and Pretoria Series) formed around and upon these even more ancient nuclei—possibly in rifts—and these sediments survive to the present day. Imagine the lottery of these scraps surviving all earthly pot-shots of potential obliteration through 2900 million years. As D. H. Lawrence said of the tortoise, they are

> *Fulfilled of the slow passion of pitching through immemorial ages*
> *Your little round house in the midst of chaos.*

By now, we might be able to make some sense of the rocky mounds and small plateaux that rise out of the plains over which we are passing. Then the Great Rift Valley appears, sheer sided, let down as suddenly as a gutter: it comes as a surprise after all that near-horizontal emptiness, so definite are its lines scored across the wilderness. There is Lake Albert below us, pointing southwards to Lake Edward and Lake Tanganyika, clear as a signpost saying: "This way, follow the geology." Those obvious volcanic rocks at the edge of the rift may be as young as the Precambrian cratons are old, so

NASA photograph of the Red Sea and the Gulf of Aqaba; it is easy from this image to imagine that this is a young ocean.

here we have the primordial ends of the earth rubbing shoulders with what could be described as a geological afterthought. We think of our distant ancestors cowering as one of the volcanoes erupted—perhaps even the one we can see on the horizon. Now we can see this valley for what it is: an attempt to break up Africa one last time, just another episode in an old, old story. Another Wilson cycle struggling to be born. To the south, there is the shimmering sheet of Lake Victoria: no rift lake this, but a huge, shallow slop accumulated in a downwarp of the ancient African crust. And then out across the desert of Somalia to the Indian Ocean, where vast breakers draw a white line at the edge of the great continent, as if to mark out the boundary along which it was once dismembered, a memory of Gondwanan disruption now drawn out in dancing and ephemeral foam.

Onwards, onwards, we follow the coast to the Horn of Africa. From the advantage of our great height, we can see that the Red Sea and Gulf of Aqaba resemble a hinge that has opened. It is hard to look at this elbow of sea, this dogleg of a crack between Africa and Arabia, without imagining a kind of creaking noise as the two continental blocks prise apart, like the opening of an old oak door in a haunted-house tale. Arabia, we can clearly see, is nothing more than part of the old African shield, and much of it was formed in the late Precambrian by the accretion of island arcs. The rocks that overlie and cover this deep foundation are the source of much of the world's oil, and no small measure of its problems. Oil is ultimately derived from the chemical transformation of the energy preserved in minute plant fossils, concentrated by natural geological processes into reserves. It just so happens that there is a disproportionate amount of both source and reserves in this desert appendage of Africa. But the truth is that there is nothing much to see from above except stretches of stony desert and fleets of dunes between bare scabs of rock, for we are skirting the edge of the *Rub al Khali,* the empty quarter. A dark and scrofulous-looking patch nearer the coast is a stretch of *sabkha*—a salt flat of a special and unpleasant kind. It is the most treacherous place I have ever visited, a darkly crusty, boiling-hot wasteland, sprouting gypsum crystals like perverse jewels. Wander off the track and the superficial crust can break, letting you and your vehicle bog down into a hot, noxious, and embracing glue. Some bacteria thrive in this environment, I am told, which proves that bacteria can live just about anywhere. As for us, we shudder and pass on.

We speed across the Arabian Sea towards the north-west coast of India. We have no time to linger on the subcontinent, but we know now that much of what has been said about Africa applies here, too: it carries the same legacy of old Gondwana dressed in the newer clothes of younger sediments. In our mind's eye we can picture the triangular continent moving northwards from its starting position as part of Pangaea, eventually to engage with Asia. John Dewey and his colleagues have calculated and mapped its trajectory step by step through 84 million years, so that its passage looks like one of those photomontages of human movement with each frame frozen in time: India kicks the backside of Asia, in living motion. Had it not completed its journey, India would now be another Australia, isolated within a great ocean, and who knows what strange animals might have lived there? It would certainly not have been reached by the polyglot profusion of peoples that makes India such an exhilarating and confusing place. Look: below us there is a stepped plateau landscape that seems half familiar. We must be passing over the Deccan Traps: one of those caves decorated with a thousand human details lies beneath us. But from above, it is possible to imagine the whole area black and slick with fresh lava boiled out of the earth's interior, one layer after another, spilling out over onto another part of the old Precambrian core of Gondwana.

The great river system of the Ganges and its many tributaries lies before us; to the east, the Brahmaputra is nearly as great a waterway. We can see the infinitely complex, but infinitely repeated pattern of tiny farms, growing rice to feed the burgeoning millions on the subcontinent; in Bangladesh, there are even three different kinds of rice to be grown according to the season, and not a day wasted in the business of cultivation. During the monsoon season, floods are both a boon and a curse: what can fertilize one area can devastate another. Vast quantities of sediment are shifted then, hundreds of millions of tons every year, much of which finishes up as an apron of sediment blanketing the sea floor of the Bay of Bengal. We can see that the Ganges drains the southern flank of the mighty Himalayan range: its tributaries cut back high into the mountains, and we can trace some of the headwaters to the glacier line. So the Himalayas are being wasted away, and ultimately ground down to silt that washes or slumps into the ocean. The Ganges Plain, flat and wide below us, is a basin that receives the debris from the tremendous range to the north, sinking to accommodate more and more sediment—ten kilometres of it: it is a molasse basin, and we might, for a moment, recall those reddish rocks that cropped out among the orchards on the flanks of the Rigi in Switzerland. The upper part of the Brahmaputra cuts through the *whole* Himalayan range, in a series of spectacular gorges. The river evidently antedated the rise of the mountains, working feverishly over millions of years to maintain its position and cut through what tectonics had raised aloft: the Via Mala raised to the *n*th degree. The mountains are still rising.

Vegetation cloaks the foothills: nothing to see from above except a green swathe tucked into rocky ridges. This is the natural habitat of the deodar and rhododendron species that make groves on the mountainsides. The mountains rise so steeply in Nepal that they seem like a wall dividing the world. However we may remind ourselves that this is just the outcome of the way plates collide, there remains something obstinately purposeful about such a massive geographical and cultural barrier—no matter if geophysics tells us that the mountain front follows the Main Boundary Thrust, which can be traced to depth on seismic profiles. There are glaciers below us now, streaked and blue, and closely following valleys down from the heights, as if they had been coloured in by opaque crayon to cover up the brown rocks that surround them. The streaked lines on their surfaces are rocky waste—yet more erosional detritus. And there is Mount Everest itself, 8848 metres high on the "roof of the world"—there is no avoiding the obvious phrase, and no way to know whether it is the highest mountain there has *ever* been. Was it topped in some other Wilson cycle long ago, in the Caledonides, perhaps, or the Hercynides? Any such peak will always remain Mount Invisible, for even its memory has been eroded away. It is unusual to get such a clear view of Mount Everest, for these altitudes are frequently wrapped in storms. The mountain was named for Sir George Everest, superintendent of the trigonometrical survey of India in 1823 and Surveyor General in 1830. He was also the first to prove by measurement* that the unexpectedly low gravitational attraction of the Himalayas showed that there had to be "roots" of lighter rock beneath the great mountains—the first indication of the crustal thickening and shortening that had happened along the tectonic line where the two plates collided. The appropriate dip in the Mohorovičíc Discontinuity marking the base of the crust was subsequently recognized from seismic evidence.

As to the how and when: India moved northwards at a rate of about fifteen centimetres a year until it collided with Asia about 45 million years ago—after which it still continued to move northwards at five centimetres annually. This was the "big push," a heave that continued long into Tertiary times and spread its influence northwards into Asia. The Karakoram and the high plateau of Tibet, the "celestial mountains" of Tien Shan, and the Altai Mountains—all these are a consequence of the persevering shove of one great continent against another. So the wrinkles on the face of the

* He followed the example of Pierre Bouguer, leader of the Andes Expedition of 1735, who had noticed that the deflection of a plumb line towards the volcanic peak of Chimborazo was less than it should have been considering its size. The measurements of variations in the gravity field due to density variations in the subsurface are still called Bouguer anomalies.

earth in Asia are not the furrowed frown of one confrontation, but phalanxes of massive corrugations spread across 1000 kilometres. The evolution of the Himalayan orogen was evidently long and complex. As push piled on push, a total of perhaps 1000 kilometres of crustal shortening was accumulated. As we have seen in the Appalachian–Caledonian chain—buried among the conifers of Newfoundland and halfway distant around the world—collision was not simply a question of continental "crunch to crunch." There were earlier "dockings" of terranes that preceded the main continent-to-continent event. "Pips" of ophiolites were squeezed out as nappes, to preserve fragments of vanished ocean floors high in the Himalayas: it is like the Oman Mountains promoted to the gods. Then came the confrontation of the prodigious protagonists: the Brobdingnagian broadside, the Titan's tussle, as the two continents, India and Asia, finally collided. A series of mighty thrusts directed broadly northwards mark the boundaries of parallel zones running all through the Himalayas. At depth, the crust of the Indian subcontinent dives beneath the Tibetan crust: the seismic profiles tell us so. In the face-off, the southerly giant ultimately cringed before its northern antagonist. Slices are piled atop this deep confrontation, each bounded by thrusts; granites were melted out and mobilized between 12 and 23 million years ago, ultimately to rise skywards. They make some of the most jagged peaks. Old geology was revivified, punched upwards, as deeper, ancient rocks were exhumed from their buried redoubts: there are even Ordovician fossils—old friends that I could recognize—on the summit of Mount Everest itself. Tethyan sediments were scrunched and metamorphosed, belt after belt. And these thrusts, as you would expect, are younging from north to south, as the Indian Plate pushed onwards, onwards.

The mighty range might seem all too complicated for the human mind to grasp, and from our height we can see none of its details; but, at the end of the last nappe, it is only a compendium and concatenation of everything we have met before in this book: a combination of Newfoundland, the Andes, and the Alps, with a soupçon of South-east Asia thrown in for good measure. Sometimes it seems as if nothing is so complex on the face of the earth that it cannot be explained in Lyellian mode: match your mountain, fix your fault, gauge your geochemistry. Does the fact that Mount Everest can be rationally explained diminish its splendour? Would it be more impressive if we still believed it to be the outcome of battle royal among the gods? Quite the contrary, I believe. Our high prospect over the "roof of the world" is enough to prove that the majesty of mountains endures even when its crags and folds have been rationalized. Far from whittling it away, understanding only increases our awe. As John Ruskin wrote in 1856: "Mountains are the beginning and the end of all natural scenery."

As we move northwards, we pass over the Tibetan Plateau, "the loftiest of the world's plateau, with an average height of 16,000 feet," as Arthur Holmes described it; and, he might have added, one of the toughest of human habitats—a cold brownish waste, a vaster version of the Altiplano, where morose yaks take the place of alpaca. There's little enough sign of human habitation from above. But what caused Tibet's elevation? Arthur Holmes seemed to have no trouble in believing that "the ancestral Indian shield . . . instead of staging a head-on collision was drawn downwards to continue its unfinished journey beneath the area that has become Tibet," thus doubling the thickness of the light continental crust. "The isostatic uplift was the simplest of the many tectonic consequences of this stupendous encounter," he continued: it floated up, if you like. Tibet does, indeed, show crustal thickening, but here the problems begin. The rate of uplift of Tibet to its present height about 8 million years ago was calculated as just too fast to be accountable to "bouncing up": something else must be involved. That something else has proved controversial. One theory, evolved by Professors Platt and England (geologists at the universities of London and Oxford, respectively), has found favour recently. They proposed that at a critical stage in crustal thickening, a deep convection current in the asthenosphere removed, or polished away, part of the thickened, and relatively cool, continental "root." The cooler volume was thus replaced by hot asthenosphere, and this is what gave a thermal boost to the elevation, a helping hand to Tibet's ascent. It is all so deep and far away, how could you ever know? Some of the predictions of the effects on the basin far above, such as faults around its perimeter, proved consistent with the theory. If correct, it should change our understanding of certain ancient mountain belts, too, and I should be obliged to rewrite some of what lies above. Time's tapestry is, after all, woven to a common design.

As a postscript to the story of Tibet, it has recently been suggested that the elevation of the high plateau 8 million years ago was responsible for modifying the climate. The atmospheric circulation was changed sufficiently to produce the current Indian Summer Monsoon. The knock-on effect was to increase the rate of erosion of the Himalayan flanks—and this is ultimately reflected in the Bengal Fan extending oceanwards 2500 kilometres from the subcontinent, where the mineralogy of the deep-sea sediments changed at exactly this time. Geological processes make moun-

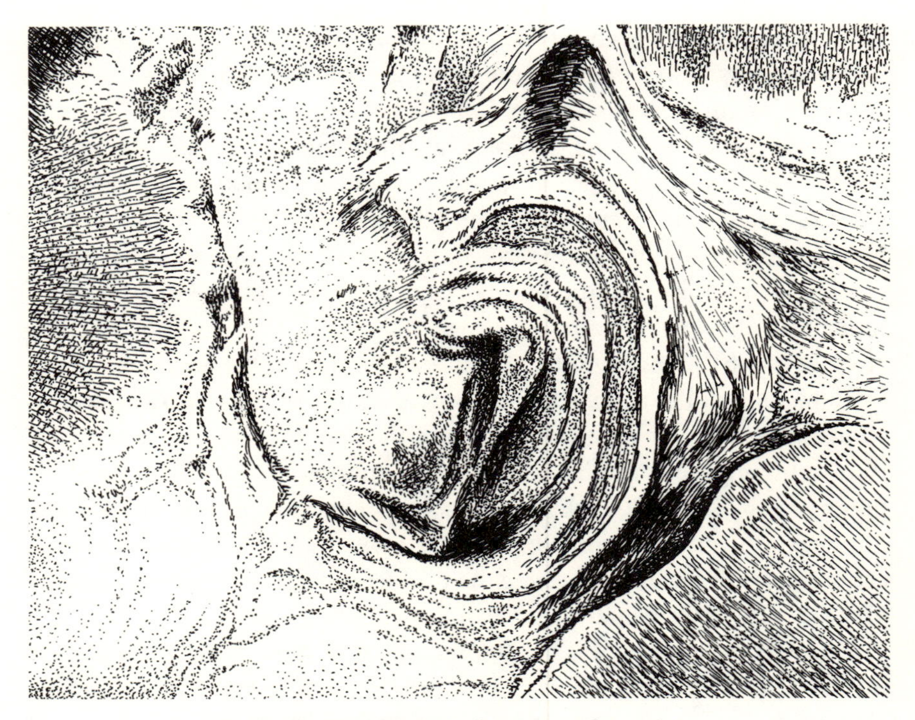

The "Great Ear of China," Lop Nor, from above.

tains, and mountains make weather patterns. Weather patterns influence erosion. Erosion produces the rocks of the next Wilson cycle. Everything connects.

We have dallied too long. We must move northwards as fast as we can across the terrible Takla Makan Desert. The Himalayas have stolen all the water that might have irrigated this arid region. As a result, it is virtually lifeless. We can make out the "Great Ear of China," Lop Nor, to the east: from above, this salt lake does indeed closely resemble an ear eighty kilometres (fifty miles) across. Its outlines are sculpted in ridges of crystallized salt and sand. The Chinese use this desolate place to test nuclear weapons. Further to the north lies another gigantic ridge, the Tien Shan Range; below it is the Turfan Depression, the second lowest place on earth at 150 metres below sea level. Cultivation of this remote area has been happening since the time of Christ, using water derived from the melting snows in the "Celestial Mountains" beyond, streams that all too soon drain down into the subsurface. The local farmers tap into the aquifers using horizontal wells called *karez.* The Great Silk Road* ran along the north side of the Turfan Depression. Camel caravans once carried bales of precious stuff to Europe in the shadow of the Tien Shan: there was good money to be made in the middle of nowhere. A railway line follows the same route today. Marco Polo passed this way on his journey to China from Venice in 1271–75, flashing the seal of the Great Khan to ensure he would receive hospitality everywhere he went. We see that the fretted surface of Tien Shan spreads out into great alluvial fans both to the north and south as if its burden of rocky outcrop were too much for the range to bear and it wished to shrug off as much as possible. The Tien Shan Range formed *within* the Asian Plate as a series of mighty thrusts following an old line of weakness. These faults, like everything else, run east–west; they bring ancient Ordovician rocks to the surface, which even carry fossils of my own animals—trilobites. Ultimately, the range was yet another consequence of the continuing northward push of India, another spasm of the crust. What seems mighty on a human scale is but a shrug of the lithosphere, a tic in time.

We must turn westwards to follow the old Silk Road through the mountains of Tadjikistan and Uzbekistan. The geology of this area is immensely complicated: a collage of bits and pieces of geology sewn together over hundreds of millions of years. The geological map is a kaleidoscope of different colours, patterned like a tie-dye T-shirt. We pass above Samarkand, the "Rome of the East," capital city of Tamerlane the Great (1336–1405) and where glazed tiles reach their geometrical apotheosis. Before we pass further westwards, it is worth reflecting that the great warrior's grandson, Ulugbek, made Samarkand into one of the scientific centres of the Middle Ages, at a time when much western intellectual endeavour was still devoted to theological minutiae. Ulugbek himself taught mathematics. If it had not been for these Middle Eastern bridges between classical and Renaissance science, who knows if geology as a discipline would have been born at all?

Westwards again, and what might be an ocean basin appears to the north. The Caspian Sea is, in fact, the most extensive freshwater body in the world. It, too, owes its existence to the heave of tectonics, for it was once connected southwards to the ocean until the elevation of the land to the south landlocked it. Seals trapped there evolved into an endemic species

* There were eventually three silk roads running from west to east, of which the one running by Turfan was the most important.

and there are hundreds of species of crustaceans found nowhere else. Thus geology and biological evolution collaborated in adding richness to our planet. Everything connects. We can see where the Volga River empties into the north of the Caspian, clouding the clear water with sediment. One day, even the Caspian, mightiest of ponds, will silt up. And northwards again a glimpse of the end of the Ural chain comes almost as a shock after the transverse grain of the world we have followed ever since the Himalayas; for this range points north–south. It is the line marking a dead Wilson cycle, the coming-together of Eurasia prior to the full assembly of Pangaea. If Europe had a geological definition, this would mark the line of its eastern boundary, where a vanished ocean still leaves a legacy of copper mines and smashed remnants of mid-ocean ridges 400 million years old. How different a journey around the world would have been that long ago, how unfamiliar the geography; only the motor of the plates would have worked in the same fashion.

By now, we have left the influence of India, only to fall under the sway of Arabia and Africa. There seems to be no end to the mountains. The northward movement of Africa—or its outriders—brings back memories of the Alps. We are now moving towards the shuffling and grinding edges of the plates that shaped Europe. As we cross in the shadow of the Caucasus between the Caspian and the Black Sea, Mount Ararat, or Agri Dagi (5165 metres), lies to the south. In the biblical account, the Ark came to shore here after the Flood, and we may remember how Eduard Suess spent so much of his great book providing a historical (and geological) explanation for the Flood as a massive incursion of the sea, garnering his evidence from the Epic of Gilgamesh. Only ten years ago, the remains of the Ark were supposed to have been discovered on Mount Ararat: to a geologist, however, the claimed "Ark" was obviously a syncline composed of well-bedded sedimentary rocks. And now before us lies that most perverse of oceans, the Black Sea, which narrowly avoids being landlocked and whose shimmering surface belies its poisoned depths deprived of oxygen, so poor is its circulation. A much more recent explanation of the biblical Flood comes to mind—the breaching of the narrow entrance into the Black Sea that dramatically filled it with salt water, raising its surface level with that of the Mediterranean and drowning shores once rich with life. This is an intellectual territory as murky as the dark depths of the Black Sea itself, a place where history, mythology, and geology compete for the hearts and minds of humankind. To some, a cherished myth is still worth more than an unvarnished fact. Time has not yet delivered its judgement on the Black

Sea flood, let alone whether it can really be identified with the Flood of Gilgamesh and the Bible. We know that there *were* catastrophes in the geological past that belie Charles Lyell's most extreme view of uniformity of process through geological time; equally, we know that sifting the evidence to understand the past requires rigour and experiment that are unquestionably Lyellian in spirit. Black Sea flood or not, we know only too well that great faults take up the stresses in the crust as it moves along the southern shore of the Black Sea; it is shifting in juddering, and sometimes catastrophic, steps eastwards relative to the rest of Turkey to the south.

Let us speed above the Dardanelles into the Mediterranean Sea and southwards through the Aegean. A pattern of islands set in a sparkling sea seems so intimate compared with what we have seen in the great ranges and wastes of the world, yet it was here that the adventurers of classical times faced monsters and tempests. Fossil bones became the limbs of giants: part of the appeal of mythic explanations lies in their proximity to our human concerns, our intimate history. It might seem to rob these islands of their individuality to learn that they are tipped blocks of limestone strata, which, in turn, are part of a response to north-dipping subduction along the Hellenic Arc lying to the south of Crete.* On the contrary, when you clamber along shore-lines shaped by the sea, looking down through the clear water at rocks that are bored by clams, like those that violated the columns at the "Temple of Serapis," you rejoice in the intimate variations upon general geological processes. Geology may provide the foundations and the structure, but there are infinite possibilities for weather and vegetation to supply the decoration. Further to the east, we remember that Cyprus carries obducted ocean floor running along the length of the island; but does it not add to our enjoyment of the richness of the world to know that Mount Olympus itself is a mass of ultramafic rocks transported from the depths of the crust, that the home of the gods is truly a production of Hades?

As for the Mediterranean Sea itself, we know enough by now to be aware that the cradle of western civilization is really an unreliable, shifting nursery. Other seas have come and gone here, Suess' Tethys most notable among them, and one day the current sea will die. In a reverse of the Black Sea scenario, the Mediterranean almost dried out in the Miocene, when the narrow link to the Atlantic Ocean was temporarily sealed by tectonics. The sea evaporated under the fierce sun, throwing down huge quantities

* In the jargon, a back-arc basin.

of salt, which still remain locked away under the current sea floor. Imagine if we had taken our journey then: the Takla Makan desert would have been as nothing to the cracked and glistening wasteland beneath us, with patches of treacherous *sabkha* waiting to engulf any beast that dared to try its luck in crossing. The flood of floods would have happened when the dam broke, the force of a thousand Niagaras crashing through the Straits of Gibraltar, tumbling fish and mollusc willy-nilly into what would become their new home. Then the waters settled; what had been mountains not many months before took on a new life as islands; and the scene was laid for proud Odysseus and all the wrangling gods on their Olympian heights.

Nor can we see the two plates comprising the eastern part of the Mediterranean region—or microplates as they should be called, because they are so much smaller than the vast slabs in the Atlantic and Pacific Oceans. They are termed the Adriatic and Anatolian Microplates, the Adriatic Sea being immediately east of Italy, and Anatolia comprising a large part of Asia Minor. Their current movements are still dictated by the northward movement of Africa towards Europe, variously twisting and thinning the crust. The African Plate continues northwards beyond its continental coastline, feeding into a subduction zone along the Cyprian and Hellenic arcs south of Turkey and Greece and the Calabrian arc south of Italy's "sole." The latter supplies magma to the active volcanoes in the Aeolian islands, which lie, where they should, above the subduction zone. Meanwhile, the Adriatic Plate is rotating anticlockwise relative to the European Plate and, as one consequence, Italy is still dangerously seismically active: frescos tumble, mudslides bury villages. The collision front between the Adriatic Plate and "Europe" lies under the eastern part of Italy, running along the length of the "boot." The relative movement between the Adriatic Plate and Europe has also caused deep fault systems to migrate, which in turn has made the volcanoes related to them progressively appear to the south over the last 5 million years. It is a complicated piece of the world. You have to envisage masses of the earth jostling one another, twisting and sliding, as well as subducting; even in some places—like the Tyrrhenian Sea—locally stretching, and all within the crushing embrace of the great continents to north and south.

The time approaches to complete our Puckish aerial circumnavigation. So much there was to see, so much remains unseen. We turn northwards towards Calabria, the toe of Italy; the elongate triangle of the island of Sicily lies to our left. That plume of smoke streaming out like a wispy tress of dark hair must be emanating from Mount Etna—the first volcano actually in process of erupting that we have encountered on our long journey. Etna celebrated the new millennium by springing vigorously to life, something it has done with infuriating unpredictability since historical records began, and many millennia before that. In 1669, it destroyed the western part of the large town of Catania, whose castle was moated by erupting lava. In 1971, the volcano even destroyed the observatory built to predict its behaviour, an example of nature cocking a snook if ever there was one. During the 1992 eruption it was estimated that 240 million cubic metres of lava were spewed forth between the months of March and May. In 2001, it was the turn of the village of Sapienza. Fortunately, Etna's lava moves slowly enough for sensible evacuation procedures to be observed and also allow attempts to divert it from its apparently inexorable course. In the midst of the latest eruption, an eyewitness recorded the smell of brimstone mixing with the aromatic whiff of burning pine resin, like a medical treatment from the time of Paracelsus.

Northwards along the Italian coast towards Naples, lower now, and to the east we see the wooded slopes of the Apennines, our last mountain chain. In spirit, we have almost come back to Glarus, for the spine of Italy is a mass of piled nappes, composed overwhelmingly of Mesozoic and Tertiary strata. The nappes came from what is now the west, while to the north-east the strata are comparatively undisturbed. Italy somehow seems out of place on the map of Europe—less a boot, more a sore thumb sticking out, away from the east–west trend of the Alps. How could this have happened as a result of an African–European rapprochement? Perhaps it is fitting to end with a puzzle. One theory promulgated by Italian geologists generates the orogen by swivelling a "proto-Italy" outwards from a "normal" Alpine position. You have to imagine the "boot" starting adjacent to France, on a line with the Alpine chain, then swinging out at its "knee" from this original position before the Oligocene, and finally, gradually, attaining its current position, all the while piling up nappes, with Corsica and Sardinia being left behind in the process. However it formed, the Italian peninsula is a scrambled mass of geology, and a rich resource. We think of churches floored with slabs of *ammonitico rosso*—a stone both warmly red and mottled with a subtlety no industrial process could duplicate, with every now and again the spiral ghost of the ammonite fossil that gave it its name. We think of marble- and limestone-clad side-chapels dedicated to the saint of choice; and, from Carrara in the north of the country, the white marble baked from the purest limestone from which Michelangelo made his unequalled sculptures: the progeny of tectonics transformed into art by the hand of genius.

See, there is Vesuvius below us now. The fug of Naples must have been blown clear away today. The wrecked courses of old lava flows can be seen tumbling and skulking among the trees. It is not difficult to imagine another one—perhaps a pyroclastic surge—wiping away urban man's impertinent roads and buildings. And there is Pompeii, looking from this height like some unfathomable board game, exhumed proof of our vulnerability. The duration of our present version of civilization, and all the versions that preceded it, is as one day in the life of the volcano, and the life of that volcano but a single breath in the life of the earth. Look, the Phlegraean Fields lie beyond the city (can you feel the magma rising deep beneath them?): that tranquil crater-lake was once the entrance to Hades. With what you now know, your mind's eye might follow Persephone downwards to the seat of the tectonic plates. And there is Baia at the end of the bay, where, Eduard Suess tells us, "Nero attempted to drown his mother in a treacherous boat" and Pliny made the observations in A.D. 79 that marked the beginnings of geology and the death of his uncle. Earthwards now, and we can see the seafront at Pozzuoli. There are some local fishermen dining alfresco on fusilli dressed with tiny squid and such fish as the nets brought in. And beyond them is the "Temple of Serapis," where my story began. A few tourists wander about the ancient market-place: perhaps one or two of them may wonder about the pitted discoloration of the great columns that dominate the square. We have come a long way.

Life on a plate.

We shall not cease from exploration
And the end of all our exploring
Will be to arrive where we started
And know the place for the first time.

T. S. Eliot, *Little Gidding*

Picture Credits

The author and publishers gratefully acknowledge the following sources for permission to reproduce illustrations:

INTEGRATED ILLUSTRATIONS

4 The Bay of Naples, engraved by E. Benjamin from an original by G. Arnald, c. 1830. *Mary Evans Picture Library.*

10 Inside the cone of Vesuvius. *Author's own collection.*

19 The "Temple of Serapis," from the frontispiece of Lyell's *Principles of Geology,* first edition, 1830.

22 "*The Light of Science Dispelling the Darkness which Covered the World.*" Drawn by Henry de la Beche.

32 Plinian-style eruption of Vesuvius in 1822. Illustration from *Considerations on Volcanos* by George Poulett Scrope.

36–37 View of Honolulu, from Manley Hopkins, *Hawaii: The Past, Present, and Future of Its Island-Kingdom* (London, 1862). *Bishop Museum, Honolulu, Hawai'i.*

42 Sketch map of the Hawaiian chain. *Ray Burrows.*

47 Frozen lava: Pele's tears and Pele's hair, resting on pumice. *Photo © Rob Francis.*

70 HMS *Challenger. Author's own collection.*

86 The Lochseiten section. *Ray Burrows.*

87 Pebbles elongated by Alpine deformation. *Ray Burrows.*

89 Fossil fish from the Flysch at Engi, as described by Louis Agassiz. *Photo courtesy of Peter Forey.*

91 Albert Heim with his mountain models, 1905. *Photo Image Archive of the ETH-Bibliothek, Zürich.*

94 The Glarus Nappe. Illustrated in *Principles of Physical Geology* by Arthur Holmes.

95 Drawings of limestones on the shore of the Urnersee by J. J. Scheuchzer. *Courtesy of Geoff Milnes.*

98 Albert Heim with folded strata in the Verrucano at Mels in St. Gallen, Switzerland. *Photo Image Archive of the ETH-Bibliothek, Zürich.*

268–69 The aftermath of the San Francisco earthquake, 1906. *Courtesy Karl V. Steinbrugge Collection, Earthquake Engineering Center, University of California, Berkeley.*

286 The Pony Express Trail, Great Basin, Nevada. *Author's own collection.*

306 Fossil stromatolite from Precambrian rocks of eastern Siberia. *Photo © Natural History Museum, London.*

312 Rodinia, late Precambrian supercontinent. *Author's own collection.*

334 The Kolb family in Rust's cable car. *Photo © Emery Kolb collection, Cline Library, North Arizona University.*

335 Hut designed by Mary Colter at the Phantom Ranch. *Ray Burrows.*

342 The deep structure of the earth. *Ray Burrows.*

355 The atomic structure of perovskite. *Ray Burrows.*

372 Cretaceous Chalk, magnified. *Ray Burrows.*

378 Crater Lake, Oregon. *Ray Burrows.*

381 Alfred Russel Wallace. *Courtesy of Sandra Knapp.*

389 The Amazon. *Photo © NASA.*

392 The Red Sea. *Photo © NASA.*

398 The "Great Ear of China," Lop Nor, from above. *Ray Burrows.*

404 Life on a plate. *Ray Burrows.*

COLOUR PLATES

I

1 Satellite view of the Bay of Naples. *Photo © NASA.*
Label showing view of Vesuvius from a Sorrento hotel. *Mary Evans Picture Library.*

2 Limestone cliffs, Capri. *Author's own collection.*
Figures overwhelmed by the eruption of Vesuvius in A.D. 79 at Pompeii. *Author's own collection.*
Mosaic floor at Pompeii, Casa del Fauno. *Photo © AKG London/Eric Lessing.*

3 Lake Agnano, Phlegraean Fields, from *Campi Phlegraei* by William Hamilton, illustrated by Piero Fabris, 1776. *Science Museum/Heritage Images.*
The Lyell Medal of the Geological Society of London. *Author's own collection.*
Arthur Holmes. *Photo courtesy of the Geological Society of London.*
The earliest seismograph, designed by Ascanio Filomarino, from the old observatory on Vesuvius. *Author's own collection.*

4 Feathered cloak, probably belonging to King Kalani'opu'u, in the collection of the Bishop Museum, Honolulu, Hawai'i. *Photo by Ben Patnoi.*
Pahoehoe lava, the Big Island, Hawai'i. *Photo © Rob Francis.*
The Lava Rock Café in Volcano, the Big Island, Hawai'i. *Author's own collection.*
A house engulfed by lava flows, the Big Island. *Photo © Rob Francis.*

5 Crusty a'a lava by a roadside on the Big Island. *Author's own collection.*
Thurston Lava Tube, Volcanoes National Park, the Big Island. *Author's own collection.*

6 Recent lava flows in Volcanoes National Park, the Big Island. *Photo © Rob Francis.*
Black sand beach, with Hawaiian green turtles. *Author's own collection.*

7 Na Pali cliffs on Kaua'i, Hawai'i. *Photo © Rob Francis.*
Waimea Canyon, Kaua'i. *Author's own collection.*

8 Coral reef, Fiji. *Photo © Rob Francis.*
Carbonate chimney, Mid-Atlantic Ridge. *Photo © Mitch Elend/University of Washington.*
The Blue Lagoon, Svartsengi, Iceland. *Photo © Rob Francis.*

II

1 Weissbad, Canton of Appenzell, eastern Switzerland. *Photo Image Archive of the ETH-Bibliothek, Zürich.*
The base of the Glarus Nappe at the Lochseiten section in Switzerland. *Author's own collection.*

2 Slate quarry at Engi, near Elms, in the Sernft Valley, Switzerland. *Author's own collection.*
Bridge crossing the Insubric Line, near Pianazzo, Switzerland. *Author's own collection.*

3 The Glarus "Thrust," from Flimerstein. *Photo courtesy of Geoff Milnes.*
Summit of the old San Bernardino Pass, Switzerland. *Author's own collection.*

4 Satellite photo of the Jura Mountains. *Image courtesy of Earth Sciences and Image Analysis Laboratory, NASA Johnson Space Center (http:eol.jsc.nasa.gov).*

5 The Oman ophiolite. *Photo courtesy of Mike Welland.*
South Wind, Clear Sky, Red Fuji, from *36 Views of Mount Fuji* by Hokusai, c. 1830. *Photo courtesy of British Library/Heritage Images.*

6 The Narrows, St. John's, Newfoundland. *Author's own collection.*
Almandine garnets. *Photo © Geoscience Features.*
Bell Island, Newfoundland. *Photo courtesy of Reg Durdle.*

7 Western coast of Newfoundland. *Author's own collection.*
Rocks at Cow Head, western Newfoundland. *Photo courtesy of Robin Cocks.*

8 The Blue Ridge Mountains, United States. *Photo © Jason Hawkes Library.*
An Ordovician trilobite fossil, *Pricyclopyge. Author's own collection.*

III

1 View of Snowdon. *Photo © National Trust Photo Library/Joe Cornish.*
Belvedere Castle, Central Park, New York. *Photo © DK Images.*

2 Silver thaler. *Photo © Kunsthistorisches Museum, Vienna.*
Marie Curie in her laboratory. *Photo: Mary Evans Picture Library.*
3 Uncut ruby. *Photo © GeoScience Features Picture Library/A. Fisher.*
Wollastonite crystals. *Photo © Natural History Museum, London.*
The Portland Vase. *Photo © Natural History Museum, London.*
The Wicklow Nugget. *Photo © Natural History Museum, London.*
4 The Ajanta temples, the Deccan Traps, India. *Photo © Rob Francis.*
Temple at Ellora, Deccan Traps, India. *Photo © Rob Francis.*
Sculpture at Ellora. *Photo © Rob Francis.*
Murals in the Ajanta Caves. *Photo © Rob Francis.*
5 Bowerman's Nose, Dartmoor. *Photo © National Trust Photo Library/David Dixon.*
Polished Dartmoor granite. *Photo © John Moorby.*
6 St. Michael's Mount, Cornwall. *Photo © Caroline Jones.*
Lava lamp. *Photo © John Moorby.*
Wheal Coats, north Cornwall. *Author's own collection.*
7 Mount Kinabalu, Indonesia. *Photo © Rob Francis.*
The collapsed Hanshin expressway, Kobe. *Photo © Topfoto.*
8 San Andreas Fault, California. *Photo courtesy Earthquake Engineering Center, University of California, Berkeley.*

IV

1 Death Valley, California. *Photo © Rob Francis.*
Eureka, Nevada. *Author's own collection.*
Basin and Range, Nevada. *Photo © Rob Francis.*
2 Lake Natron, African Rift Valley. *Photo © Rob Francis.*
2–3 Great Rift Valley in Kenya. *Photo © Corbis.*
4 Scourie Dyke. *Photo courtesy Graham Park.*
Kinlochewe Thrust. *Photo courtesy Graham Park.*
Torridonian rocks, Liathach, northwest Scotland. *Photo courtesy Graham Park.*
5 Crofter's cottage, South Harris. *Photo courtesy Graham Park.*
Lewisian gneiss, south of Loch Tollie. *Photo courtesy Graham Park.*
Precambrian banded iron formation. *Photo © Geoscience Features Picture Library.*
6 The Grand Canyon. *Photo © Rob Francis.*
Mules on Bright Angel Trail, the Grand Canyon. *Photo courtesy Jerry Ortner.*
Rock formations in the Grand Canyon. *Ray Burrows.*
7 The Inner Canyon, showing the Colorado River. *Author's own collection.*
Winter at Phantom Ranch. *Author's own collection.*

High Coast World Heritage Site, Sweden. *Photo courtesy County Administration of Vasternorrland/Metriä.*

8 Sulphur miner, Kawah Ijen volcano, Java. *Photo © Rob Francis.*
Blue Mountains, Australia. *Photo © Rob Francis.*
Mount Cook, New Zealand. *Photo © Rob Francis.*
Atacama Desert, Chile. *Photo © Rob Francis.*

Further Reading

Aubouin, J. *Geosynclines.* Elsevier, 1965.

Bancroft, P. "Gem and Mineral Treasures." *Western Enterprises/Mineralogical Record* (1984).

Beus, S. S., and M. Morales, eds. *Grand Canyon Geology.* 2nd ed. Oxford University Press, 2003.

Blundell, D. J., and A. C. Scott, eds. "Lyell: The Past Is the Key to the Present." Special Publications of the Geological Society of London 143 (1998).

Campbell, W. H. *Earth Magnetism: A Guided Tour Through Magnetic Fields.* Academic Press, 2000.

Condie, K. C. *Plate Tectonics and Crustal Evolution.* 4th ed. Heinemann, 1997.

Craig, G. Y., and J. H. Hull. "James Hutton—Present and Future." Special paper of the Geological Society of London 150 (1999).

Dolnick, E. *Down the Great Unknown: John Wesley Powell's Journey of Discovery and Tragedy Through the Grand Canyon.* HarperCollins, 2002.

Drury, S. *Stepping Stones: The Making of Our Home World.* Oxford University Press, 1999.

Du Toit, A. L. *The Geology of South Africa.* 3rd ed. Oliver and Boyd, 1954.

Eide, E. A., ed. *Batlas: Mid Norway Plate Reconstruction Atlas with Global and Atlantic Perspective.* Geological Survey of Norway, 2002.

Ernst, W. G., ed. *Earth Systems: Processes and Issues.* Cambridge University Press, 2000.

Fiero, W. *Geology of the Great Basin.* University of Nevada Press, 1986.

Fisher, R. V., and G. Hecken. *Volcanoes: Crucibles of Change.* Princeton University Press, 1997.

Fortey, R. A. *Life: An Unauthorised Biography.* Flamingo, 1998.

Glen, W. *Continental Drift and Plate Tectonics.* Merrill, 1975.

Grayson, D. K. *The Desert's Past: A Natural Prehistory of the Great Basin.* Smithsonian Institution, 1993.

Greene, M. T. *Geology in the Nineteenth Century.* Cornell University Press, 1982.

Hallam, A. *Great Geological Controversies.* 2nd ed. Oxford University Press, 1989.

———. *A Revolution in the Earth Sciences: From Continental Drift to Plate Tectonics.* Oxford: Clarendon Press, 1973.

Hazlett, R. W., and D. W. Hyndman. *Roadside Geology of Hawai'i.* Mountain Press, 1996.

Hill, M. *Gold: A Californian Story.* University of California Press, 1999.

Holmes, A. *Principles of Physical Geology.* 2nd ed. Nelson, 1964.

Jacobs, J. A. *Deep Interior of the Earth.* Chapman and Hall, 1992.

Keary, P., ed. *The Encyclopedia of the Solid Earth Sciences.* Blackwell, 1993.

Keary, P., and F. J. Vine. *Global Tectonics.* 2nd ed. Blackwell, 1996.

Kilburn, C., and W. McGuire. *Italian Volcanoes.* Terra, 2001.

Koyi, H. A., and N. S. Mancktelow, eds. "Tectonic Modelling: A Volume in Honour of Hans Ramberg." Memoir of the Geological Society of America 193 (2001).

Kunzig, R. *Mapping the Deep.* Sort of Books, 2000.

Lewis, C. *The Dating Game: One Man's Search for the Age of the Earth.* Cambridge University Press, 2000.

McDonald, G. A. *Volcanoes in the Sea: The Geology of Hawaii.* University of Hawaii Press, 1979.

McGuire, W., and C. Kilburn. *Volcanoes of the World.* Thunder Bay Press, 1997.

Menard, H. W. *Geology, Resources and Society.* W. H. Freeman, 1974.

Mussett, A. E., and M. A. Khan. *Looking into the Earth.* Cambridge University Press, 2000.

National Museum of Australia. *Gold and Civilisation.* National Museum of Australia Press, 2000.

Nisbet, E. G. *The Young Earth: An Introduction to Archaean Geology.* Allen and Unwin, 1987.

O'Donoghue, M., and L. Joyner. *Identification of Gemstones.* Butterworth-Heinemann, 2002.

Oreskes, N., ed. *Plate Tectonics: An Insider's History of the Modern Theory of the Earth.* Westview Press, 2001.

Peltier, W. R., ed. *Mantle Convection: Plate Tectonics and Global Dynamics.* Gordon and Breach, 1989.

Penhallurick, R. D. *Tin in Antiquity.* Institute of Metals, 1986.

Pfiffner, O. A., et al., eds. *Deep Structure of the Swiss Alps.* Birkhauser, 1995.

Powell, R. E., R. J. Weldon, and J. C. Matti, eds. *The San Andreas Fault System.* Geological Society of America, 1993.

Repchek, J. *The Man Who Found Time: James Hutton and the Discovery of the Earth's Antiquity.* Simon and Schuster, 2003.

Sparks, R. S. J., et al. *Volcanic Plumes.* John Wiley & Sons, 1997.

Suess, E. *The Face of the Earth.* Clarendon Press, 1904–24.

Walker, G. *Snowball Earth.* Bloomsbury, 2003.

Windley, B. F. *The Evolving Continents.* 3rd ed. Wiley, 1995.

Index

Page numbers in *italics* refer to illustrations.

A NOTE ABOUT THE AUTHOR

Richard Fortey is a senior palaeontologist at the Natural History Museum in London. His previous books include the critically acclaimed *Life: A Natural History of the First Four Billion Years of Life on Earth,* short-listed for the Rhône-Poulenc Prize in 1998, *Trilobite! Eyewitness to Evolution,* short-listed for the Samuel Johnson Prize in 2001, and *The Hidden Landscape,* which won the Natural World Book of the Year in 1993. He was awarded the Lewis Thomas Prize for science writing by Rockefeller University in 2004. He was Collier Professor for the Public Understanding of Science at the University of Bristol in 2002 and is a Fellow of the Royal Society.

A NOTE ON THE TYPE

This book was set in Minion, a typeface produced by the Adobe Corporation specifically for the Macintosh personal computer and released in 1990. Designed by Robert Slimbach, Minion combines the classic characteristics of old-style faces with the full compliment of weights required for modern typesetting.

Composed by North Market Street Graphics,
Lancaster, Pennsylvania
Printed and bound by Berryville Graphics
Berryville, Virginia
Designed by Virginia Tan

MaBP*	ERA	MAJOR DIVISIONS	
(0.01)	QUATERNARY	HOLOCENE	Ice Ages: northern hemisphere greatly affected, specially adapted mammals
1.64		PLEISTOCENE	
5.2	TERTIARY	PLIOCENE	
23.3		MIOCENE	
35.4		OLIGOCENE	Alpine evolution
56.5		EOCENE	
65		PALAEOCENE	Mammals and birds diversity
			Great Extinction 'End of the dinosaurs'
145	MESOZOIC	CRETACEOUS	Chalk deposited widely
208		JURASSIC	Modern oceans widen
245		TRIASSIC	
			Great Extinction
290	UPPER PALAEOZOIC	PERMIAN	Pangea Supercontinent
362.5		CARBONIFEROUS	'Ice Age' Coal Swamps
408		DEVONIAN	Fish and amphibia
438	LOWER PALAEOZOIC	SILURIAN	Caledonian mountains at zenith, colonization of the land
505		ORDOVICIAN	The ancient ocean Iapetus at its widest
543		CAMBRIAN	Trilobites and other marine animals appear
2500	PRECAMBRIAN	PROTEROZOIC	'Snowball Earth' Multicellular life
			Oxygenated Atmosphere develops
3500		ARCHAEAN	Life – traces in the rocks
4550		(HADEAN)	

MaBP* Millions of years before present